IUTAM Symposium on Elastohydrodynamics and Micro-elastohydrodynamics

SOLID MECHANICS AND ITS APPLICATIONS
Volume 134

Series Editor: G.M.L. GLADWELL
Department of Civil Engineering
University of Waterloo
Waterloo, Ontario, Canada N2L 3GI

Aims and Scope of the Series

The fundamental questions arising in mechanics are: *Why?, How?,* and *How much?* The aim of this series is to provide lucid accounts written by authoritative researchers giving vision and insight in answering these questions on the subject of mechanics as it relates to solids.

The scope of the series covers the entire spectrum of solid mechanics. Thus it includes the foundation of mechanics; variational formulations; computational mechanics; statics, kinematics and dynamics of rigid and elastic bodies: vibrations of solids and structures; dynamical systems and chaos; the theories of elasticity, plasticity and viscoelasticity; composite materials; rods, beams, shells and membranes; structural control and stability; soils, rocks and geomechanics; fracture; tribology; experimental mechanics; biomechanics and machine design.

The median level of presentation is the first year graduate student. Some texts are monographs defining the current state of the field; others are accessible to final year undergraduates; but essentially the emphasis is on readability and clarity.

For a list of related mechanics titles, see final pages.

IUTAM Symposium on Elastohydrodynamics and Micro-elastohydrodynamics

Proceedings of the IUTAM Symposium held in Cardiff, UK, 1–3 September 2004

Edited by

R. W. SNIDLE

University of Cardiff, U.K.

and

H. P. EVANS

University of Cardiff, U.K.

Springer

A C.I.P. Catalogue record for this book is available from the Library of Congress.

ISBN-10 1-4020-4532-8 (HB)
ISBN-13 978-1-4020-4532-5 (HB)
ISBN-10 1-4020-4533-6 (e-book)
ISBN-13 978-1-4020-4533-2 (e-book)

Published by Springer,
P.O. Box 17, 3300 AA Dordrecht, The Netherlands.

www.springer.com

Printed on acid-free paper

Printed in the Netherlands.

CONTENTS

CONTENTS

PREFACE

This volume contains the proceedings of the IUTAM Symposium on Elastohydrodynamics and Microelastohydrodynamics held in Cardiff from 1st to 3rd September 2004. The symposium focused on theoretical, experimental and computational issues in elastohydrodynamic lubrication (EHL) both in relation to smooth surfaces and in situations where the film is of the same order or thinner than the surface roughness (micro-EHL). The last IUTAM Symposium in this general area of contact of deformable bodies was in 1974. The emphasis in the Symposium was upon fundamental issues such as: solution methods; lubricant rheological models, thermal effects; both low and high elastic modulus situations; human and replacement joints; fluid traction; dynamic effects, asperity lubrication and the failure of lubrication; surface fatigue and thermal distress under EHL conditions.

Delegates were welcomed to Wales and the Cardiff School of Engineering by the head of the School, Professor Hywel Thomas. The opening lecture was given jointly by Professor Duncan Dowson, FRS and Sir Gordon Higginson, the distinguished partnership which produced some of the most important numerical solutions to the fundamental EHL problem which led to the first reliable film thickness formula for isothermal, Newtonian conditions. Their presentation reviewed the early developments in the subject and included some fascinating details of the difficulties overcome and the scientific personalities involved.

A total of 33 presentations were given over a period of three days. A particularly thought-provoking presentation was that by Dr Scott Bair (Georgia Institute of Technology, USA) on the ongoing challenges posed by lubricant rheology. If we are to understand lubricant behaviour under the very severe conditions of a real EHL contact we need to draw upon sources outside conventional tribology such as molecular dynamics simulations, and pursue improved techniques for high pressure measurements. Other papers were on state of the art developments in rough surface lubrication, journal bearing solution techniques, and natural and replacement human joints. The importance of experimental validation techniques was emphasised in a number of papers on thin film methods.

This Symposium on the specialised topic of EHL was attended by leading experts in the field and was judged by the delegates to have succeeded in: attracting stimulating presentations on basic EHL research, encouraging lively and informative discussion, and identifying future goals for the subject. We hope that the Symposium has renewed interest in basic EHL and that a further IUTAM gathering in this field will be held in the future.

The meeting attracted 45 participants from ten countries (China, Czech Republic, Egypt, France, Germany, Italy, Japan, The Netherlands, USA, UK).

The International Scientific Committee responsible for the Symposium comprised the following:

R.W. Snidle (Chair, UK), L. Chang (USA), H.P. Evans (UK), M. Kaneta (Japan), T. Knudsen (Denmark), A.A. Lubrecht (France), F. Sadeghi (USA), C.H. Venner (The Netherlands), K. Walters, FRS (UK) and D.H. van Campen (The Netherlands).

The Committee gratefully acknowledges financial support for the Symposium from the International Union of Theoretical and Applied Mechanics, and the School of Engineering, Cardiff University. The smooth running of the Symposium owes much to the efforts of Cherrie Summers, Aderyn Reid, Chris Davies, Ajay Dhulipalla, Mark Holmes and Kayri Sharif. To all of them our sincere thanks.

R.W. Snidle
H.P. Evans
Cardiff, March 2005

IN MEMORIAM

Ian Lee-Prudhoe
1970–2003

Ian had infectious enthusiasm that came across as soon as people met him – his patience with people was remarkable. He was always a very hard worker and really loved his job at PCS Instruments. Everyone who came into contact with him, found him very approachable, and nothing was too much trouble for him.

His love of everything in his life, his family, friends, his work, his house, his numerous cars and bikes. He loved music, socialising, running (he ran two marathons and numerous half-marathons), helping others whenever possible, whether it be helping someone to move house, fixing a computer problem or just having a quiet chat over a pint. Somehow you got energy from him when he was around.

He was never sad, well he never showed it anyway, never argued with people, always let them have their say, but he was always honest.

Our thoughts are still with his family and close friends – you should all be very proud of what he achieved in such a short time.

He lived, he laughed, he loved, he left.

SESSION 1

REFLECTIONS ON EARLY STUDIES OF ELASTO-HYDRODYNAMIC LUBRICATION

Duncan Dowson[1] and Gordon Robert Higginson[2]

[1] *"Ryedale", 23 Church Lane, Adel, Leeds LS16 8DQ, U.K.*

[2] *61, Albany Park Court, Westwood Road, Southampton SO17 1LA, U.K.*

Abstract It is almost fifty years since theoretical work on elastohydrodynamic lubrication commenced in the Department of Mechanical Engineering of The University of Leeds. Details of the development of numerical solutions to the line contact problem during the six year period (1956–1962) that the authors worked together on the problem will be outlined. The computing aids available during the eighteen month period involved in generating the first solution consisted of two hand operated mechanical calculating machines, with the first digital computer at Leeds being installed in 1959. The general research environment during the period will be recalled and a number of significant events recorded. It is appropriate to record at this Symposium aspects of these initial developments in a subject which has dominated research in tribology throughout the latter part of the 20th and into the early years of the 21st centuries. The excitement of being involved in taking some of the first steps in a field described by the late Professor F.T. Barwell (1970) as "*the major event in the development of lubrication science since Reynolds's own paper*", will be recalled.

Keywords: elasto-hydrodynamic lubrication (EHL), elastic deformation, pressure viscosity, dimensionless groups (G-materials, H-film thickness, U-speed, W-load), line contact, film thickness equation.

1. INTRODUCTION

Significant progress in plain bearing design, manufacture and operation was evident within a quarter of a century of the publication by Osborne Reynolds (1866) of his classical paper exposing the fundamentals of fluid-film lubrication. However, the mechanism of gear lubrication remained a mystery for a further fifty years or so. Martin (1916) applied Reynolds equation to the counter-formal profiles presented by gear teeth but his calculated film thicknesses for rigid teeth lubricated by an incompressible, iso-viscous fluid were considerably smaller than the surface roughnesses generated by contemporary manufacturing techniques. This presented a quandary, since hydrodynamic

R.W. Snidle and H.P. Evans (eds), IUTAM Symposium on Elastohydrodynamics and Microelastohydrodynamics, 3–21.

theory failed to predict that the teeth could be separated by a fluid-film, yet as Martin observed "... *The absence of wear must be attributed to the presence of an oil film between the teeth*".

In the 1930s and 1940s attempts were made to extend the analysis by incorporating separately the effects of the very high pressure in the gear contacts upon elastic distortion of the solids and the viscosity of the lubricant (see Dowson and Higginson, 1966, 1977, for an account of these developments). The calculated film thicknesses were increased by these extensions to the theory, as expected, but only at most by about 150%. This was still much too small to bring hydrodynamic predictions of film thickness into accord with experimental observations and gear operating experience.

The synergistic effects of incorporating both elastic deformation and pressure-viscosity effects simultaneously into solutions of the Reynolds equation were truly amazing (Ertel, 1945; Grubin, 1949; Petrusevich, 1951; Dowson and Higginson, 1959). Predicted film thicknesses were not simply increased by a factor of about two from the classical rigid solids, iso-viscous lubricant solutions, but by one or two orders of magnitude. The predictions were at long last broadly consistent with practical experience and the fundamentals of elasto-hydrodynamic lubrication (EHL) had been firmly established and well recognized. Indeed, it was to dominate much of the literature on tribology for the next half century and, as this Symposium demonstrates, the subject flourishes still.

The inspired approximate analytical solution derived by Ertel (1945) and presented in his PhD thesis, emerged during the traumatic period of World War II and remained in relative obscurity until Grubin promoted its publication in Moscow (1949). Some time later a valuable English translation was produced by the DSIR in the U.K. Alastair Cameron's (1985) account of the story behind the link between Grubin and Ertel represents a remarkable feature of EHL history. Alastair recommended that the approximate analytical method and solution of the line-contact problem previously associated with Grubin (1949) should henceforth be attributed to Ertel (1945). Later it was suggested that Grubin's initiative in publishing the work (1949) justified joint recognition as the *Ertel Grubin solution*.

Work on EHL in Leeds University commenced in 1956 when two contemporary former undergraduate (1947–1950) and PhD (1950–1952) students of Professor Derman Guy Christopherson were re-united as young lecturers in Mechanical Engineering. One (D.D.) had worked on aerodynamics in the the guided missile division of an aircraft manufacturer and the other (G.R.H.) on projectile stability at Fort Halstead. It is appropriate to recall the years in which their first steps were being taken in the then new field of EHL, on an occasion held almost half a century later, at which significant modern developments in the subject are being presented to this IUTAM Symposium on *Elastohydrodynamic and Microelastohydrodynamic Lubrication*.

2. DEPARTMENT OF SCIENTIFIC AND INDUSTRIAL RESEARCH (DSIR) INITIATIVE (1956)

In setting the scene for the development of EHL studies at Leeds the paucity of external funding for University research almost half a century ago should be recalled. This made the 1956 initiative by F.T. Barwell of the highly respected Lubrication and Wear Division of the DSIR Mechanical Engineering Research Laboratory, Thorntonhall, East Kilbride, Glasgow, truly inspired and unusual.

The DSIR proposal for extra-departmental research into elasto-hydrodynamic lubrication problems was prompted by a recognition that "*conventional hydrodynamic lubrication theory and contact stress analysis are inadequate for full understanding of conditions of lubrication and surface failure in the thin film range*". Practical manifestations of this situation were related to the pitting and surface rippling observed on surfaces in Hertzian contact, including the distribution and orientation of fatigue cracks in pitting of ball bearings and the plastic flow of hypoid gear teeth. The proposed contract was expected to promote "*... a synthesis of the hydrodynamic and elastic approaches, taking into account also such relevant work as may be done on the properties of lubricants under high pressure, the influence of steep-fronted pressure waves on materials, fatigue etc. ...*"

In May 1956 expressions of interest and definite proposals were invited from four institutions where members of staff were thought to be interested in undertaking such investigations. These were:

- Imperial College (Professor Christopherson and Dr. Cameron).
- Cambridge (Dr. K.L. Johnson).
- Royal Technical College, Glasgow (Professor Thompson).
- Leeds University (Professor D.C. Johnson).

Dan Johnson had been appointed Head of the Department of Mechanical Engineering at Leeds in 1955 (to 1962), having already established a distinguished standing in the fields of dynamics/vibrations and gearing at Cambridge. He readily recognized the significance of the proposal and invited two of his young lecturers (Duncan Dowson – fluid mechanics/lubrication-appointed December 1954; Gordon Robert Higginson – solid mechanics/stress analysis-appointed September 1956) to consider it. DSIR sponsorship was quickly agreed and a start was made on the development of a line contact numerical solution in the autumn of 1956.

The authors initially undertook the work alone, since it was unusual for junior members of staff to supervise PhD students in those days.

Furthermore they somewhat underestimated the time and effort required to solve EHL problems, and fully expected to complete the bulk of the work in a matter of months, if not weeks! An interesting cameo from this period is recorded in Note 1.[1]

An outline of some of the joys and frustrations experienced in developing solutions to the line contact EHL problem in the following months and years is presented in Section 3.

3. DEVELOPMENT OF NUMERICAL SOLUTIONS TO THE EHL LINE CONTACT PROBLEM

The first paper we published together (Dowson and Higginson, 1959) was "A numerical solution to the elasto-hydrodynamic problem". It was accepted for publication in the *Proceedings of the Institution of Mechanical Engineers*, but after a long delay it appeared in the first issue of the *Journal of Mechanical Engineering Science*, 1959.[2] The new journal (JMES) became known affectionately as James. This paper took much longer to prepare than any other we wrote together.

We fell into the young man's trap of trying to solve the problem using a method which had already failed in the hands of others – the straightforward iteration of

$$\text{pressure} \rightarrow \text{elastically deformed film profile} \rightarrow \text{pressure}.$$

We were teased by the process, because each time-consuming step seemed to be leading to a neat convergence until suddenly a huge discrepancy opened between successive steps. The steps were time- consuming because we had between us only two small hand operated mechanical calculators, a Facit and a Brunsviga.

We were driven to seek a different approach, using the same simplifying assumptions of plane strain in the elastic solids and two-dimensional incompressible flow in the fluid. Starting from an assumed pressure distribution, calculating the elastic distortion of the solids and comparing the corresponding shape with the geometrical profile demanded by an inverse hydrodynamic solution of the Reynolds equation for the fluid: the two profiles were then used

[1] Professor Ken Johnson (private communication) reminded us that he was unable to undertake work on the topic in 1956, but when a potential research student came forward in 1958 they visited Leeds and Jim Crook in AEI Aldermaston and decided that "both the theoretical and experimental aspects of the problem were being well taken care of, and we retreated back to 'dry' contacts".

[2] The quaint units of tons and inches used in this first paper were derived from an idiosyncratic disc machine designed for the Department by a research student studying Novikov gears.

to determine the next guess of the pressure distribution:

$$\text{pressure} \rightarrow \begin{array}{c}\text{elastic distortion} \rightarrow \text{film shape} \\ \text{inverse hydrodynamic film shape}\end{array} \rightarrow \text{pressure}$$

On hearing about this, Dan Johnson described it as "*relaxation proper*", but we preferred to call it "*relaxation improper*".

Much practice, and the arrival of a Ferranti Pegasus University mainframe computer, installed in a disused church adjacent to the campus, allowed us to produce enough results to prepare our initial paper outlining the method of solution for EHL line contacts. The calculation procedures are described in detail in Dowson and Higginson (1959).

The initial converged EHL solution had been achieved with the aid of the hand operated mechanical calculators in about eighteen months; a figure that may interest experts attending this Symposium to discuss details of their sophisticated programs developed for much more complex problems and with run times measured in minutes. After attending a one day Autocode Course on January 14th 1959, we felt quite exhilarated by the relatively rapid solution process on Pegasus.

Figure 1 shows the separate effects of elasticity and pressure-dependent viscosity and their huge effect in combination, while Figure 2 illustrates the near-Hertzian pressure distribution at higher loads and the very slow variation of minimum film thickness with load.

In the second paper (Dowson and Higginson, 1960), published under the title "The effect of material properties on the lubrication of elastic rollers", some rather spectacular pressure spikes emerged, whereas none were evident for the conditions considered in the 1959 paper. The role of lubricant and solid properties, together with external operating conditions (speed and load), in determining not only the film thickness but also the presence or otherwise of the spike and its location and height evoked much interest. With great foresight the need to have a film restriction and a pressure spike in the outlet region had been predicted in Grubin's (1949) paper and confirmed numerically by Petrusevich (1951).

Petrusevich was presenting papers on fatigue and the design of gears when the authors were but six years old. During the second world war he was responsible for research on gear materials at the Central Scientific Research Institute of Technology and Engineering (Moscow) and he later became deputy director at the Institute of Machine Science. In 1955 he was appointed Counselor at the Soviet Embassy, 13, Kensington Palace Gardens, London W8. In 1959 we somewhat brashly invited him to visit us in Leeds, since we were anxious to learn more of his numerical approach to line contact EHL problems and to discuss various features of his solutions. The curious absence of elastic modulus in his published film thickness formula was resolved when it

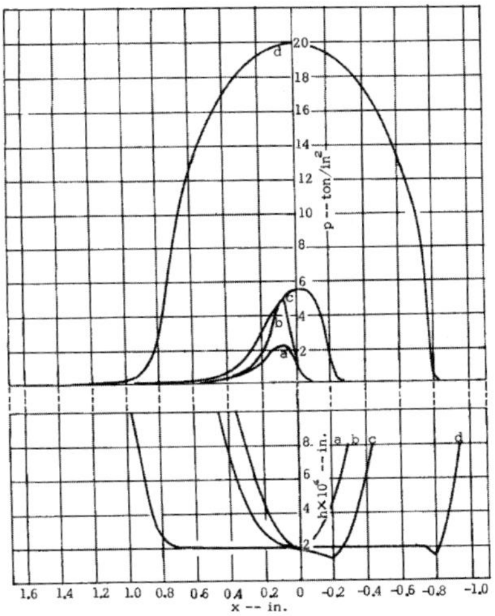

Figure 1. Pressure distributions and film shapes with the same central film thickness; (a) constant viscosity-rigid cylinders; (b) pressure-dependent viscosity-rigid cylinders. (c) constant viscosity-elastic cylinders; (d) pressure-dependent viscosity-elastic cylinders.

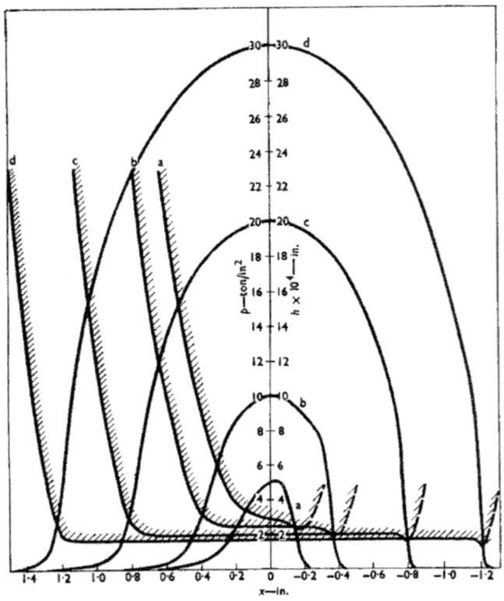

Figure 2. Pressure distributions and film shapes with pressure-dependent viscosity and elastic cylinders. P_{max} tons/in^2; (a) 5; (b) 10; (c) 20; (d) 30.

was realized that a specific value of 2,150,000 kg/cm^2 (210.84 GPa) had been built into it. The indication that film thickness increased as load and hence Hertzian pressure increased, remained a puzzle.

We first met Petrusevich at a Research Symposium on the Relaxation of Oils on Monday 16th February, 1959 and he graciously agreed to visit us. After a delay caused by a ban on Soviet diplomats leaving London, Petrusevich made a memorable visit to Leeds. He was very charming and open in conversation. He surprised us greatly by not being able to remember how he did his calculations, but on reflection we did not fully appreciate that they had been completed almost ten years earlier. We now entirely understand his problem!

The 1960 paper confirmed the dominance of changes in (U) and the relative unimportance of changes in (W) in determining the film thickness in lubricated, highly-loaded contacts. It also illustrated, for the first time, the influence of the parameter (G) upon the shape of the pressure distribution at the outlet end. High (G), for example steel and mineral oil, gave a Petruevich spike; lower (G) bronze and mineral oil, no spike. It was also demonstrated that the spike draws the maximum principal stress difference in the solid towards the surface in quite a spectacular manner, which is important in surface fatigue.

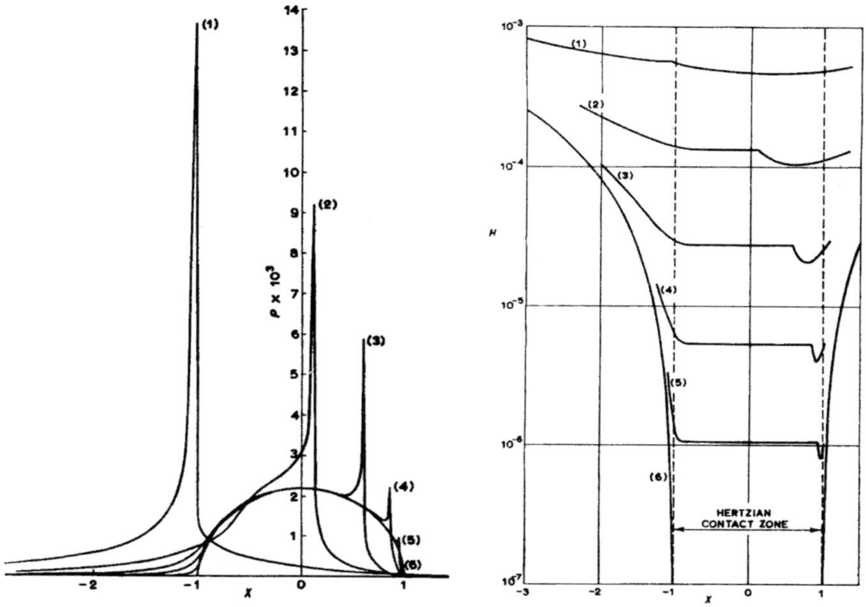

(I) (II)

Figure 3. Pressure distributions (I) and film shapes (II) for an incompressible lubricant. $W = 3 \times 10^{-5}$; $G = 5000$; $U = 0$ (dry contact), (1) 10^{-13}, (2) 10^{-12}, (3) 10^{-11}, (4) 10^{-10}, (5) 10^{-9}.

Figure 3 shows the huge variation with (U) of the height of the theoretical pressure spike and the corresponding film shapes from dry contact to near rigid solid profiles.

Our third paper was "New roller bearing lubrication formula" (Dowson and Higginson, 1961). It was published in Engineering, a highly regarded journal carrying research and industrial articles, on 4th August 1961. The attraction for us was that it was widely read by engineers and that it published accepted articles very quickly, on the timescale of a newspaper or weekly magazine. A drawback was that the editor wrote the title, which struck us as rather "sensational". We chose this rapid publication route to announce our minimum film thickness formula for line contacts and to bring the findings to the largest possible readership. We also thought, wrongly as it happened, that we were to be beaten to it by a rival group in the U.K.

In general terms:

$$H = f(WUG) \quad \text{or} \quad H = kW^a U^b G^c,$$

where the dimensionless groups are written as

$$H_{\min} = \frac{h_{\min}}{R} \quad \text{minimum film thickness},$$

$$W = \frac{w}{E'R} \quad \text{load parameter,}$$
$$U = \frac{\eta_0 u}{E'R} \quad \text{speed parameter,}$$
$$G = \alpha E' \quad \text{materials parameter.}$$

Some chose to reduce the number of dimensionless groups from four to three, but the pragmatic attraction to machine designers of the separate groups related to operating conditions caused us to retain the full list of variables. An analysis of the numerical solutions obtained by the beginning of the 1960s enabled the following dimensionless relationship between (W), (U), (G), and minimum film thickness $H_{\rm min}$ to be established

$$H_{\rm min} = 1.6\frac{G^{0.6}U^{0.7}}{W^{0.13}}. \tag{1}$$

Or, in terms of individual physical quantities:

$$h = 1.6\alpha^{0.6}(E')^{0.03}R^{0.43}\frac{(\eta_0 u)^{0.7}}{(w)^{0.13}}. \tag{2}$$

The corresponding central film thickness expression published by Grubin (1949) but probably derived by Ertel (1945), can be written as

$$H_{\rm cen} = 1.95\frac{(GU)^{8/11}}{W^{1/11}} \quad \text{or} \quad H_{\rm cen} = 1.95\frac{G^{0.73}U^{0.73}}{W^{0.091}}. \tag{3}$$

The powers on the independent variables in equations (1) and (3) are encouragingly similar, while the predicted minimum film thicknesses ((1) DD/GRH) are about 75–80% of the central ((3) Ertel/Grubin) film thickness for typical line contact elastohydrodynamic conjunctions in many engineering components.

Equations (1) and (3) have been widely adopted in the calculation of minimum and central film thickness in elastohydrodynamic line contacts. It is therefore of some interest to recall the small number of solutions then available for the derivation of equation (1).

While the range of independent variables considered was quite extensive, the number of solutions within each range was modest. For example, for incompressible lubricants, solutions were obtained for values of the materials parameter (G) of 2,500 and 5,000 at two loads. The computed power (c) on (G) varied from 0.55 to 0.63 and so a representative value of 0.6 was adopted. The power on (U) was more firmly based with no less than seven different dimensionless speeds between 10^{-13} and 10^{-9} being considered at a load (W) of 3×10^{-5} and a value of (G) of 5,000. The load (W) range considered was

3×10^{-5} to 3×10^{-4} and the power on this less influential parameter (a) was found to be 0.13 from some four solutions. At a later stage the powers were adjusted (Dowson, 1968) to satisfy the overall requirements of dimensional analysis, but the process had little effect upon accuracy. The revised equation was

$$H_{\min} = 2.65 \frac{G^{0.54} U^{0.70}}{W^{0.13}}. \tag{4}$$

In this regard it should be noted that there was more uncertainty about the power on the materials parameter (G) than on speed. However, the range of (G) encountered in engineering is restricted compared to the speed range. The well-known stiffness of the film with changing load is also apparent and thus errors associated with load changes are generally negligible.

If consideration is given to metallic machine components lubricated by mineral oils, a further useful simplification can be applied (Dowson and Higginson, 1966, 1977), since

$$h = k(\eta_0 u R)^{1/2}. \tag{5}$$

For SI units the value of (k) is 1.6×10^{-5}.

A less audacious approach to the construction of our minimum film thickness equation (1) over forty years ago might have been appropriate if not only the subsequent heavy reliance upon it but also its extrapolation to conditions well outside the range considered in the solution domain had been anticipated! The agreement between the individually computed points and the predictions of equation (1) was nevertheless most encouraging. Furthermore, the close accord between the predictions based upon equation (1) and experimental measurements by capacitance (Crook, 1961a); and X-ray transmission (Sibley and Orcutt, 1961) shown in Figure 4 provided welcome encouragement.

Crook (1961a) measured film thickness by a capacitance method and in his experiments he varied load and speed. Sibley et al. used an X-ray transmission technique to measure directly the minimum film thickness at various loads, speeds, viscosities and pressure-viscosity indices.

The 1962 paper on "The lubrication of elastic rollers" (Dowson and Higginson, 1962) offered the first opportunity to present our results overseas. It was presented to the Tenth International Congress of Applied Mechanics in Stresa, Italy. This was an exciting event for two young men and a sight of the other side of academic life. Opportunities to travel were still greatly restricted in those far off days, and very different from the present situation. A rather long but cheap train journey from Leeds to Stresa, Italy was followed by a traumatic experience with visual aids. As the box of old fashioned glass slides was handed to the projectionist it fell onto the marble floor and all were shattered! There were however, pleasant compensations in attending a scientific meeting in Stresa! It was particularly gratifying and enjoyable to spend some time with

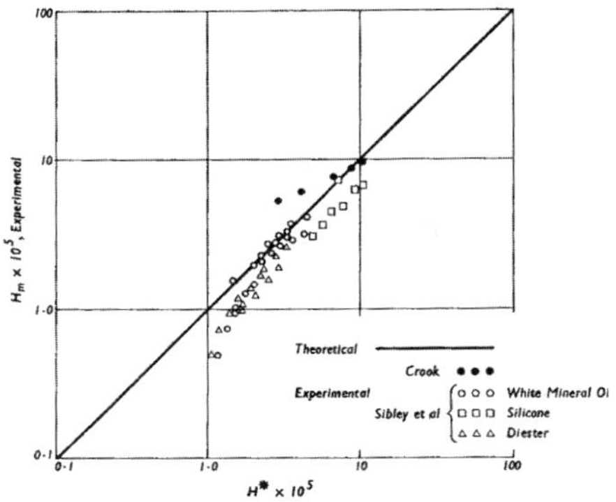

Figure 4. Comparison of experimental minimum film thicknesses and predictions of equation (1).

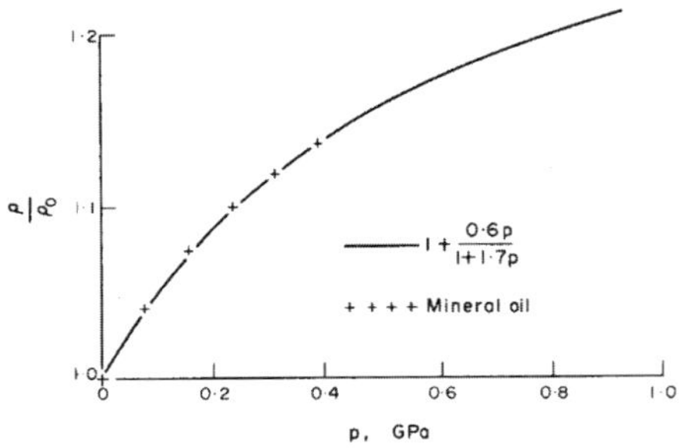

Figure 5. Density-pressure relationship (note: in this expression (p) is in GPa, whereas the original expression was based upon ton/in^2).

the late Alan Milne, who had played a central role in the promotion of the DSIR initiative outlined in Section 2.

Once the solution procedure had been established, the DSIR agreed further support for a research assistant dedicated to EHL computation. Alan Whitaker was appointed to rewrite and develop our computer programs and to tidy up the numerical procedures. At about this time we introduced lubricant compressibility into the Reynolds equation, using a simple expression (6) to describe the variation of density with pressure. The Thornton Research Centre (Shell) provided experimental data for a particular mineral oil and the five measured values, together with the expression adopted to fit them (note (p) in GPa), are shown in Figure 5. The limit of compression of mineral oils is about 25%, giving a maximum density increase of 33%. This representative expression has been widely used by many investigators, which is perhaps surprising in view of the limited data upon which it was based!

$$\frac{\rho}{\rho_0} = 1 + \frac{0.6p}{1 + 1.7p}, \tag{6}$$

where (p) is in GPa.

The full range of solutions previously obtained for incompressible lubricants was reviewed and new solutions for compressible lubricants added in the 1962 paper on "Elasto-hydrodynamic lubrication: A survey of isothermal solutions" (Dowson et al., 1962).

The introduction of compressibility had little effect on minimum film film thickness, although the central film thickness decreased as the lubricant was compressed by near Hertzian pressures (Figure 6). The pressure distribution

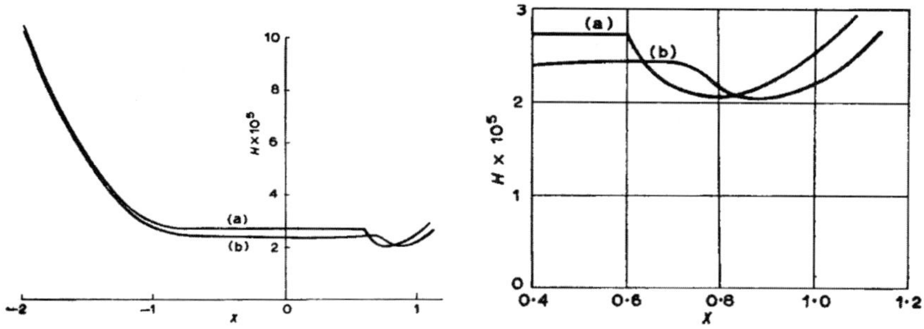

(I) (II)

Figure 6. Conjunction film shapes (I) and details of film shapes at outlet (II). $W = 3 \times 10^{-5}$; $G = 5000$; $U = 10^{-11}$. (a) Incompressible lubricant. (b) Compressible lubricant.

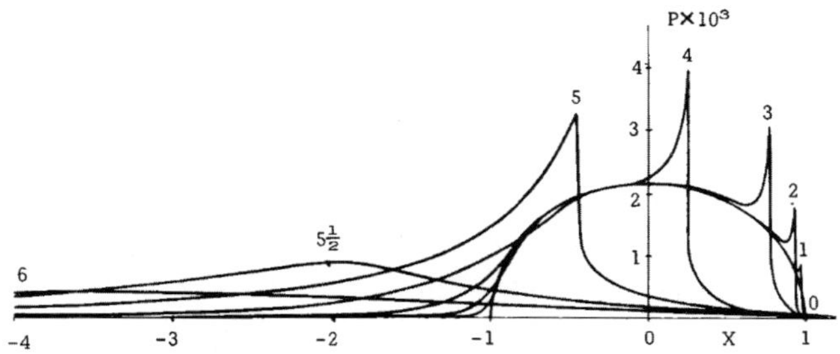

Figure 7. Pressure distributions for a compressible lubricant. $U =$ (0) dry contact, (1) 10^{-13}, (2) 10^{-12}, (3) 10^{-11}, (4) 10^{-10}, (5) 10^{-9}, ($5\frac{1}{2}$) $10^{-8.5}$, (6) 10^{-8}). $W = 3 \times 10^{-5}$; $G = 5000$.

was, however, altered appreciably, especially the height and location of the Petrusevich pressure spike, giving it a more feasible appearance (Figure 7).

In 1963 the Institution of Mechanical Engineers ran a major Symposium on Fatigue in rolling contact; a topic then, as now, of some interest and importance. Our paper "Stress distribution in lubricated rolling contacts" (Dowson et al., 1964) appeared in the Symposium Proceedings. Attention was focused upon the stress distributions in EHL line contacts and contours of maximum shear stress in the solids as they passed through the loaded conjunction were recorded. Particular attention was drawn to the modification to the Hertzian stress distribution, then extensively used in design and analysis of many highly stressed machine components such as gears, rolling bearings and cams-followers, by the presence of an EHL film of lubricant. In the symposium itself many variables were examined in addition to lubrication and eventually much of the mystery and inconsistency in practical performance was removed by the use of vacuum remelting of steels.

Once the film thickness formula (1) had been established with fair support from experimental studies carried out elsewhere, attention was increasingly focused upon its applications to established machine components.

The last paper we wrote together before GRH left Leeds in 1962, just six years after we started work on the EHL problem, was entitled "Theory of roller-bearing lubrication and deformation". It was presented in the highly successful series of annual Conventions arranged by the Lubrication and Wear Group of the Institution of Mechanical Engineers. A number of these meetings were held at seaside resorts on the south coast of England and our paper was presented at the 1963 Convention held in Bournemouth (Dowson and Higginson, 1964). The behaviour of cylindrical roller bearings with variable radial clearance and a range of loads and speeds was examined theoretically and, for the first time, by incorporating elastohydrodynamic lubrication concepts. Apart from the initial displacement the stiffness was found to be only slightly affected by radial clearance, but the stiffness was increased by small interferences. The EHL analysis showed that under substantial load the motion of the rollers was epicyclic, even when a significant film separated the solids.

Our next "applications" paper addressed the long-standing problem of gear lubrication. The Mechanical Tests of Lubricants Panel of the Institute of Petroleum held its first Symposium on the general subject of gear lubrication in London in 1952. During the next decade there was much pressure to increase specific loadings on gear trains and a growing recognition that this could not be achieved without full attention to lubricant development and the incorporation of new understanding of fundamental lubrication mechanisms into gear design and operation.

The second Gear Lubrication Symposium was held in Brighton in 1964, with the Proceedings being published in 1966. In this we presented a paper on "A theory of involute gear lubrication" (Dowson and Higgginson, 1966) in which guidance was offered to gear designers on the calculation of EHL film thicknesses in spur gears. Carpet graphs enabled (hmin) to be readily determined for given wheel speeds and centre distances at a gear ratio of unity, a load of 1 ton/inch and a viscosity of 0.75 poise. Graphs of correction factors for alternative loads, gear ratios and lubricant viscosities were also offered.

A further jointly authored paper was published eight years after the hectic period of working together in Leeds. This was prepared on the occasion of the award of the British Society of Rheology Gold Medal. This comprehensive paper (Dowson and Higginson, 1970) represented an interpretation of the state of EHL some sixteen years after our joint work on the subject began.

4. EXPERIMENTAL CONFIRMATION OF BASIC FEATURES OF EHL LINE CONTACTS

It was most exciting to work on EHL when the initial comparative theoretical and experimental data was emerging. Such comparisons emerged from many sources, particularly during the 1960s, but three deserve special mention.

The general and encouraging agreement between the predicted film thickness and experimental measurements based upon capacitance (Crook, 1961a) and X-ray transmission techniques (Sibley and Orcutt, 1961), which emerged during the period of our studies has been shown in Figure 4.

Crook determined by capacitance measurements not only the mean film thickness in EHL line contacts, but also the film shape (Crook, 1961b). He inserted a central glass disc in his remarkable four disc machine and evaporated a chromium electrode upon its surface. As the electrode traversed the EHL conjunction an outstanding portrayal of the predicted outlet restriction emerged to confirm one of the most distinctive features of line contact EHL (Figure 8). The film thickness reduction in the outlet constriction was about 10% of the central film thickness and the slope in the centre of the conjunction about 0.03 degree.

The first Symposium on Elastohydrodynamic Lubrication was organized by the Lubrication and Wear Group of the Institution of Mechanical Engineers and held in Leeds in September 1965. It turned out to be a landmark event in the development of EHL studies. Two adjacent papers by Dyson et al. (1966) and Kannel (1966) attracted much attention. In the former, interpretation of capacitance measurements on a two disc machine enabled convincing correlation with theoretical predictions to be demonstrated for a range of lubricants. The agreement for mineral oils was particularly impressive and the results for a medium viscosity-index mineral oil are shown in Figure 9.

In the following paper, Kannell (1966) presented the first experimental confirmation of the second remarkable feature of EHL line contacts: the Petrusevich pressure spike. A thin narrow strip of manganin was evaporated onto the surface of one of two quartz discs mounted in a two disc machine.

The change in resistance of the 0.002 in (50 μm) wide manganin strip as it traversed the EHL conjunction was related to the film pressure to yield traces of the form shown in Figure 10. The distinctive rise in pressure in the outlet zones was the first experimental evidence of the pressure concentrations predicted theoretically (Petrusevich, 1951; Dowson and Higginson, 1960). The experimental measurements were, of course, much less spectacular than the theoretical predictions, since the manganin strip was a few times wider than the expected width of the pressure spike. As Kanel (1966) noted "... *since the pressures detected by the transducer are actually the average pressures over*

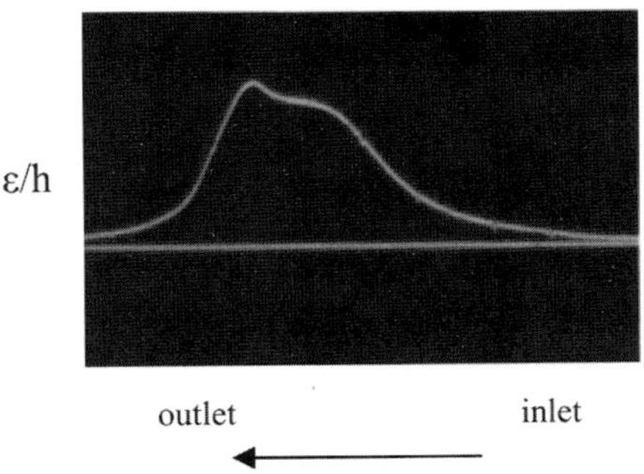

Figure 8. Crook's (1961b) oscilloscope trace of electrical potential across an EHL film (*h* film thickness; *e* dielectric constant for lubricant).

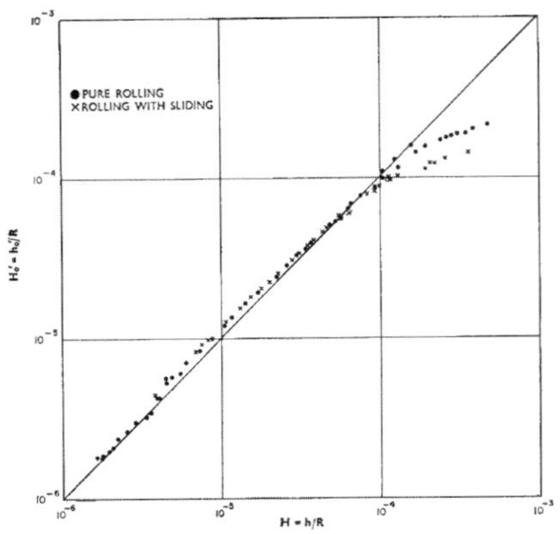

Figure 9. Dyson et al.'s (1966) comparison of measured non-dimensional film thickness with predicted values ($H = h/R$). (Medium viscosity-index mineral oil to OM100 specification).

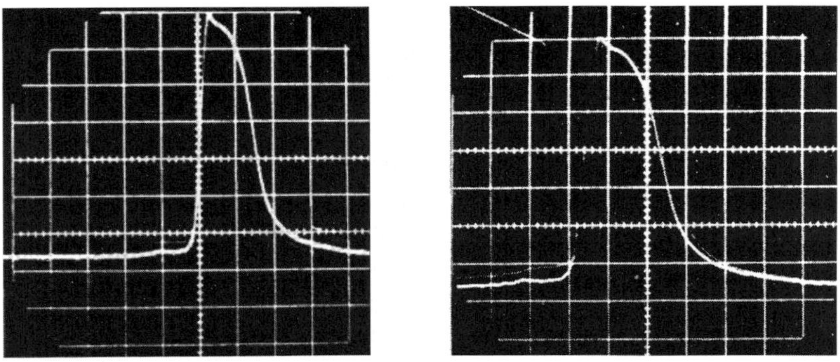

Figure 10. Kannel's (1966) pressure measurements using manganin strip transducers.

the width of the transducer, any irregularity in the data, such as a pressure spike, would be expected to be smoothed".

If the agreement between measured and predicted film thicknesses (Crook, 1961a; Sibley and Orcutt, 1961; Dyson et al., 1966) was most encouraging, Crook's (1961b) experimental film shape caused a flutter of the heart, while Kannel's pressure profiles amounted to a mystical experience.

A fourth experimental study reported in the early 1960s was particularly significant, since it related failure by pitting to film thickness. Dawson (1962) found convincing evidence that such failure was related to the ratio of surface roughness to theoretical EHL film thickness. This significant observation by Peter Dawson not only enhanced the value of EHL film thickness predictions, but also linked lubrication to surface failure in highly stressed machine elements in a most direct manner.

5. SHORT COURSE AND BOOK ON ELASTOHYDRODYNAMIC LUBRICATION

5.1 Short Course (1962)

Once a numerical procedure for the solution of line contact problems had been established, attention was focused upon the physical explanation of the emerging characteristics of EHL conjunctions. As the distinctive roles of the three dimensionless parameters (G, U, W) upon the dimensionless film thickness (H) emerged, machine designers, manufacturers and plant operators showed no reluctance to become familiar with the practical implications and the potential of this newly revealed aspect of fluid film lubrication. It was therefore decided, early in the 1960s, to prepare and present a short course of lectures on the subject. In later years short courses related to research activities became an established and almost essential activity for University Departments, but forty years ago it was still something of a risk and a novelty. The Short Course extended over two and a half days from Monday 26th to Wednesday 28th March 1962. The Course Fee was £5 and accommodation in one of the new University Halls of Residence cost about £1–5s (£1.25p) for one day and night. The authors shared nine of the ten hour-long lectures and were pleased to receive Dr. A.W. Crook of the AEI Research Laboratories, Aldermaston Court, to present an account of his impressive experimental investigations and confirmation of the essential features of EHL film shape.

Examination of the set of notes provided for the delegates shows that they were typed on foolscap sheets, with all equations being inserted in distinctive hand writing! A Notation was provided together with a list of 39 references.

The attendance list makes interesting reading. There were 84 registered delegates, all but three (H. Christensen, Norway; G.G. Hirs and F.H. Thyse, The Netherlands) being from the U.K. The Course was also attended by 13 members of the Department of Mechanical Engineering from the home University and there were three course lecturers. A pleasing feature of the list of 84 registered delegates was that 70% were from industry, with the rolling element bearing and oil companies and leading manufacturers from the aircraft, engine, textile, power generating and motor car industries being well represented. Many of the delegates subsequently played major roles in the development of EHL studies and tribology in general.

5.2 Book on Elastohydrodynamic Lubrication (1966)

Notes from the 1962 short course provided the foundation for a research monograph (Dowson and Higginson, 1966). When invited to publish this book the notes were enlarged considerably and the visiting lecturer on the short course, Jim Crook, and his colleague at AEI, Jack Archard readily agreed to prepare chapters on experimental work dealing essentially with "film thickness and film shape" and "friction and viscosity" respectively. The division of labour was not as neat as it appeared, since although the four authors were working in two pairs at Leeds University and AEI at the time the work was undertaken, at the time of writing they were in four different places.

The 235 page book was published by the Applied Mechanics Division of the Commonwealth and International Library, of the publisher Robert Maxwell's Pergamon Press. It was widely used and referenced and requests for copies are still received! The initial volume adopted the contemporary British yard-pound-second system of units, but long after it went out of print a second edition in SI units was published (Dowson and Higginson, 1977). A translation appeared in China in 1982.

6. CLOSURE

In presenting this story some of the background to early work on EHL in the U.K. almost half a century ago, or as much of it as we can remember, has been recalled. While aware of the general significance of the subject, we certainly did not envisage that the topic would prove to be so central to studies of lubrication into the 21st century. Our joint work in Leeds was carried out between 1956 and 1962 and it is interesting to recall that this was prior to the formation of the Jost Committee and four years before the publication of the Report establishing the "new" word tribology.

Several international symposia on elastohydrodynamics have been arranged over the past fifty years and the present event demonstrates that there is continuing activity and interest in the subject. There have been too many significant developments for us to outline accounts of progress in any detail, but some headings of the major fields of endeavor illustrate the trends.

- development of robust/ rapid numerical procedures for EHL problems;
- solutions for circular and elliptical point contacts;
- non-steady state conditions;
- development of experimental techniques (interferometry);
- study of very thin films (nm); molecular dynamics;
- lubricant rheology;

- thermal effects (on lubricants and solids);
- micro-EHL (asperity lubrication for "soft", "hard" and layered solids);
- friction and traction (slip; spin losses);
- analysis of real (rough) surfaces;
- integration with mixed lubrication studies;
- starvation and cavitation;
- analysis of engineering machine elements (e.g. cams and followers; piston rings; cvt's; gears; bearings);
- bio-tribology (biological systems-eyes; synovial joints; cardiovascular flows and prostheses- total replacement synovial joints).

The reader will judge, however, whether attention to the transfer of understanding of the amazing features of EHL to industry and the development of guidance for machine designers deserve as much priority today as they did fifty years ago.

ACKNOWLEDGEMENTS

The authors would like to express their appreciation to all their colleagues who have contributed to the development of EHL and promoted stimulating discussions.

They are grateful to Professional Engineering Publications for permission to reproduce Figures 1–7; 9 and 10 and to *Nature* for permission to reproduce Figure 8.

APPENDIX

Table 1. Joint publications by D. Dowson and G.R. Higginson.

Paper (No.)	Year	Title	Authors	Journal
1	1959	A numerical solution to the elasto-hydrodynamic problem	DD & GRH	*J.M.E.S.* Vol. 1 No. 1
2	1960	The effect of material properties on the lubrication of elastic rollers	DD & GRH	*J.M.E.S.* Vol. 2, No. 3
3	1961	New roller-bearing lubrication formula	DD & GRH	*Engineering* 4 August, No. 192
4	1962	Elasto-hydrodynamic lubrication: a survey of isothermal solutions	DD, GRH & AVW	*J.M.E.S.* Vol. 4, No. 2

Paper (No.)	Year	Title	Authors	Journal
5	1962	The lubrication of elastic rollers	DD & GRH	*Proceedings Tenth International Congress of Applied Mechanics*
6	1963	Stress distribution in lubricated rolling contacts	DD, GRH & AVW	*I. Mech. E. Proc. Symposium on Rolling Contact Fatigue*
7	1964	Theory of roller-bearing lubrication and deformation	DD & GRH	*I. Mech. E. Proc. Lubrication and Wear Convention*, Bournemouth
8	1966	A theory of involute gear lubrication	DD & GRH	*Institute of Petroleum. Proc. Symposium on Gear Lubrication*, Brighton
9	1966	Elasto-hydrodynamic lubrication: The fundamentals of roller and gear lubrication	DD & GRH	(Ch. 9, Crook; Ch. 10, Archard) Book, Pergamon Press, pp. 1–235
10	1970	The role of lubricant rheology in engineering applications of elastohydrodynamic lubrication	DD & GRH	British Society of Rheology
11	1977	Elasto-hydrodynamic lubrication: The fundamentals of roller and gear lubrication (SI Edition)	DD & GRH	(Ch. 9, Crook; Ch. 10, Archard), Book, Pergamon Press, pp. 1–235

REFERENCES

Barwell, F.T., 1970, The founder of modern tribology, in *Osborne Reynolds and Engineering Science Today*, D.M. McDowell and J.D. Jackson (eds), Manchester University Press, pp. 240–263.

Cameron, A., 1985, Righting a 40-year-old wrong, *Tribology International*, **18**(2), 92.

Crook, A.W. 1958, The lubrication of rollers I, *Phil. Trans. Roy. Soc.*, **A250**, 387–409.

Crook, A.W., 1961a, The lubrication of rollers II, Film thickness with relation to viscosity and speed, *Phil. Trans. Roy. Soc.*, **A250**, 223–226.

Crook, A.W., 1961b, Elastohydrodynamic lubrication of rollers, *Nature*, 190, 1182.

Dawson, P.H., 1962, Effect of metallic contact on the pitting of lubricated rolling surfaces, *J. Mech. Engng. Sci.*, **4**(1), 16–21.

Dowson, D., 1968, Elastohydrodynamics, *Proc. Instn. Mechn. Engrs.*, **182**, 151–167.

Dowson, D. and Higginson, G.R., 1959, A numerical solution to the elasto-hydrodynamic problem, *J. Mech. Engng. Sci.*, **1**(1), 6–15.

Dowson, D. and Higginson, G.R., 1960, The effect of material properties on the lubrication of elastic rollers, *J. Mech. Engng. Sci.*, **2**(3), 188–194.

Dowson, D. and Higginson, G.R., 1961, New roller-bearing lubrication formula, *Engineering*, **192**(4972), 158–159.

Dowson, D. and Higginson, G.R., 1962, The lubrication of elastic rollers, in *Applied Mechanics, Proc. Tenth Intl. Congress of Appl. Math.*, F. Rolla and W.T. Koiter (eds), Elsevier Publishing Company, Amsterdam, pp. 137–139.

Dowson, D. and Higginson, G.R., 1964, Theory of roller bearing lubrication and deformation, in *Instn. Mechn. Engrs., Proc. 1963 Lubrication and Wear and Wear Group Convention*, Bournemouth, pp. 216–227.

Dowson, D. and Higginson, G.R., 1966, A theory of involute gear lubrication, in *The Institute of Petroleum, Proceedings of a Symposium on Gear Lubrication-1964*, Brighton, The Elsevier Publishing Company, Amsterdam, pp. II8–II15.

Dowson, D. and Higginson, G.R., 1966, *Elasto-hydrodynamic Lubrication; The Fundamentals of Roller and Gear Lubrication*, Pergamon Press, Oxford, pp. 1–235.

Dowson, D and Higginson, G R., 1970, The role of lubricant rheology in engineering applications of elastohydrodynamic lubrication, *British Society of Rheology*, Bulletin No. 4, Vol. 12, (Supplement), 1–22.

Dowson, D. and Higginson, G.R., 1977, *Elasto-hydrodynamic Lubrication SI Edition*, Pergamon Press, Oxford, pp. 1–235.

Dowson, D., Higginson, G.R. and Whitaker, A.V., 1962, Elastohydrodynamic lubrication: A survey of isothermal solutions, *J. Mechn. Engng. Sci.*, **4**(2), 121–126.

Dowson, D., Higginson, G.R. and Whitaker, A.V., 1964, Stress distribution in lubricated rolling contacts, in *Instn. Mechn. Engrs., Proc. Symposium on Fatigue in Rolling Contact*, pp. 66–75.

Dyson, A., Naylor, H. and Wilson, A.R., 1966, The measurement of oil-film thickness in elastohydrodynamic contacts, in *Proc. Instn. Mechn. Engrs., 1965–1966*, 180, Part 3B, "Elastohydrodynamic Lubrication", pp. 119–134.

Ertel, A.M., 1945, *Die Berechnung der hydrodynamischen Schmierung gekrummter oberflaschen unter hoher Belastung und Relativbewegung*, Fortschr. Ber. VDIZ, Reihe 1, No. 115, 1984.

Grubin, A.N., 1949, Fundamentals of the hydrodynamic theory of lubrication of heavily loaded cylindrical surfaces, in *Proceedings of Symposium on "Investigation of the Contact of Machine Components"*, by A.N. Grubin and I.E. Vinogradova, edited by Kh.F. Ketova, Central Scientific Research Institute for Technology and Mechanical Engineering (TsNIITMASh), Book No. 30, Moscow, pp. 115–166 (DSIR Translation).

Kannel, J.W., (1966), Measurements of pressures in rolling contact, in *Proc. Instn. Mechn. Engrs., 1965–1966*, 180, Part 3B, "Elastohydrodynamic Lubrication", pp. 135–142.

Martin, H.M., 1916, Lubrication of gear teeth, *Engineering* (London), **102**, 119–121.

Petrusevich, A.I., 1951, Fundamental conclusions from the contact-hydrodynamic theory of lubrication, *Izv. Uzbekist, Fil. Acad. Nauk. SSSR (OTN)*, **2**, 209–223.

Reynolds, O., 1886, On the theory of lubrication and its application to Mr. Beauchamp Tower's experiments, including an experimental determination of the viscosity of olive oil, *Phil. Trans. R. Soc.*, **177**, 157–234.

Sibley, L.B. and Orcutt, F.K., 1961, Elastohydrodynamic lubrication of rolling-contact surfaces, *ASLE Transactions*, **4**(2), 234–245.

RHEOLOGICAL CHALLENGES AND OPPORTUNITIES FOR EHL

Scott Bair[1] and Peter Gordon[2]
[1]*Center for High Pressure Rheology, George W. Woodruff School of Mechanical Engineering, Georgia Institute of Technology, Atlanta, GA 30332-0405, U.S.A.*
[2]*Corporate Strategic Research, ExxonMobil Research and Engineering Company, Annandale, NJ 08801, U.S.A.*

Abstract There can be no doubt that the assumptions used in modeling elastohydrodynamics for the temperature, pressure and shear rate dependence of viscosity are at odds with empirical fact. As a result, there has been relatively little progress since the classic Newtonian film thickness solutions toward relating film thickness and traction to the properties of individual liquid lubricants and it is not clear at this time that full numerical solutions can even be obtained for heavily loaded contacts using accurate models. One central issue is the validity of Reynolds equation, derived under the isoviscous assumption, for conditions where the local pressure-viscosity coefficient can approach 100 GPa^{-1}. Some pressing problems are reviewed in this paper, including the effects of shear-thinning on film thickness and traction, and the proper definition of the pressure-viscosity coefficient for film thickness calculations. Wherever possible, for credibility, sources from outside of Tribology will be used. Then some opportunities for the field will be discussed. These opportunities result from advances in molecular dynamics simulations and improved techniques for high-pressure measurements and should shed light on the relationship between molecular structure and performance in contacts.

Keywords: elastohydrodynamic, pressure-viscosity coefficient, shear-thinning, molecular dynamics, non-Newtonian, traction, film thickness.

1. INTRODUCTION

Even a casual survey of the rheology literature from outside of the field of tribology should lead to the conclusion that the assumed temperature, pressure and shear rate dependence of viscosity used in the study of elastohydrodynamic lubrication, EHL, cannot be supported by empirical measurement. The local values of the temperature-viscosity coefficient and the pressure-viscosity coefficient begin to increase rapidly with pressure as the glass transition is ap-

R.W. Snidle and H.P. Evans (eds), IUTAM Symposium on Elastohydrodynamics and Microelastohydrodynamics, 23–43.

proached from lower pressure. These features are generally ignored in the EHL rheological description of liquids. Compared to EHL the journal bearing community seems to have a more accurate description of shear-thinning. Where viscosity depends upon shear rate, there is a substantial interval of shear rate for which the viscosity varies with shear rate raised to a fixed exponent and often a second regime of nearly constant viscosity appears. This paper challenges the EHL field to assume accurate rheological models and property relations so that the effects of the individual rheological properties of different liquids on lubricated contact performance may be understood. Then it should be possible to capitalize on improvements in molecular dynamics modeling and high pressure viscometry to explain the differences in traction and film thickness responses of the different chemical structures of base oils and additives. It would be unfortunate if these opportunities were ignored.

2. RHEOLOGICAL CHALLENGES FOR EHL

Some apparently unresolved issues regarding the descriptions used in EHL of the effect of pressure and shear on viscosity are outlined in this section.

2.1 The Pressure-Viscosity Coefficient

The classical Newtonian film thickness solution [1] requires as a rheological description of the liquid, the ambient viscosity, μ_0, and a pressure-viscosity coefficient, α, that was originally defined as the coefficient multiplying the pressure in a purely exponential pressure-viscosity relation [1]. For the liquids typically used as lubricants, such a relation is accurate for ordinary temperatures only at high pressure near the inflection in the log viscosity versus pressure curve. Various definitions, then, of the pressure viscosity coefficient have evolved to quantify the pressure-viscosity response of lubricants within the relatively low pressure EHL inlet zone where the film thickness is established.

The authors regard the definition by Blok [2] as the most accurate for the calculation of film thickness for general pressure-viscosity response. His reciprocal asymptotic isoviscous pressure is defined [2] as

$$\alpha^* = \left[\int_0^\infty \frac{\mu(0)dp}{\mu(p)}\right]^{-1}. \tag{1}$$

Another common definition is

$$\alpha_0 = \frac{1}{\mu_0}\left.\frac{d\mu}{dp}\right|_{p=0}. \tag{2}$$

To obtain an estimate of α_0, many use the Roelands definition [3] of α_0,

$$\alpha_R(p) = \frac{z\ell n(\mu_0/\mu_R)}{p_R}, \tag{3}$$

Table 1. Viscosity and pressure-viscosity coefficients for a diester at 218°C.

p/MPa	μ/mPa·s	z	α_R/GPa^{-1}	α_B/GPa^{-1}
0.1	0.734	–	–	–
68	1.453	0.824	10.3	10.0
142	2.40	0.738	9.24	8.34
202	3.473	–	–	–
268	4.758	0.678	8.49	6.97
335	6.89	–	–	–
423	9.08	–	–	–
549	14.43	0.618	7.74	5.43
690	23.97	0.603	7.55	5.05

$\alpha^* = 6.90$ GPa^{-1}, $\alpha_0 = 11.6$ GPa^{-1}.

which has been described [3] as the "conventional pressure viscosity coefficient". Equation (3) requires that the viscosity up to some pressure, p, be fitted to the isothermal Roelands equation to find z, a dimensionless pressure-viscosity index. Yet another definition is a kind of secant coefficient,

$$\alpha_B(p) = \frac{\ell n(\mu(p)/\mu(0))}{p}, \tag{4}$$

where p has assumed values as great as 0.545 GPa [4].

For pure exponential response, the four definitions above are equivalent. The greatest differences for typical lubricants occur at high temperature. See for example the viscosity and pressure-viscosity coefficients in Table 1 for a diester, di (2-ethyl-hexyl) sebacate. The viscosity values were obtained from the 1953 ASME Pressure Viscosity Report [5]. These data were repeated at Georgia Tech at four elevated pressures to with ±3% and should be considered accurate. In addition, the data of Table 1 are very similar to the viscosity of a jet oil meeting specifications of Mil-L7808, measured in this laboratory at 220°C so that the conclusions reached here are very relevant to film forming in high temperature gas turbine bearings. For the evaluation of the integral in equation (1), a pressure of 549 MPa was sufficient to calculate α^* using a previously published algorithm [6]. The extremes of the values of the different pressure-viscosity coefficients in Table 1 vary by more than two-to-one, the highest being α_0 and the lowest being α_B at high pressure. The Roelands pressure-viscosity coefficient, α_R, varies from 10.3 to 7.6 GPa^{-1} as the pressure range is increased due to the inaccuracy of the Roelands model. It is interesting that this diester was one of the liquids investigated by Roelands [7].

Since all of the definitions used to calculate the pressure-viscosity coefficients in Table 1 are in current use, several interesting questions arise:

1. This flexibility in defining α calls into question the accuracy with which film thickness is being calculated today. Are these definitions being ma-

nipulated to make inaccurate calculations appear more accurate or to adjust for the neglect of ordinary shear-thinning?

2. Can an analysis based on general pressure-viscosity and Newtonian behavior lead to a single definition of α that can be accepted by the entire EHL community?

The failure of α_0 to serve as a universal measure of the strength of the pressure-viscosity response for film forming can be illustrated by referring to the viscosity measurements in Figure 1 for a polyol ester at 100°C and a polyphenyl ether at 83°C. These temperatures were selected because they result in values of μ_0 that are the same within experimental error ($\pm 3\%$). Interestingly, any difference in the values of α_0, the initial slope, is imperceptible as can be seen from the inset graph in Figure 1 where measurements were obtained at pressures of 0.1, 12, 25 and 50 MPa. The broken curve is Roelands' model extrapolated from α_0. If α_0 were the operative definition of pressure-viscosity coefficient, then the classical Newtonian formulae would yield the same prediction of film thickness for these two very different liquids. The film thicknesses have been measured in this laboratory [6] and the polyphenyl ether generated about 30% thicker films than the polyol ester at temperatures of Figure 1. Clearly, the definition of equation (2) is insufficient to quantify the pressure-viscosity effect for the purpose of film thickness prediction.

2.2 General Pressure-Viscosity Response at High Pressure

Little needs to be said concerning this challenge to EHL. The view of the EHL community regarding the pressure variation of low shear viscosity seems to have been that the exponential pressure dependence decreases with pressure [8]. This view is squarely at odds with measurements, some of which have been summarized in the physics literature as shown in Figure 2, reproduced from Brazhkin and Lyapin [9], in which the response is always faster than exponential at high pressure. This discrepancy between the assumed pressure-viscosity behavior and empirical reality must be central to the lack of progress in modeling traction behavior.

Another challenge regarding pressure-viscosity response aside from the numerical stability of solutions to Reynolds equation, is the validation of the Reynolds equation, derived from the isoviscous assumption, for conditions where the local pressure viscosity coefficient, α_L,

$$\alpha_L = \frac{1}{\mu}\frac{\partial u}{\partial p} \tag{5}$$

can reach values [10] near 100 GPa^{-1}. The viscosity gradients that are neglected in the Reynolds derivation must become significant at high pressure.

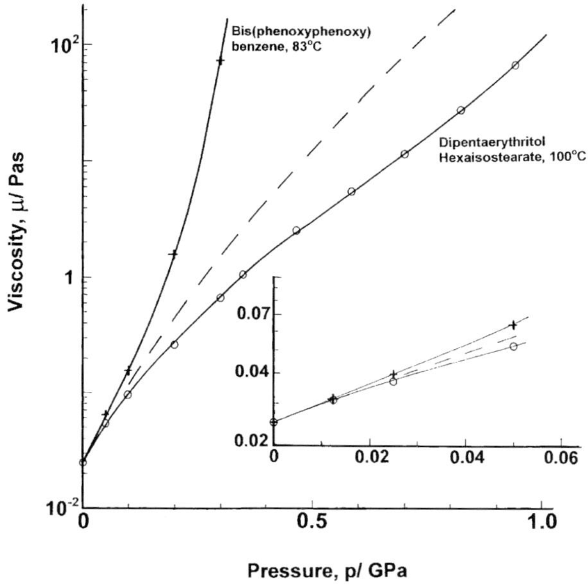

Figure 1. The low shear viscosity as a function of pressure for a polyol ester at 100°C and a polyphenyl ether at 83°C.

Renardy [11] showed that the Navier–Stokes equations for an incompressible piezoviscous Newtonian liquid may undergo a change of character for principal shear rate equal to $1/(\mu\alpha_L)$. Bair and co-workers [12] were apparently first to suggest that the Reynolds equation accurately describes the flow of liquids with pressure dependent viscosity only when the product of the shear stress and α_L are much less than one.

This limitation for the Reynolds equation was investigated further by Schafer et al. [13] and Almqvist and Larsson [14] and confirmed by Rajagopal and Szeri [15]. Bair et al. [12] were also apparently first to discover that unidirectional flow in incompressible liquids is not possible for exponentially pressure dependent viscosity; secondary flows develop. This result was confirmed by Hron et al. [16]. These secondary flows that seem to develop in response to emerging cross-film pressure gradients [17] should be important to EHL because they provide a basis for the nucleation of the shear bands that have been associated with the limiting stress effect [17]. Comparison of solutions of the Navier–Stokes equations with solutions of Reynolds equation for EHL conditions would provide a validation of the Reynolds equation for certain limits of liquid behavior; however, to date, the only such comparisons [13–15] have invoked the same unrealistic pressure-viscosity response that has become commonplace for numerical simulations of EHL.

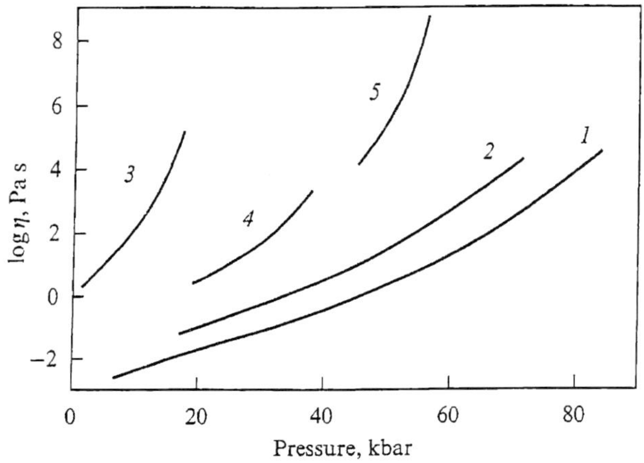

Figure 2. A review of the pressure dependence of viscosity for organic liquids showing faster than exponential response for: 1. methanol; 2. mixture of methanol and ethanol; 3. toluene; 4. butyl chloride and 5. ethyl ether. Reproduced from Brazhkin and Lyapin, *Physics Uspekhi* **43**(5), 2000, 493–508.

2.3 Shear-Thinning

For the journal bearing community, the shear-thinning model often used for multigrade motor oil is the Cross equation [18] written for viscosity as a function of shear rate, $\dot{\gamma}$, as

$$\eta = \mu_2 + (\mu - \mu_2)[1 + (\lambda\dot{\gamma})^{1-n}]^{-1}, \tag{6}$$

where μ_2 is the viscosity of a second Newtonian regime, λ is a characteristic time and n is the power-law exponent. There are many models [19] with properties similar to equation (6). As $\dot{\gamma} \to 0$, $\eta \to \mu$ and if $\mu_2 = 0$ or $\mu_2 \ll \mu$, a well-defined regime exists for which η varies with $\dot{\gamma}^{n-1}$. EHL lubricants are often multigrade motor oils and under EHL conditions the film thickness is generally less than predicted by Newtonian formulae. The pressure developed in a journal bearing film may not exceed 100 MPa or so, but the high-shear rheology of journal bearing lubrication is related to the high-pressure rheology of EHL as shown by Figure 3. Flow curves for two experimental 5W-30 oils are plotted in Figure 3 from TBS (Tapered Bearing Simulator) viscometer data given in [18]. This viscometer is utilized to certify the high-temperature, high-shear requirement of the J300 standard for motor oils. Data were obtained at Georgia Tech in a pressurized Couette viscometer for a commercial 5W-30 motor oil [19] at 20°C and $p = 350$ MPa and these data were time-temperature-pressure shifted using a published [20] shifting rule to obtain the flow curve in Figure 3. Clearly, the ordinary shear-thinning behavior of multigrade oils that has been observed in high-pressure viscometers is the same behavior that

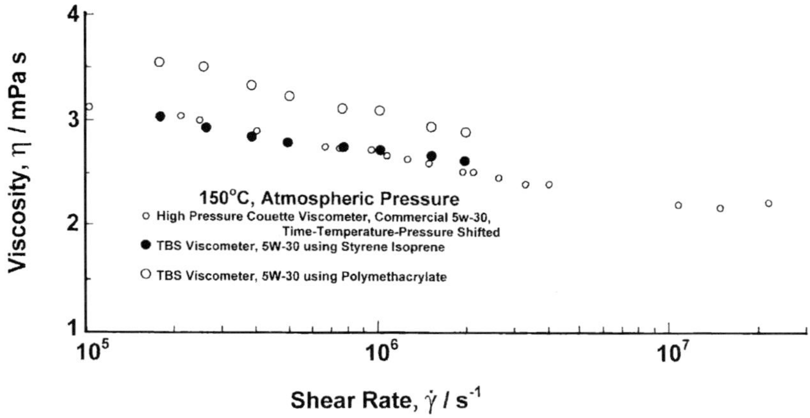

Figure 3. The viscosity as a function of shear rate for two experimental 5W-30 oils using a commercial TBS viscometer [18] compared to shifted data for a commercial 5W-30 at 20°C and 350 MPa [19] using a pressurized Couette viscometer.

is being routinely measured in high-temperature, high-shear viscometers at atmospheric pressure. Much data must be available from these routine measurements that may be used to calculate EHL film thickness and traction. Of course, this requires an accurate model such as equation (6) and possibly a molecular degradation model.

A standard reference material for shear-thinning is available from the National Institute of Standards and Technology (NIST). For this material, SRM 2490, the viscosity is certified [21] at atmospheric pressure for various shear rates and the parameters and a time-temperature shifting rule for the Cross equation (6) are provided [21]. The certified viscosity is shown in Figure 4 as the solid points and the Cross equation (6) with parameters provided with the certificate [21] is plotted as the curves for $\mu_2 = 0$. The parameters are $\lambda = 0.234$ s, $\mu_R = 100.2$ Pa·s and $n = 0.195$ and μ_R is a reference value of μ at $p = 0$, $T = 25$°C. The shifted form of the Cross equation is given [21] as

$$\eta = \left(\frac{T\rho}{T_R\rho_R}\right)\left\{\frac{\mu_R a(T)}{1 + [a(T)\lambda\dot{\gamma}]^{1-n}}\right\}, \tag{7}$$

where T_R and ρ_R are temperature and mass density at 25°C and $p = 0$ and $a(T)$ is a temperature shift factor that takes the WLF form [21] and is equal to one at $T = 25$°C. Equation (7) with the NIST parameters is plotted as the atmospheric pressure curve in Figure 4 for $\mu_2 = 0$.

The open circle data points in Figure 4 for shear rate $\dot{\gamma} > 10^2\ \text{s}^{-1}$ were obtained with the same pressurized Couette viscometer used to obtain the shifted high-pressure data in Figure 3. Both ambient and high pressure viscosities at 25°C were measured. The open circle data points for $\dot{\gamma} < 1\ \text{s}^{-1}$ were obtained

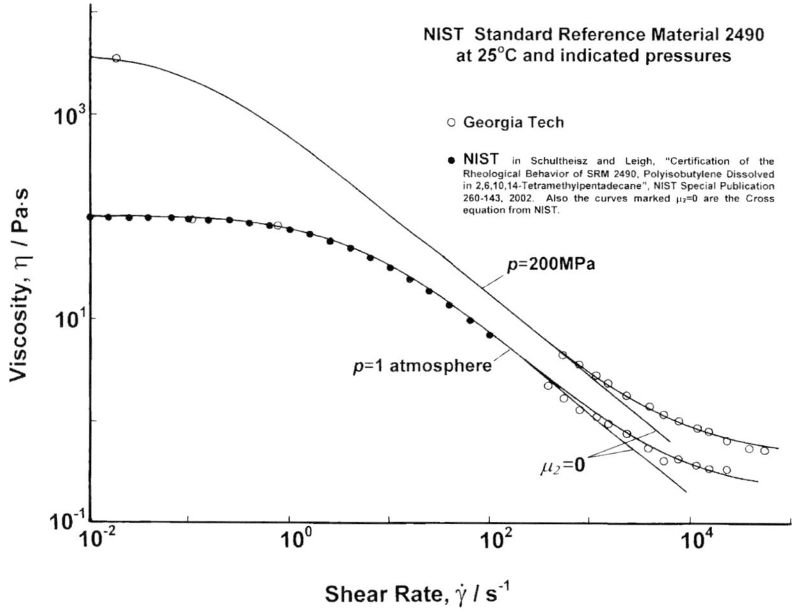

Figure 4. The certified viscosity of a standard reference material compared with measurements in a pressurized Couette viscometer.

with a falling body viscometer using two sinkers that apply a shear stress, τ, of 10 or 65 Pa.

All elevated pressure falling body measurements here utilized $\tau = 65$ Pa and these data are accurately described by a Vogel-like equation,

$$\eta(T = 25°\text{C}, \tau = 65 \text{ Pa}) = \eta_0 \exp\left(\frac{Cp}{p - p_\infty}\right), \tag{8}$$

where $\eta_0 = 87.1$ Pa·s, $C = 18.4$ and $p_\infty = -789$ MPa. Although this liquid is slightly shear-thinning for $\tau = 65$ Pa, for constant stress, the viscosity at this low stress should vary approximately with pressure in the same way as the limiting low shear viscosity, μ. The shift factor, a, should vary with pressure and temperature in approximately the same way as μ. Then a pressure shift factor can be defined as

$$a(p) = \exp\left(\frac{Cp}{p - p_\infty}\right). \tag{9}$$

The relative density was estimated for $p = 200$ MPa from data for squalane [22] to be $\rho/\rho_R = 1.10$. Substituting $a(p)$ for $a(T)$ in equation (7) results in the elevated pressure curve marked $\mu_2 = 0$ in Figure 4. The agreement of the NIST data and model with the data generated in the pressurized Couette viscometer at the lowest shear rates is good.

At the higher shear rates and more significantly, the higher shear stress, generated in the pressurized Couette viscometer, a second Newtonian begins to appear as can be seen by the leveling off of the flow curves in Figure 4. A second Newtonian and pressure shifting is accommodated here in equation (7) by rewriting it as

$$\eta = \mu_2(p) + \frac{\left(\frac{\rho T}{\rho_R T_R}\right) a(p)\mu_R - \mu_2(p)}{1 + [a(p)\lambda\dot{\gamma}]^{1-n}}, \tag{10}$$

where $\mu_2(0) = 0.2$ Pa·s and $\mu_2(p = 200\text{ MPa}) = 0.5$ Pa·s. Curves for equation (10) at $p = 0$ and 200 MPa are plotted in Figure 4. Equation (9) is then a pressure-dependent, shear-thinning model for a standard reference material for which the constituents are readily available, polyisobutylene and pristane. Another challenge then is that any successful numerical analysis of non-Newtonian EHL should, of course, be able to predict the film thickness of this liquid and in-contact methods for measuring shear-thinning should be able to recover this behavior as long as there is no mechanical degradation of the polymer.

One last comment on shear-thinning is in order. The following equation is not the Ree–Eyring model for shear-thinning [23]:

$$\eta = \frac{\tau_0}{\dot{\gamma}} \sinh^{-1}\left(\frac{\mu\dot{\gamma}}{\tau_0}\right). \tag{11}$$

Eyring found this relation to be of only limited usefulness for the description of the stress-increasing, thixotropic behavior of wax containing lubricants under pressure [24]. In that case, μ is not the same quantity as the Newtonian or limiting low shear viscosity. Ree and Eyring [23] concluded that equation (11) is not an accurate description of the constitutive behavior of a shear-thinning liquid.

The effect of ordinary shear thinning upon film thickness may be subtle and traction may be dominated by limiting stress effects, however there are instances where shear-thinning is important and it is surprising that EHL modeling has ignored this behavior of liquids for so long.

3. RHEOLOGICAL OPPORTUNITIES IN EHL

Advances in molecular dynamics simulations and improvements in high pressure viscometry have uncovered details of lubricant rheology at extremes of pressure and shear stress that may be incorporated into EHL analyses to the benefit of the field.

3.1 Additives and Mixtures

The lubricants that are used for elastohydrodynamic contacts are seldom pure liquids. Not only are base oils often blends, but a significant additive concentration is usually present. It has already been shown that the detailed structure of liquid hydrocarbons can have an extraordinary effect upon the viscosity at pressures near one gigaPascal [25] even when the low pressure viscosities are similar. Since very high pressures discriminate easily among the viscosities of chemical structures, the viscosity at these pressures may be manipulated to advantage by mixing. This would provide an inexpensive means to improve performance of lubricants at high pressures.

The study of non-Newtonian response of mixtures has generally been restricted to polymer solutions in low molecular weight solvents with the non-Newtonian effects contributed by the polymer. Some practical rules for polymer solutions have been established and a useful summary has been provided by Van Krevelen [26]. Two recent papers [20, 27] have outlined some opportunities for the manipulation of high pressure rheological properties.

Small concentrations of an additive can change in a measurable way the viscosity at high pressure without affecting the ambient viscosity and pressure-viscosity coefficient. In Figure 5 the viscosity of a mineral-based tubine oil, Shell T9, is plotted with and without 2% by weight of a long-chain fatty acid, oleic acid.

The viscosity at $p = 1$ GPa is reduced to less than half by the additive. This effect may be explained by considering the plasticizing effect of a fatty acid in reducing the glass transition temperature, or in this case, raising the glass transition pressure. The pressure variation of low-shear viscosity is probably the most important lubricant property with respect to traction. It might be fruitful to consider the pressure-viscosity behavior shown in Figure 5 in a rough surface mixed EHL analysis, such as in Holmes et al. [28], since fatty acids and esters have long been used as boundary lubrication additives.

For most EHL applications, particularly for high speed, the "best" lubricant might be one which forms a thick film while providing low friction. Then a liquid with a strong pressure-viscosity response at low pressure, for film-forming, and a weak pressure-viscosity response at Hertz pressure levels would be optimum, all other things being equal. The additive effect shown in Figure 5, then makes the turbine oil a "better" lubricant. Blending of widely different molecular weight cuts can provide similar effects [27].

Additives can be used to modify shear-thinning behavior as well. Molecular dynamics simulations [29] have shown that mixtures of two monodisperse liquid hydrocarbons of similar but sufficiently different molecular weight can display two shear-thinning transitions, one for each component occurring at a shear stress characteristic of the component. This double shear-thinning

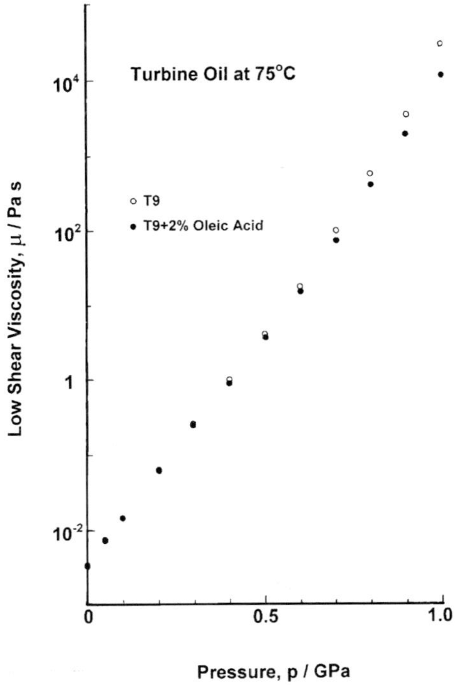

Figure 5. The effect of a fatty acid additive on the pressure dependence of viscosity for a turbine oil.

can be observed in the laboratory as well [27]. A polystyrene oligimer (PS) was added at 20% weight concentration to dibenzyl etheyl benzene (DBEB). The weight average and number average molecular weights of PS are 740 and 700 kg/kmole, respectively. The formula molecular weight of DBEB is 286 kg/kmole. Flow curves were obtained in a pressurized Couette viscometer and are plotted in Figure 6 as shear stress, τ, versus shear rate. There is a single transition in the data for DBEB at about $\tau = 6.5$ MPa and this stress is marked by the upper horizontal arrow in Figure 6. The mixture shows an additional transition marked by the lower arrow at $\tau = 1.5$ MPa. The mixture has two transitions, one for each component. If additives can modify the shear-thinning behavior in a predictable way, EHL simulations would be a guide to lubricant formulation.

3.2 Traction of Dry Contact: The Rheological Contribution of Steel

The modeling of traction has occupied a significant portion of the research effort within EHL. The isothermal, viscous regime of EHL traction is now

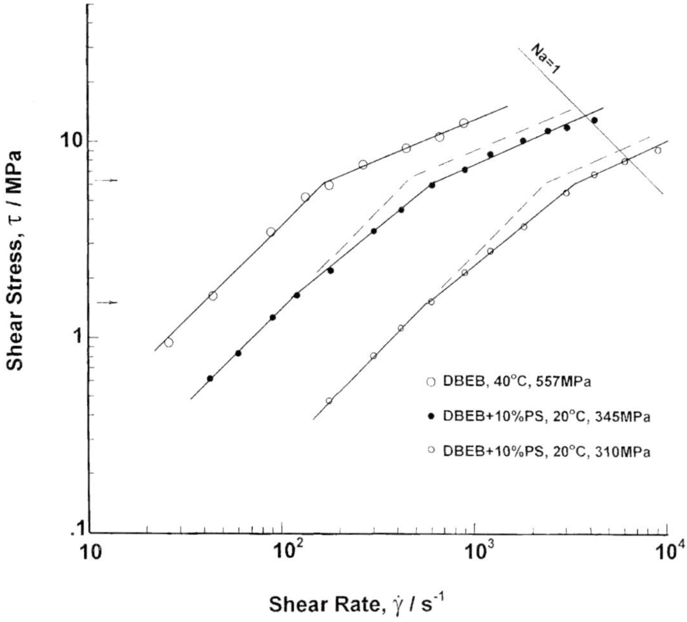

Figure 6. A flow curve for dibenzyl ethyl benzene (DBEB) and flow curves for DBEB with polystyrene (PS).

sufficiently well understood and viscometers are sufficiently advanced that accurate calculations of traction may be obtained from measured properties [22, 30] and even from molecular dynamics simulation [22].

These calculations require an accurate description of the pressure and shear rate dependence of viscosity and a limit to the shear stress that varies approximately in proportion to pressure. This limiting stress has been associated with localization of the shear deformation along intermittently operating shear bands inclined to the flow direction [17].

For very high pressures and very low slide-to-roll ratio, Σ, traction curves are better correlated with Σ than with the shear rate within the film. Since Σ is a measure of the total strain experienced by the film it has been natural to associate this regime of traction with elastic response of the film and make use of high-pressure, low Σ traction to extract values of shear modulus for the liquid. Historically, these values of shear modulus have been orders of magnitude lower than those measured by ultrasonics [31]. Johnson [32] has most thoroughly examined the difficulty of separating the elastic response of the liquid from the elastic response of the (usually steel) substrate. Recently one author (S.B.) has pointed out that unlubricated, dry traction curves have the same shape and characteristics of lubricated traction curves [33] making

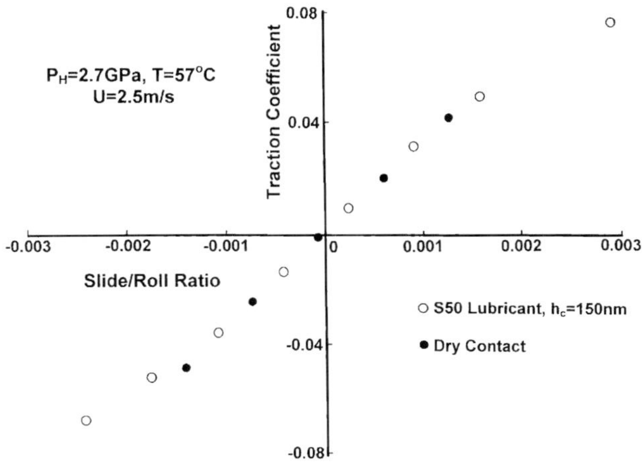

Figure 7. Traction curves for a high pressure contact with and without a liquid lubricant film.

the extraction of rheological properties of the liquid film from traction curves a dubious undertaking.

There appears to be a very simple method to model the elastic regime of traction. In Figure 7 the transverse traction coefficient is measured in point contact at maximum Hertz pressure of $p_H = 2.7$ GPa and a rolling velocity of 2.5 m/s. The solid data points indicate unlubricated rollers and the open points indicate a calculated 150 nm thick film of Santotrac 50 lubricant. For $|\Sigma| < 2 \times 10^{-3}$ the traction is essentially unaffected by the presence of the liquid film. While microslip may contribute some of the apparent compliance of the steel in dry contact, it seems that a practical solution for high-pressure, low Σ traction is to measure the linear traction gradient in dry contact and the rheological contribution of the steel can then be added to the value of Σ that has been obtained from a viscous calculation [34]. This gradient, m_s, in our traction rig, can be related to contact pressure roughly by

$$m_s = \frac{G_s}{p_H}, \tag{12}$$

where G_s is the shear modulus of the (steel) substrate.

Now, provided that the above relation is universal, there may be an opportunity to accurately calculate traction over the entire range of EHL operating conditions.

3.3 The Effect of Shear-Thinning on Scuffing and Wear in Sliding Contacts

At this time the only Reynolds equation for ordinary shear-thinning is Greenwood's [35] Reynolds–Rabinowitsch equation. With the Rabinowitsch model

$n \equiv 1/3$. For the general case of variable n, apparently the only EHL film thickness solutions available are central film thicknesses obtained from Grubin-style calculations using direct integration across the film.

These calculations [36] of central film thickness clearly show a dependence of film thinning on the slide-to-roll ratio, Σ, and the power-law exponent, n. Reducing n makes the film thickness more sensitive to Σ resulting in a faster loss of central thickness with increasing Σ for lower values of n.

To date there has been no investigation of the effect of ordinary shear-thinning on the minimum film thickness in sliding point contact. Clearly the effect of sliding should be to increase the sideways Poiseuille flow resulting from the reduction of viscosity due to shear in the sliding direction. This increased side-leakage should result in a reduced minimum film thickness at the side lobes – the reduction being greater for liquids with a lower value of power-law exponent, n, all other things being equal. Then the power-law exponent may be an extremely important parameter related to scuffing in sliding contacts. Analytical work at Cardiff [28] has begun to show the relationship between side leakage flow and scuffing in rough sliding contacts. Some evidence already exists that there is a connection between scuffing and shear-thinning. In a Ryder gear test, the scuffing load for a polyglycol was about twice that of diesters of similar viscosity [37]. In high-pressure measurements of shear-dependent viscosity [25] a polyglycol gave $n = 0.66$ and a diester, dioctylphthalate gave $n = 0.41$. Considering the economic impact of scuffing failures, the illucidation of any connection between shear-thinning and minimum film thickness would be a major success for EHL. This effect may be most pronounced in contacts that are long in the direction of motion.

3.4 Molecular Dynamics Simulations

Recent advances in computational power, algorithmic development, and intermolecular potential development have put molecular dynamics simulation techniques on the cusp of making very real contributions to the EHL community by explicitly linking lubricant structure to rheological response at the high pressures typical of EHL contacts. The key advantage to molecular simulation is that one can unambiguously define composition, allowing detailed structure-property investigations to be conducted. This is of particular utility for synthetic lubricant development, in which the molecular structure of the base stock can be controlled and modified to produce desirable rheological characteristics.

In conventional "equilibrium" molecular dynamics (EMD), the Newtonian viscosity of a system can be obtained from the appropriate Green–Kubo integral of the stress auto-correlation function,

$$\mu_{\mathrm{GK}} = \frac{V}{kT} \int_0^\infty dt \langle P_{xy}(0) P_{xy}(t) \rangle. \tag{13}$$

Viscosity as a function of shear rate can be obtained from "non-equilibrium" molecular dynamics (NEMD), in which a planar Couette flow field is imposed on the natural equilibrium dynamics of the system. In this case, the linear response of the driven system is identified as the viscosity,

$$\eta_{\mathrm{NEMD}}(\dot{\gamma}) = \frac{\langle P_{xy} \rangle}{\dot{\gamma}}. \tag{14}$$

Note that the NEMD approach yields strain rate dependent viscosities. For typical lubricant molecules, it has been found that when the applied strain rate exceeds the longest relaxation time of the molecules in the system, typically τ_{rot}, the rotational relaxation time, shear thinning is observed.

When applied to molecular systems under conditions typical of EHL contacts, both methods suffer from the computational burden associated with extracting the rheological response. For EMD, equation (13) requires simulation times on the order of several 100 intervals of τ_{rot} for a reasonable degree of convergence. For NEMD, the Newtonian plateau of the viscosity vs. strain rate curve is reached only for relatively low applied strain rates; the time averaged stress in equation (NEMD) converges slowly in the face of the poor signal to noise resolution of the weak applied strain field.

Despite these difficulties, several novel approaches offer access to the rheological response at extremes of pressure. McCabe and co-workers [38] have demonstrated that standard time-temperature superposition techniques can be applied to viscosity vs. strain rate data derived from NEMD simulations to collapse transport data over a wide range of states onto a single master-curve. From this curve, the Newtonian viscosity can be estimated from a small number of high strain simulations at the temperature and pressure of interest. At high pressures, these high strain simulations converge much more quickly than those at applied strain rates corresponding to the non-shear thinning regime. Bair et al. [39] have further demonstrated that a combination of rheological data for squalane, obtained from experiment and NEMD simulations, can be rationalized with time-temperature superposition. The rheological model obtained from this combination can be incorporated into a viscous EHL traction calculation and the resulting traction predictions compare quite favorably with experimental traction measurements [22].

In addition, Gordon et al. [40, 41] has developed an approach based on Stokes–Einstein relationships to make quantitative, extrapolative predictions of viscosity from EMD. In this method, high temperature simulations are used to establish the scaling relationship between self-diffusion and viscosity. At high temperatures, transport properties converge with relatively little computational effort. The scaling relationship can be thought of as a master curve, from which viscosity can be estimated on the basis of calculations of self-diffusion and geometric characterization, quantities that are much easier to compute from a molecular dynamics simulation.

Figure 8(a) depicts an example of the scaling of such transport data for a 5-n-propyl-5-n-butylnonane modeled with the TRAPPE-UA intermolecular potential [42]. Brute force calculations of viscosity via equation (13) at lower temperatures and high pressures confirm that transport properties obey the same general scaling relationship as those obtained at high temperatures. This analysis can lead to useful observations about the impact of molecular structure on rheological response at Hertzian pressures akin to those measured from experiment [25].

Quantitative prediction of rheological (or any other!) properties of a system from molecular dynamics requires an accurate description of the energies and forces governing molecular motion. For complex fluids representative of lubricants, this description usually includes a combination of terms governing intramolecular degrees of freedom, such as bond stretching, angle bending, and torsional rotation, and intermolecular interactions governing van der Waals and electrostatic interactions. Recent advances in force field development have produced potential models that are able to capture much of the essential physics of the molecular interactions.

An example of this is provided in Figure 9, which demonstrates that while the viscosity is uniformly underpredicted by the model compared to experiment, the pressure-viscosity response of squalane, a branched C_{30} isoparaffin, is accurately captured from simulation predictions. We note that, to date, most potential model development has emphasized describing thermodynamic behavior rather than transport properties; we believe that a significant opportunity exists for developing improved potential models capable of quantitatively estimating rheological response. Combined with the computational approaches outlined above, aimed at estimating viscosity under EHL conditions, this should be of tremendous benefit to the EHL community.

4. CONCLUSION

4.1 Challenges to the EHL Field

1. Various definitions of the pressure-viscosity coefficient are in use, their values can vary by 2:1, and it is possible that the definitions are being manipulated to improve the apparent accuracy of film calculations.

2. A single engineering definition of alpha for general pressure-viscosity response accepted by the entire field is essential.

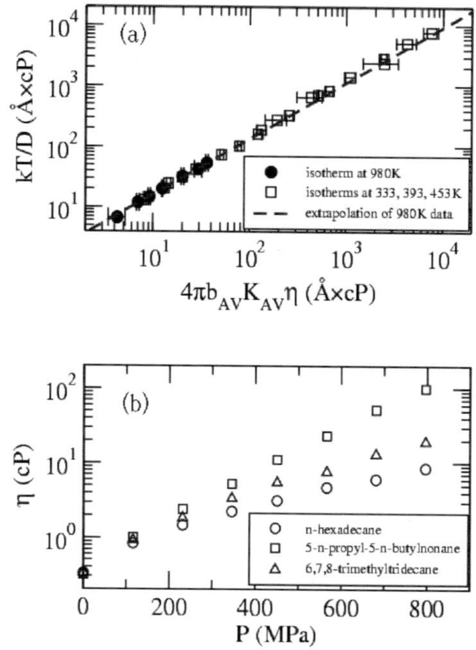

Figure 8. (a) Transport data of model 5-n-propyl-5-n-butylnonane interpreted through a Stokes–Einstein framework. High temperature simulations at 980K and a range of densities scale in the same way as transport data obtained at lower temperatures, and extrapolations of the curve can be used to estimate the viscosity from calculations of self-diffusion and geometric characterization of the molecules. (b) Predictions of viscosity vs. pressure for several isomers of C_{16} paraffins at 453K.

3. Greater than exponential pressure-viscosity response is necessary for calculating traction at high pressures from material properties and must be incorporated into traction calculations.

4. The Reynolds equation should be validated for EHL conditions with reasonable pressure-dependence of viscosity.

5. Calculate film thickness in sliding point contact for ordinary shear-thinning and validate with a non-Newtonian standard.

6. Calculate traction for ordinary shear-thinning.

4.2 Opportunities for the EHL Field

The use of an accurate description of the pressure, temperature and shear dependence of viscosity would make the following opportunities available to the EHL field.

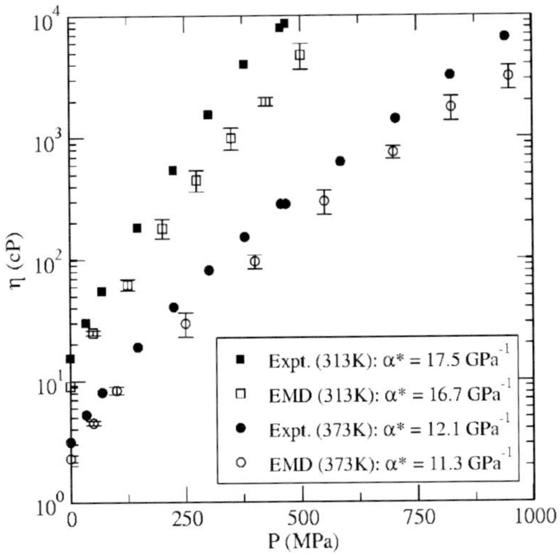

Figure 9. Viscosity vs. pressure for squalane at several temperatures. Filled symbols are experimental data [25] and open symbols are from EMD + Stokes–Einstein calculations.

1. Additives can modify the pressure dependence of viscosity and the shear dependence of viscosity. Then EHL simulations using accurate rheological models would aid in lubricant formulation.

2. Comparing dry traction with lubricated traction may reveal the rheological contribution of elastic compliance of the substrate, possibly precluding a calculation of the elastic response of the liquid.

3. The power-law exponent may be an essential property of liquid lubricants related to scuffing of rough sliding surfaces.

4. Molecular simulation is on the verge of providing a sufficiently accurate description of the pressure and shear dependence of viscosity of low-molecular-weight liquids that the EHL performance of speculative structures can be tested before they are synthesized.

NOMENCLATURE

μ	low shear viscosity, Pa·s
η	generalized viscosity, Pa·s
τ	shear stress, Pa
$\dot{\gamma}$	shear rate, s^{-1}
n	power-law exponent
G	modulus or critical stress, Pa
u	absolute value of rolling velocity, m/s

μ_0	low shear viscosity at ambient pressure, Pa·s
h	film thickness, m
T	temperature, °K
λ	characteristic time, s
Σ	slide-to-roll ratio
μ_2	viscosity of a second Newtonian, Pa·s
α	pressure-viscosity coefficient, Pa^{-1}
p	pressure, Pa
b_{AV}	small axis of molecular ellipsoid, m
K_{AV}	average resistance of molecular translation
z	pressure-viscosity index
a	shift factor
ρ	mass density, $\mathrm{kg/m^3}$
C	constant
p_H	Hertz pressure, Pa
G_s	shear modulus of substrate, Pa
m_s	linear traction gradient in dry contact
P_{xy}	instantaneous microscopic shear stress, Pa
k	Boltzmann's constant (1.38×10^{23} J/atom.K)
μ_{GK}	Newtonian viscosity obtained from equilibrium molecular dynamics, mPa·s
μ_{NEMD}	strain rate-dependent viscosity obtained from non-equlibrium molecular dynamics, mPa·s
τ_{rot}	rotational relaxation time of molecules, ps

ACKNOWLEDGEMENT

One of the authors (Scott Bair) was supported by a grant from the Timken Company.

REFERENCES

[1] D. Dowson and G.R. Higginson, *Elasto-Hydrodynamic Lubrication*, Pergamon Press, Oxford, 1966, p. 69.

[2] H. Blok, Inverse problems in hydrodynamic lubrication and design directives for lubricated flexible surfaces, in *Proc. Intern. Symp. Lubrication and Wear*, Muster and Sternlicht (eds), 1963, p. 74.

[3] J. Zhao and F. Sadeghi, Analysis of EHL circular contact shut down, *ASME J. Tribology*, **125**, 2003, 76–90.

[4] R.P. Glovnea and H. A. Spikes, Elastohydrodynamic film collapse during rapid deceleration. Part I: Experimental results, *ASME J. Tribology*, **123**, 2001, 254–261.

[5] R.V. Kleinschmidt, D. Bradbury and M. Mark, *Viscosity and Density of over Forty Lubricating Fluids of Known Composition at Pressures to 150,000 psi and Temperatures to 425 F*, ASME, New York, 1953.

[6] S. Bair, An experimental verification of the significance of the reciprocal asymptotic isoviscous pressure, *STLE Tribology Transactions*, **36**, 1993, 153–162.

[7] C.J.A. Roelands, Correlational aspects of viscosity-temperature-pressure relationship of lubricating oils, Thesis, Delft, 1966, p. 106.

[8] A.J. Moore, The behavior of lubricants in elastohydrodynamic contacts, *Proc. Instn. Mech. Engrs.*, **211**(J), 1997, 91–106.

[9] V.V. Brazhkin and A.G. Lyapin, Universal viscosity growth in metallic melts at megabar pressures: The vitreous state of the earth's inner core, *Physics Uspekhi*, **43**(5), 2000, 493–508.

[10] A. Dandridge and D.A. Jackson, Measurements of viscosity under pressure: A new method, *J. Physics, D: Appl. Physics*, **14**, 1981, 829–831.

[11] M. Renardy, Some remarks on the Navier-Stokes equations with a pressure-dependent viscosity, *Comm. in Partial Diff. Eqn.*, **11**(7), 1986, 779–793.

[12] S. Bair, M. Khonsari and W.O. Winer, High-Pressure rheology of lubricants and limitations of the Reynolds equation, *Trib. Intern.*, **31**(10), 1998, 573–586.

[13] C.T. Schaefer, P. Giese, W.B. Rowe and N.H. Woolley, elastohydrodynamically lubricated line contact based on the Navier–Stokes equations, in *Proc. 26th Leeds-Lyon Symp. Trib.*, Elsevier, Amsterdam, 2000, pp. 57–59.

[14] T. Almqvist and R. Larsson, The Navier–Stokes approach for thermal EHL line contact solutions, *Trib. Intern.*, **35**, 2002, 163–170.

[15] K.R. Rajagopal and A.Z. Szeri, On an inconsistency in the derivation of the equations of elastohydrodynamic lubrication, *Proc. Roy. Soc. Lond. A*, **459**, 2003, 2771–2786.

[16] J. Hron, J. Malek and K.R. Rajagopal, Simple flows of liquids with pressure-dependent viscosities, *Proc. Roy. Soc. Lond. A*, **457**, 2001, 1603–1622.

[17] S. Bair and C. McCabe, A study of mechanical shear bands in liquids at high pressure, *Trib. Intern.*, **37**(10), 2004, 783–789.

[18] R.B. Rhodes, Development of ASTM Standard Test Methods for measuring engine oil viscosity using rotational viscometers at high-temperature and high shear rates, in *High-Temperature, High-Shear (HTHS) Oil Viscosity: Measurement and Relationship to Engine Operation*, ASTM STP 1068, James A. Spearot (ed.), American Society for Testing and Materials, Philadelphia, 1989, pp. 14–22.

[19] S. Bair, A more complete description of the shear rheology of high-temperature, high-shear journal bearing lubrication, *STLE Trib. Trans.*, accepted 2004.

[20] S. Bair and F. Qureshi, The high-pressure rheology of polymer-oil solutions, *Trib. Intern.*, **36**, 2003, 637–645.

[21] C.R. Schultheisz and S.D. Leigh, Certification of the rheological behavior of SRM 2490, Polyisobutylene dissolved in 2,6,10,14-Tetramethylpentadecane, *NIST Special Publication 260-143*, 2002, pp. 2–27.

[22] S. Bair, C. McCabe and P.T. Cummings, Calculation of viscous EHL traction for Squalane using molecular simulation and rheometry, *Trib. Lett.*, **13**(4), 2002, 251–254.

[23] F.H. Ree, T. Ree and H. Eyring, Relaxation theory of transport problems in condensed systems, *Ind. Eng. Chem.*, **50**, 1958, 1036–1038.

[24] S.J. Hahn, H. Eyring, I. Higuchi and T. Ree, Flow properties of lubricating oils under pressure, *NLGI Spokesman*, **21**(3), 1958, 123–128.

[25] S. Bair, The high pressure rheology of some simple model hydrocarbons, *Proc. Instn. Mech. Engrs.*, **216**(J), 2002, 139–149.

[26] D.W. Van Krevelen, *Properties of Polymers*, 3rd Edition, Elsevier, Amsterdam, 1990, pp. 475–486.

[27] S. Bair, The high-pressure rheology of mixtures, *ASME J. Tribology*, Paper 1040, 2004.

[28] M.J.A. Holmes, H.P. Evans, T.G. Hughes and R.W. Snidle, Transient elastohydrodynamic point contact analysis using a new coupled differential deflection method. Part 2: Results, *Proc. Instn. Mech. Engrs.*, **217**(J), 2003, 305–321.

[29] L.I. Kioupis and E.J. Maginn, Rheology, dynamics, and structure of hydrocarbon blends: A molecular dynamics study of n-Hexane/n-Hexadecane mixtures, *Chem. Eng. J.*, **74**, 1999, 129–146.

[30] S. Bair, Complete isothermal solution for the viscous regime of concentrated contact traction, *Intl. J. Appl. Mech. Engr.*, **7**(3), 2002, 719–727.

[31] J.F. Hutton and M.C. Phillips, Shear modulus of liquids at elastohydrodynamic pressures, *Nature Phys. Sci.*, **238**, 1972, 141–142.

[32] K.L. Johnson and J.L. Tevaarwerk, Shear behavior of elastohydrodynamic oil films, *Proc. Roy. Soc. Lond. A*, **356**, 1977, 215–236.

[33] S. Bair, The nature of the logarithmic traction gradient, *Trib. Intern.*, **35**, 2002, 591–597.

[34] S. Bair, Ordinary shear-thinning behavior in liquids and its effect upon EHL traction, in *Tribology Research: From Model Experiment to Industrial Problem*, G. Dalmaz et al. (eds), Elsevier, 2001, pp. 733–742.

[35] J.A. Greenwood, Two-dimensional flow of a non-Newtonian lubricant, *Proc. Instn. Mech. Engrs.*, **214**(J), 2000, 29–41.

[36] S. Bair and F. Qureshi, Ordinary shear-thinning behavior and its effect upon EHL film thickness, in *Proc. 29th Leeds-Lyon Symp.*, Tribology, Elsevier, Amsterdam, 2003, pp. 693–699.

[37] R.C. Gunderson and A.W. Hart, *Synthetic Lubricants*, Reinhold, New York, 1962, pp. 87–88.

[38] C. McCabe, C.W. Manke and P.T. Cummings, Predicting the Newtonian viscosity of complex fluids from high strain rate molecular simulations, *J. Chem. Phys.*, **116**(8), 2002, 3339–3342.

[39] S. Bair, McCabe and P.T. Cummings, Comparison of non-equilibrium molecular dynamics with experimental measurements in the nonlinear shear-thinning regime, *Phys. Rev. Lett.*, **88**(5), 2002, 058302.

[40] P.A. Gordon, Characterizing isoparaffin transport properties with Stokes–Einstein relationships, *Ind. Eng. Chem. Res.*, **42**(26), 2003, 7025–7036.

[41] P.A. Gordon, Extrapolation of rheological properties for lubricant components with Stokes–Einstein relationships, to be submitted.

[42] M.G. Martin and J.I. Siepman, Novel configurational-bias Monte Carlo method for branched molecules. Transferable potentials for phase equilibria. 2. United-atom description of branched alkanes, *J. Phys. Chem. B*, **103**(21), 1999, 4508–4517.

SESSION 2

ADJOINT ERROR ESTIMATION AND SPATIAL ADAPTIVITY FOR EHL-LIKE MODELS

Dan Hart[1], Christopher E. Goodyer[1], Martin Berzins[1,2], Peter K. Jimack[1] and Laurence Scales[1,3]

[1]*University of Leeds, Leeds LS2 9JT, U.K.;* [2]*SCI Institute, University of Utah, Salt Lake City, Utah, U.S.A.;* [3]*Shell Global Solutions, Cheshire Innovation Park, Chester, U.K.*

Abstract The use of adjoint error estimation techniques is described for a model problem that is a simplified version of an EHL line contact. Quantities of interest, such as friction, may be dependent upon the accuracy of the solution in some parts of the domain more than in others. The use of an inexpensive extra solve to calculate an adjoint solution is described for estimating the inter-grid error in the value of friction calculated, and as a basis for local refinement. It is demonstrated that this enables an accurate estimate for the quantity of interest to be obtained from a less accurate solution of the model problem.

Key words: error estimation, discrete adjoint, EHL, friction calculation, mesh adaptation.

1. INTRODUCTION

Numerical simulations of lubrication problems are of great benefit to industry for the design and evaluation of both oils and contacts since laboratory experiments are costly in terms of both time and money. Simulations enable a much wider range of cases to be evaluated provided that appropriate physical and mathematical models are used, and that the software is fast, robust and accurate.

One potential vehicle for allowing these requirements to be met is the use of *a posteriori* error estimation combined with spatial mesh adaptation, where more points are placed in regions where the solution is sensitive to the local resolution and fewer points where it is insensitive. These ideas are not new to computational engineering and have been used in problems such as

R.W. Snidle and H.P. Evans (eds), IUTAM Symposium on Elastohydrodynamics and Micro-elastohydrodynamics, 47–58.

elastohydrodynamic lubrication (EHL) by a number of authors[1,2,3,4,5,6]. In this paper we note that, when solving such problems, the goal of the engineer may typically be to obtain knowledge about some quantity that may be derived from the solution, such as the total friction through the contact for example. For this reason it may not always be most efficient to develop the mesh adaptation strategy so as to maximise the overall increase in the accuracy of the computed solution, since this may have relatively little effect on the quantity of interest.

In this work we consider an alternative mechanism for controlling mesh adaptivity for EHL-like problems, based upon controlling the error in the quantity of interest, the calculated friction for example, rather than the solution as a whole. This is achieved at the relatively small expense of solving an additional problem which yields an approximation to an adjoint solution. Using an adjoint system to gain extra information about a solution has long been exploited in optimal shape design[7], and recently these ideas have become common in aerospace engineering[8,9].

In the case of EHL an additional complicating factor is the presence of cavitation which leads to a free boundary in the mathematical model. An initial investigation into the application of adjoints to this type of problem has been made by Hart et al.[10], where the free boundary problem was considered in the context of uniform mesh refinement only. In this work we extend the approach to demonstrate that the adjoint solution can also yield local information that may be used to drive non-uniform refinement in an effective manner. Furthermore it is shown how varying the quantity of interest results in different local refinement patterns.

The following section provides an outline of the theory that lies behind adjoint error estimation whilst Section 3 describes the nonlinear, free-boundary, model problem being solved. Results are then shown in Section 4 illustrating the accuracy of the error estimator that has been implemented, as well as the performance of mesh adaptation based upon this. The paper concludes with a short discussion of the material presented and of the directions in which ongoing research is headed.

2. ADJOINT ERROR ESTIMATION THEORY

In this section, the abstract background to the adjoint estimation of an error is introduced. The starting point is to define two meshes with spacing $h = \Delta x$ and $H = \Delta X = m \times \Delta x$ (m some integer > 1). The idea is that mesh size H is fine enough to capture the features of the problem being solved, and coarse enough to be solved in a reasonable time, while the fine mesh size h would give the solution to a greater accuracy but in an unacceptable time.

Consider an arbitrary problem whose discrete form may be represented as $A_h u_h = f_h$ on the finer mesh, and $A_H u_H = f_H$ on the coarser mesh. Let u_h^H be an approximation to u_h obtained by interpolation of the coarse mesh solution: $u_h = I_h^H u_H$. A Taylor series expansion for the fine grid residual function $R_h(u_h) = f_h - A_h u_h$ as explained by Darmofal and Venditti[9], shows that

$$
\begin{aligned}
0 = R_h(u_h) &= R_h(u_h^H + (u_h - u_h^H)) \\
&= (f_h - A_h u_h^H) + \left[\left. \frac{\partial R_h}{\partial u_h} \right|_{u_h^H} \right] (u_h - u_h^H) + h.o.t.
\end{aligned}
\tag{1}
$$

Neglecting higher order terms, (1) may be written as

$$
(u_h - u_h^H) = -\left[\left. \frac{\partial R_h}{\partial u_h} \right|_{u_h^H} \right]^{-1} (f_h - A_h u_h) \tag{2}
$$

Suppose that the quantity of interest for this problem is a functional which may be expressed as $F_h(u_h)$ on the finer grid. This can also be expanded about the interpolated coarse mesh solution to give

$$
F_h(u_h) = F_h(u_h^H) + \left(\left. \frac{\partial F_h}{\partial u_h} \right|_{u_h^H} \right)^T (u_h - u_h^H) + h.o.t.
$$

Expression (2) may be substituted into this and, again discounting higher order terms, this becomes

$$
F_h(u_h) = F_h(u_h^H) - \left(\left. \frac{\partial F_h}{\partial u_h} \right|_{u_h^H} \right)^T \left[\left. \frac{\partial R_h}{\partial u_h} \right|_{u_h^H} \right]^{-1} (f_h - A_h u_h) \tag{3}
$$

By setting

$$
\Psi_h = \left(\left. \frac{\partial F_h}{\partial u_h} \right|_{u_h^H} \right)^T \left[\left. \frac{\partial R_h}{\partial u_h} \right|_{u_h^H} \right]^{-1}
$$

it follows that Ψ_h must satisfy

$$
\left[\left. \frac{\partial R_h}{\partial u_h} \right|_{u_h^H} \right]^T \Psi_h = \left(\left. \frac{\partial F_h}{\partial u_h} \right|_{u_h^H} \right)^T \tag{4}
$$

Throughout this paper we will refer to (4) as the adjoint system and to Ψ_h as the adjoint solution. Note that if Ψ_h is known then (3) may be expressed as $F_h(u_h) = F_h(u_h^H) - (\Psi_h)^T (f_h - A_h u_h^H)$ and a value for $F_h(u_h)$ may be found without having calculated u_h explicitly. Of course solving (4) exactly to obtain Ψ_h is likely to be as expensive as obtaining u_h exactly. Hence, instead of solving the adjoint system on this fine mesh, an approximation to it is solved for on the coarse mesh. This is given by

$$\left[\frac{\partial R_H}{\partial u_H}\right]^T \Psi_H = \left(\frac{\partial F_H}{\partial u_H}\right)^T \tag{5}$$

This coarse grid adjoint solution is then interpolated onto the fine grid to give $\Psi_h^H = I_h^H \Psi_H$. The correction operation for the functional is given by

$$F_h(u_h) \approx \widetilde{F}_h(u_H) = F_h(u_h^H) - (\Psi_h^H)^T R_h(u_h^H) \tag{6}$$

An estimation of the error in the functional between the two grids has now been obtained simply by solving an additional (adjoint) problem on the coarser grid. Note that we will refer to the expression $(\Psi_h^H)^T R_h(u_h^H)$ in (6) as the "correction" to the functional $F_h(u_h^H)$. A summary of this error estimation procedure is presented in Figure 1.

3. MODEL PROBLEM

In order to help understand how the above adjoint theory may be applied to a full EHL problem, in this work we focus on a slightly simplified model problem. This is designed to capture the key nonlinear features of the force balance equation and the free boundary for the cavitation position. The elastic deformation is therefore neglected at this stage.

Pressure is solved from a hydrodynamic "Reynolds" equation

$$\frac{d}{dX}\left(H^3 \frac{dP}{dX}\right) - \lambda \frac{dH}{dX} = 0 \tag{7}$$

with the film thickness given by

$$H(X) = H_0 + \frac{X^2}{2} \tag{8}$$

1. Solve the system of equations $A_H u_H = f_H$ on the coarser grid
2. Solve the adjoint system (5) on the coarser grid
3. Interpolate u_H and Ψ_H onto the finer grid
4. Calculate the functional $F_h(u_h^H)$ using the interpolated solution
5. Calculate the fine grid residuals $R_h(u_h^H)$ using the interpolated solution
6. Calculate the correction (product of adjoint and residuals)

Figure 1. Summary of the adjoint error estimation procedure.

The deformation term is omitted, leaving the separation to be governed only by H_0. The values of H_0 and the cavitation position X_c must be found such that both the force balance equation

$$\int_{X_{-\infty}}^{X_c} P(X)dX = L \tag{9}$$

and the cavitation condition

$$P'(X_c) = 0 \tag{10}$$

are satisfied. In equation (9) $X_{-\infty}$ is defined to be equal to X_c minus a given, constant, domain size (chosen to be sufficiently large as to not influence the solution in the contact region significantly).

3.1 Solution Procedure

For a given value of X_c the Reynolds equation, (7), may be discretized using a second order finite order finite difference scheme, as previously described[10], for example. For simplicity the mesh spacing may be considered to be fixed, at ΔX say, although in practice it will be allowed to adapt in different regions of the domain. In order to satisfy (10) it is necessary to allow X_c to vary until both the discrete forms of (7) and (10) are satisfied together (achieved via an iterative procedure). Other approaches to dealing with the cavitation region may be found elsewhere[11,12,13], however it is important for this work that the cavitation position, X_c, be treated as a continuous variable, and as such the region beyond X_c need not be considered. This allows it to fit more naturally into the theoretical framework introduced above. Finally note that by defining $X_{-\infty}$ to be X_c minus some fixed value, as X_c varies within the solution procedure $X_{-\infty}$ varies too, so that the size of the computational domain remains fixed. This removes the problem of having ΔX depend on X_c (which would make the sparsity pattern of the Jacobian matrix in (5) dense) and also, with the extension to deforming contacts in mind, eliminates the need to recompute the kernel function. The fact that the left-hand boundary moves is countered by the fact that it is chosen to be sufficiently far to the left to make no significant difference to the calculated value of the friction. In summary, the discrete problem that must be solved is to find P_1 to P_{n-1}, H_0 and X_c such that the following residuals are zero:

$$R_i = \Delta X\left(\lambda X_i - \left(\frac{H^3_{i+½}P_{i+1} - (H^3_{i+½} + H^3_{i-½})P_i + H^3_{i-½}P_{i-1}}{(\Delta X)^2}\right)\right) \tag{11}$$

for $i = 1, \ldots, n-1$,

$$R_{H_0} = L - \sum_{i=0}^{n-2} \frac{P_i + P_{i+1}}{2} \Delta X \tag{12}$$

and

$$R_{X_c} = \frac{3P_{n-1} - 4P_{n-2} + P_{n-3}}{2\Delta X} \tag{13}$$

Having fully specified the model problem, attention can be turned to the discrete adjoint formulation.

3.2 Adjoint Problem

As stated in (5) the adjoint system is based upon the transpose of the Jacobian of the residual equations. The Jacobian entries are defined by $\frac{\partial R_i}{\partial P_j}$, $\frac{\partial R_i}{\partial H_0}$, $\frac{\partial R_i}{\partial X_c}$, $\frac{\partial R_{H0}}{\partial P_i}$, $\frac{\partial R_{H0}}{\partial H_0}$, $\frac{\partial R_{H0}}{\partial X_c}$, $\frac{\partial R_{X_c}}{\partial P_i}$, $\frac{\partial R_{X_c}}{\partial H_0}$, $\frac{\partial R_{X_c}}{\partial X_c}$, for $i, j \in 1, \ldots, n-1$. Based on (11) - (13), the sparsity pattern of the Jacobian can be easily obtained and is shown in Figure 2.

The functional that we are interested in for this work is the friction, which for this model problem is given by

$$F(P) = \int_{-\infty}^{X_c} \left(-\frac{\partial P}{\partial X} \frac{H}{2} + \frac{\overline{\eta}}{H} \right) dX \tag{14}$$

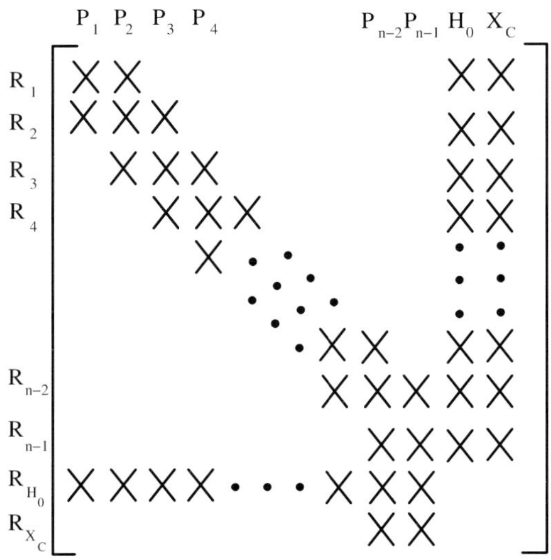

Figure 2. Sparsity pattern for the Jacobian of the residuals.

This may be calculated using a simple quadrature scheme, such as the compound midpoint rule, whereby the integrand (by physical analogy we refer to this as the shear stress) is estimated on each cell of the mesh using the pointwise values of P. To form the right-hand side of the adjoint system (5) the derivative of this quadrature expression with respect to each P, along with H_0 and X_c must be obtained: these are easily calculated analytically. Once a solution has been found, the right-hand side of (5) may then be evaluated and Ψ_H may be obtained.

The solution Ψ_H of the adjoint system represents the sensitivity of the friction to each of the residuals. Hence multiplying each of these adjoint variables by the corresponding residual and taking the sum allows an estimate to be obtained of the error in the friction between the two grids. Note that although Ψ_H and u_H are found on the coarser grid the best approximation to the error is obtained by interpolating these to the finer grid before finding the residual $R_h(u_h^H)$ and its product with Ψ_h^H. This is the last expression of equation (6).

4. RESULTS

In this section a small number of sample results are presented. In all cases, unless stated otherwise, the value of L in (9) is taken as 5.0 and the value of $\overline{\eta}$ in (14) is taken as 20.0.

4.1 Uniform Mesh Results

Before considering a sequence of locally refined meshes, we begin by presenting results on uniform meshes. Table 1 shows the performance of the predicted error in the friction, as calculated using the adjoint approach, by comparing it with the true error when solving on the next mesh. Note that in this context we use the term error to mean $F(u_h^H)-F(u_h)$ (as opposed to $F(u_h^H)-F(u)$ where u is the exact solution of the continuous problem).

The first column of the table shows the grid level for the coarser of the two grids, and has a number of points equal to $2^{g+1}+1$. Using the solution from this grid, interpolated onto grid $g+1$, a friction value is calculated which is shown in the second column. Column three shows the correction to this friction, as calculated using the adjoint system solved on the coarse grid g. The corrected friction is shown in column 4, with the "true" friction value for grid $g+1$ shown in column 5. The measured error between columns 2 and 5 is shown in column 6. The final column shows the ratio of the measured error to the estimated error (known as the Effectivity Index) and can be seen to approach unity with increasing mesh resolution, which indicates that for uniform grids the friction error estimate is remarkably accurate. If the adjoint solution were not available, it would be necessary to keep computing on finer and finer grids until the friction changed by less than ε, at which point the last (and most expensive) solution does not yield a friction value that is significantly more accurate than the previous. By using this strategy but based upon the adjoint estimate the same accuracy will be achieved but we will have saved the cost of computing a solution on the finest mesh in this sequence.

4.2 Adaptive Mesh Results

Table 2 shows that with non-uniform meshes the adjoint error estimation approach is still reliable, in the sense that the ratio of the predicted correction to the actual difference in friction on consecutive meshes still tends to one as the meshes are refined. Note that in order to obtain these results global mesh refinement, based upon element bisection, has still been used, but now the initial mesh (and hence all subsequent meshes) is non-uniform. Clearly the residual equation (7) has to be modified to reflect this non-uniformity in ΔX and the grid level referred to in the first column of the table indicates the finest level that is present in the mesh.

Having demonstrated that the predicted error is still reliable on non-uniform meshes it is now possible to use these values as the basis for local, rather than global, mesh refinement. It should be noted, however, that the correction value given by the last term in equation (6) is just a single number indicating the current error in the friction and so further information is required in order to determine where the contribution to this error is the greatest. In the following example we base the local refinement on the magnitude of $(\Psi_h^H)_i \times (R_h(u_h^H))_i$ locally, and refer to this as the correction component of mesh point i. Figure 3 shows the computed correction components across the domain after a number of local refinements have been undertaken. Starting from the left it may be seen that the contribution to the estimated friction error gradually increases until the first region of local refinement is reached, whereupon it drops suddenly. The contribution to the error then grows again until the next region of local refinement is reached, and so on. The contribution to the error is always kept below 10^{-7} in this particular example.

Table 1. Adjoint based inter-grid friction error on non-uniform meshes, each with the same refinement.

Grid (g)	Interpolated Friction (g)	Calculated Correction	Corrected Friction (g)	Friction (g+1)	Measured Error	Effectivity Index
5	87.95668	15.27956	72.67711	68.02241	19.93427	1.304
6	68.67781	2.64095	66.03686	66.37680	2.30101	0.871
7	66.52919	0.31057	66.21862	66.31442	0.21477	0.691
8	66.35216	0.01504	66.33712	66.34818	0.00398	0.264
9	66.35761	-0.00255	66.36015	66.36125	-0.00365	1.432
10	66.36361	-0.00124	66.36484	66.36496	-0.00135	1.094
11	66.36555	-0.00037	66.36592	66.36593	-0.00038	1.035
12	66.36608	-0.00010	66.36618	66.36618	-0.00010	1.016

Table 2. Adjoint based inter-grid friction error on non-uniform meshes, each with the same refinement.

Grid (g)	Interpolated Friction (g)	Calculated Correction	Corrected Friction (g)	Friction (g+1)	Measured Error	Effectivity Index
5	87.66173	-4.56301	92.22474	67.98769	19.67404	-4.311
6	68.65988	2.62800	66.03187	66.36806	2.29181	0.872
7	66.52502	0.30892	66.21611	66.31220	0.21282	0.688
8	66.35111	0.01462	66.33649	66.34762	0.00348	0.238
9	66.35734	-0.00266	66.36000	66.36111	-0.00377	1.417
10	66.36354	-0.00127	66.36481	66.36492	-0.00138	1.093
11	66.36553	-0.00038	66.36591	66.36592	-0.00039	1.035
12	66.36608	-0.00010	66.36618	66.36618	-0.00010	1.016

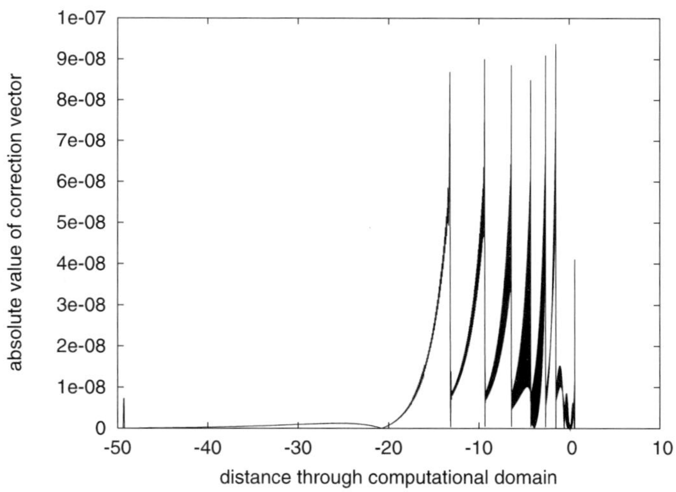

Figure 3. Plot showing the absolute value of the correction vector, and how it is distributed through local mesh refinement.

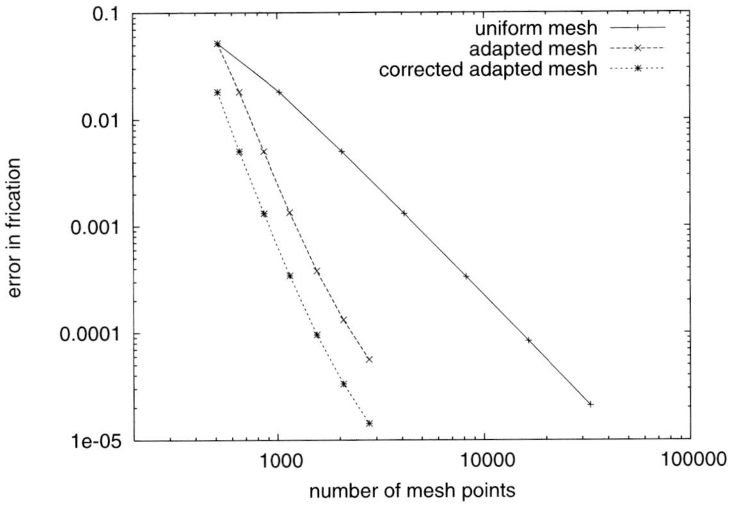

Figure 4. Plot showing error reduction for uniform and adaptive grids.

Figure 4 shows the overall effectiveness of this strategy compared to the use of uniform mesh refinement. In this case the plot is of the error in the friction (as compared against a friction value calculated on a so-called "truth mesh" containing approximately 250,000 equally spaced points) versus the total number of nodes present in the mesh. Unsurprisingly the uniform refinement strategy converges most slowly, the next curve shows the error in the friction on the locally refined (adapted) mesh, whilst the final curve shows the error in the corrected friction value on the adapted mesh.

5. DISCUSSION

Results have been presented which show that the adjoint error estimation approach may be used effectively for a non-linear EHL-like model problem containing a free boundary due to the cavitation condition. The effectivity of this estimate on uniformly refined meshes may be used to provide a reliable stopping criterion without the need to solve on the finest mesh. Furthermore, the components of the correction term are shown to provide an appropriate basis for determining where to refine locally. The resulting meshes can yield solutions of a considerably greater accuracy (in terms of friction, for example) than obtained on correspondingly sized uniform grids.

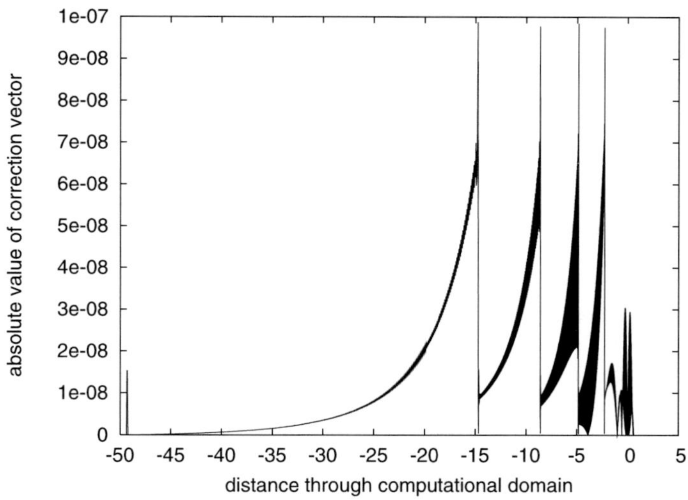

Figure 5. Plot showing the absolute value of the correction vector, and how it is distributed through local mesh refinement.

It should be noted that the adjoint procedure provides an estimate that may be used to control mesh adaptivity based upon a particular quantity of interest. When this quantity is altered (changing the value of $\overline{\eta}$ in (14) for example) then the resulting refinement of the mesh will be altered accordingly. This behaviour is illustrated in Figure 5 which shows the equivalent plot to that of Figure 3 but for the case $\overline{\eta} = 0.0$. Again a sequence of local refinements have been undertaken so as to keep the local contribution to the error below 10^{-7} everywhere but the resulting grid is very different (the regions of change in refinement level can clearly be seen to differ between Figures 3 and 5.

The next stage in this work is to move on to a true EHL line contact problem by introducing elastic deformation and more realistic viscosity and density models. Being highly localised computations, neither of the rheological equations should adversely affect the sparsity of the Jacobian. However, the introduction of the deflection calculation will clearly make the Jacobian dense. One possible solution may be to use the differential deflection method proposed by Evans and Hughes[14] to compute the elastic deformation of the solid. One of the longer term goals of this work is to solve problems with real surface roughness, where it is hoped that adaptation can take place to resolve only those features of most importance to the friction calculation.

REFERENCES

1. C. E. Goodyer, R. Fairlie, D. E. Hart, M. Berzins and L. E. Scales, Calculation of Friction in Steady-State and Transient EHL Simulations, in: *Transient Processes in Tribology: Proceedings of the 30th Leeds-Lyon Symposium on Tribology*, A. A. Lubrecht and G. Dalmaz (eds), Elsevier, 2004.
2. C. E. Goodyer, R. Fairlie, M. Berzins and L. E. Scales, Adaptive Mesh Methods for Elastohydrodynamic Lubrication, in *ECCOMAS CFD 2001: Computational Fluid Dynamics Conference Proceedings*, Institute of Mathematics and Its Applications, 2001.
3. S. R. Wu and J. T. Oden, A Note on Application of Adaptive Finite Elements to Elastohydrodynamic Lubrication Problems, *Comm. Appl. Num. Meth.*, **3**, 485–494 (1987).
4. A. A. Lubrecht, Numerical solution of the EHL line and point contact problem using multigrid techniques, PhD Thesis, University of Twente, Enschede, The Netherlands, 1987.
5. C. H. Venner, Multilevel Solution of the EHL Line and Point Contact Problems, PhD Thesis, University of Twente, Enschede, The Netherlands, 1991.
6. C. H. Venner and A. A. Lubrecht, *Multilevel Methods in Lubrication*, Elsevier, 2000.
7. O. Pironneau, *Optimal Shape Design for Elliptic Systems*, Springer-Verlag, 1984.
8. N. A. Pierce and, M. B. Giles, Adjoint Recovery of Superconvergent Functionals from PDE Approximations, *SIAM Review* **42**(2), 247–264 (2000).
9. D. A. Venditti and D. L. Darmofal, Grid Adaptation for Functional Outputs: Application to Two-Dimensional Inviscid Flows, *J. Comp. Phys.*, **176**, 40–69 (2002).
10. D. E. Hart, M. Berzins, C. E. Goodyer, P. K. Jimack and L. E. Scales, Adjoint Error Estimation for EHL-like Models, *Int. J. Numer. Meth. Fluids*, **47**, 1069–1075 (2005).
11. S. R. Wu, A Penalty Formulation and Numerical Approximation of the Reynolds-Hertz Problem of Elastohydrodynamic Lubrication, *Int. J. Eng. Sci.*, **24**, 1001–1013 (1986).
12. D. Dowson and C. M. Taylor, Cavitation in Bearings, *Ann. Rev. Fluid Mech.*, **11**, 35–66 (1979).
13. D. G. Christopherson, A New Mathematical Method for the Solution of Film Lubrication Problems, *Proc. Inst. Mech. Eng.*, **146**, 126–135 (1942).
14. H. P. Evans and T. G. Hughes, Evaluation of deflection in semi-infinite bodies by a differential method, *Proc. Inst. Mech. Eng., Part C*, **214**, 563–584 (2000).

SURFACE ROUGHNESS ATTENUATION IN EHL LINE AND POINT CONTACTS UNDER CONDITIONS OF STARVED LUBRICATION

C.H. Venner[1] and C.J. Hooke[2]
[1]*Faculty of Engineering Technology, University of Twente, Enschede, The Netherlands*
[2]*School of Manufacturing and Mechanical Engineering, University of Birmingham, U.K.*

Abstract The authors have previously examined the effect of surface roughness in line and point EHL contacts and have shown that it is the ratio of the wavelength to the inlet pressure sweep that determines the degree of roughness attenuation under the contact. Because of this a single curve can be used to determine the attenuation for all contacts.

However this work was carried out for fully flooded contacts and many practical contacts operate under conditions of starved lubrication. This paper extends the analysis to this type of conjunction.

Keywords: starvation, roughness attenuation.

1. INTRODUCTION

Because the operating conditions of EHL contacts are becoming more extreme, the film thickness level at which they have to operate reliably is decreasing and the time dependent effects of the surface roughness moving through the conjunction are becoming more important. At present full scale numerical simulations for rough surfaces are feasible but apart from millions of numbers such simulations produce very little information that is actually useful in, for example, surface design. What is needed is the identification of the (dimensionless) parameters that determine the behaviour and developing simple, easy to use and reliable rules for application in engineering that accurately reflect the physical mechanisms governing the behaviour.

In the past years it has been shown that this can be achieved through the analysis of the effect of single harmonic components [1–9].

It was shown that as waviness patterns pass through an EHL conjunction they are modified with the longer wavelength components being reduced in amplitude. At the same time they produce rippling of the pressure distribution.

R.W. Snidle and H.P. Evans (eds), IUTAM Symposium on Elastohydrodynamics and Micro-elastohydrodynamics, 59–70.

In a series of papers with results for line and point contacts, and for waviness patterns with different orientation and topography it was shown that the physical mechanism of attenuation of waviness is identical for line and point contacts and depends solely on the ratio of the roughness wavelength to the length of the inlet sweep. It was shown that there is a single master curve for all contacts that predicts the amplitude reduction. This curve, when combined with Fourier analysis techniques, provides a very practical tool for the analysis of real rough surface deformation inside a contact. For example, it can help to identify those components causing a high risk of metallic contact or those which will be deformed but will cause local pressure oscillations and may reduce the surface fatigue life.

All the results previously presented applied to fully flooded conditions. However, many practical EHL contacts operate under conditions of starved lubrication due to limited lubricant supply. One of the first studies of the effects of starvation on film thickness was by Wedeven [10]. The first numerical studies of the starved problem were with a simple model varying the inlet location of the domain [11]. Recently Chevalier et al. [12] studied the steady state problem numerically with a more advanced approach. A solver for the study of the full time dependent problem needed to study waviness moving through the contact was developed by Wijnant [13]. In [14] results were presented for a circular contact. It was shown that the results for different degrees of starvation could be scaled on the mastercurve of the fully flooded contact but no theoretical foundation was given. In this paper it is shown that it is the effect of the shortening of the inlet sweep as a result of the limited lubricant supply that needs to be accounted for in starved contacts. Thus it is again shown that the basic mechanism of waviness attenuation is the same for line and point contacts. The present paper considers only transverse roughness in rolling contacts. However, the methods described can readily be extended to other, more complex patterns of roughness.

2. NOTATION

a	Hertz semi-contact width in direction of entrainment
A	amplitude of surface roughness
h	film thickness
h_{m0}	$= h_m \rho_m / \rho_0$ equivalent central film thickness
h_{oil}	total thickness of film entering contact
δh	clearance variation inside contact
L	length of contact inlet
P	pressure
P	Greenwood pressure parameter
P_H	Hertz pressure

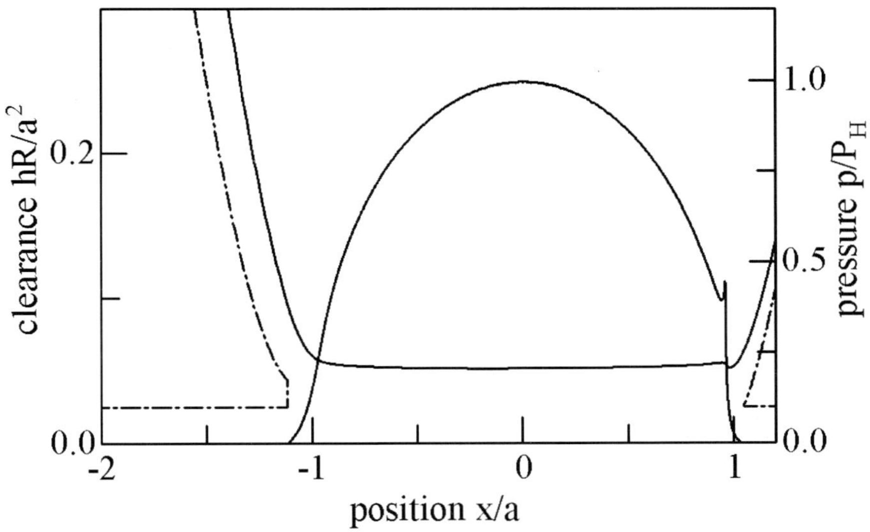

Figure 1. Starved EHL contact.

δP	pressure variation inside contact
R	effective contact radius
S	Greenwood speed parameter
u	entrainment velocity
x	distance through contact
x_{start}	distance from point of film formation to Hertz boundary
η	viscosity
λ	roughness wavelength
ρ	density

3. STARVATION

As the oil supply to an EHL contact is reduced the inlet becomes starved with the start of the pressure sweep being delayed and the amount of fluid entering the conjunction being reduced. This situation is shown, schematically, in Figure 1.

In the inlet the fluid enters the contact attached to the two surfaces. In this region the pressures are zero. This fluid is shown schematically by the chained line in the figure. At some point a full oil film forms and the inlet pressure sweep starts. The location of this point is determined by the flow balance between the oil entering the film and that being convected into the conjunction.

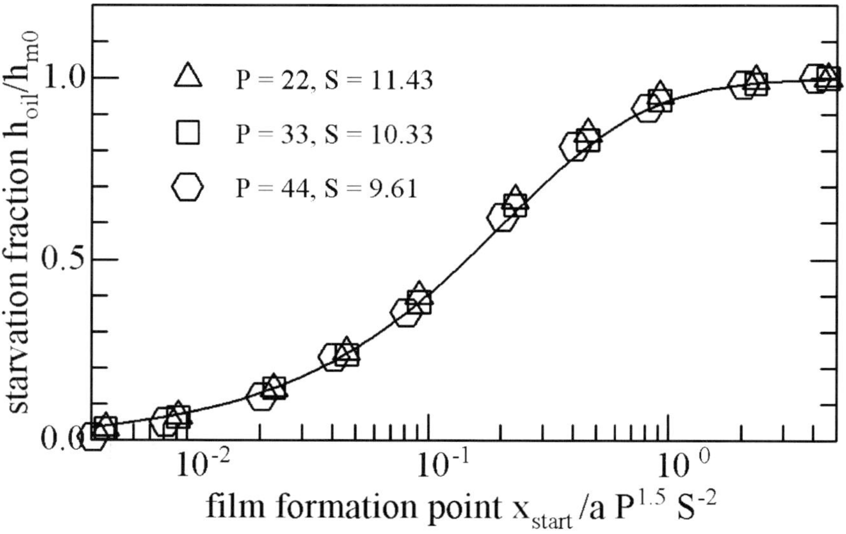

Figure 2. Relationship between degree of starvation and the start of film formation.

Defining the total thickness of the fluid entering the contact as h_{oil} the flow balance is given by:

$$uh_{\text{oil}} = uh - \frac{h^3}{12\eta}\frac{\partial P}{\partial x}. \tag{1}$$

For line contacts the easiest way to define the degree of starvation is to define the point at which the oil film forms and then, once the EHL problem has been solved, to find hoil from the central film thickness. (Flow continuity requires that the flow through the contact is identical at all points.) Figure 2 shows the result for a range of operating conditions.

It may be noted that h_{m0} is the central film thickness under fully flooded conditions scaled by the change in density at the contact centre. Thus the ratio h_{oil}/h_{m0} represents the reduction in central clearance as the degree of starvation increases. x_{start} is the distance to the left of the Hertz contact boundary at which the inlet pressure sweep starts.

Under fully flooded conditions the length of the inlet region is proportional to $aP^{-1.5}S^2$ and x_{start} has been scaled by this distance in the figure. It may be seen that there is a simple relationship between the reduction in film thickness and the location of this point relative to the fully flooded inlet length and that a single curve applies for all operating conditions.

The trend line in the figure was obtained from an analysis of the inlet of a heavily loaded contact and details are given in the Appendix.

The behaviour of point contacts is somewhat different. For the line contact problem the only steady state solution that can exist for $h_{\text{oil}}/h_{m0} \geq 1$ is

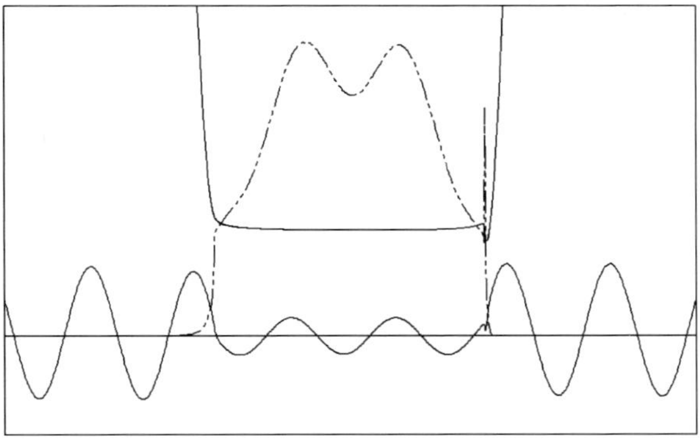

Figure 3. Sinusoidal surface passing through a starved EHL contact.

the fully flooded solution, as, by definition, all lubricant supplied has to pass through the centre. A point contact in such cases still has a starved steady state solution as lubricant also flows around the contact. However, with increasing degree of starvation this amount decreases and as was shown by Chevalier et al. [12] for heavily starved contacts almost all lubricant passes through the central region. So, for $h_{\text{oil}}/h_{m0} < 1$, the results in Figure 2 will still apply with reasonable accuracy on the contact's centre-line.

4. EFFECT OF ROUGHNESS

As a rough surface passes through an EHL contact the roughness tends to be reduced in amplitude. This is shown in Figure 3 for a surface with transverse roughness having a wavelength equal to 80% of the Hertz semi-width. Although the results in this paper are for low amplitude roughness, a relatively large amplitude has been used here in order to allow the behaviour to be shown clearly.

The roughness is partially flattened under the conjunction producing a sinusoidal clearance variation of near constant amplitude with a wavelength equal to that of the original surface. Although not obvious, there is a slight phase difference between the clearance variation and the original profile. A pressure ripple is associated with the attenuation of the roughness and this is related to the change in profile via the elasticity equations for the surfaces.

The ratio of the clearance variation to the original roughness may either be calculated from numerical solutions of the full time dependent problem with the waviness moving through the contact, see Figure 3, or by a perturbation analysis. These processes have been described in detail elsewhere [2, 6] and will not be repeated here. In these analyses care is required to ensure that flow continuity is maintained at the inlet to the conjunction and it is, perhaps, worth detailing this behaviour here.

If the location of the start of the fluid film is defined by $x = q$, where the location will vary as the rough surface enters the contact, then the equation of flow continuity becomes:

$$u(h - h_{\mathrm{oil}}) - \frac{h^3}{12\eta}\frac{\partial P}{\partial x} - (h - h_{\mathrm{oil}})\frac{\partial q}{\partial t} = 0. \tag{2}$$

Each term represents a loss of fluid from the point of film formation. The first term gives the fluid convected away by the mean surface velocity. The second the flow due to the pressure gradient while the third is the loss of fluid due to the movement of the formation point.

The continuity equation is required in this explicit form for the perturbation analysis used, in this paper, to obtain the line contact results. For the point contact problem the approach of Chevalier et al. [12] and Wijnant [13] has been used. A generalized Reynolds equation is used in which an additional variable representing the degree to which the film is filled is introduced. Subsequently the problem is solved as a complementarity problem in terms of this variable and the pressure, i.e. when the film is locally fully filled a pressure is solved. When it is only partially filled the pressure must be zero and the degree to which the film is filled is solved from flow continuity. In this process the boundary between the pressurized and cavitated/starved regions is solved implicitly. For details regarding the equations and the numerical approach the reader is referred to [13]. A typical point contact solution is shown in Figure 4.

5. RESULTS

Values of the amplitude ratio were obtained for three operating conditions for a number of roughness wavelengths and for a number of different degrees of starvation. A sample of these is shown in Figure 5. In this figure the hollow symbols represent results for line contacts, the solid symbols those for point contacts.

It may be seen that the short wavelengths roughness enters the contact unchanged while long wavelength profiles produce little change in clearance. The wavelength at which the transition from short to long wavelength behaviour occurs depends on the operating conditions and on the degree of starvation. First, as the contact becomes more heavily loaded and the length of the inlet decreases, the transition moves to shorter wavelengths. A similar effect is seen where the inlet length is reduced by starvation.

For fully flooded conjunctions it was found that it was the ratio of the roughness wavelength to the length of the contact inlet – defined by $aP^{-1.5}S^2$ – that controlled the amplitude ratio. To see whether this applies to starved contacts

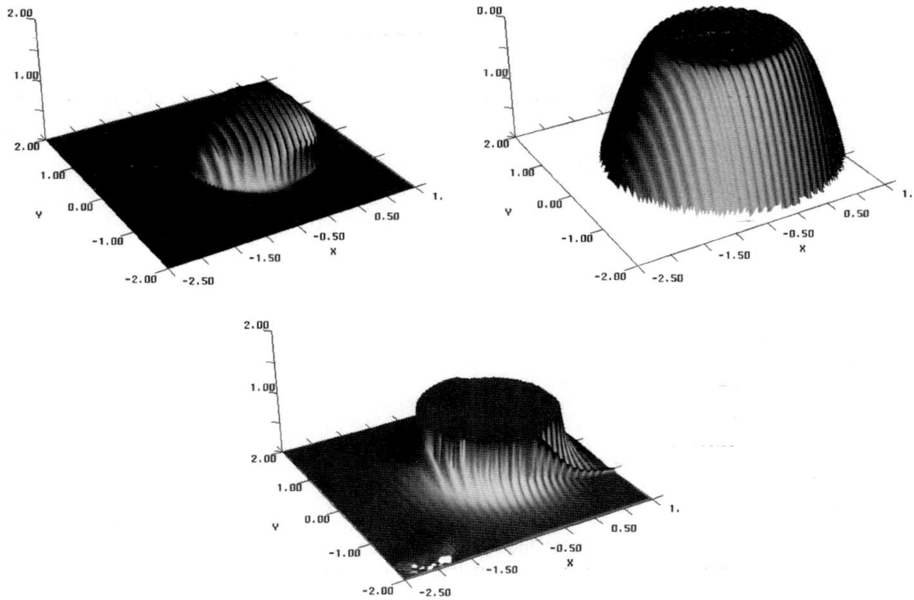

Figure 4. (Dimensionless) pressure, film thickness, and fractional film content for a starved EHL contact with transverse waviness. $P = 22$, $S = 10.3$, $\lambda/a = 0.125$, $h_{oil}/h_{m0} = 0.4$.

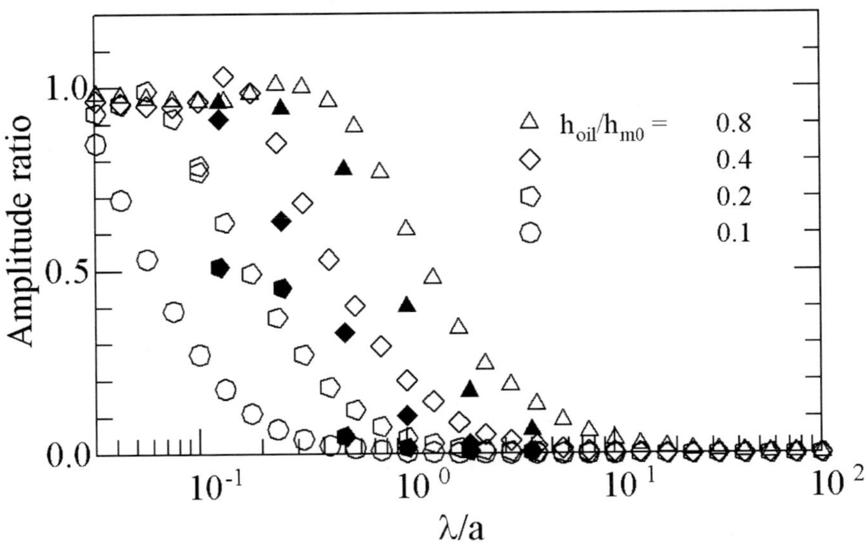

Figure 5. Amplitude ratio vs wavelength for different levels of oil starvation. Filled symbols – point contacts; hollow symbols – line contacts. $P = 22$, $S = 11.4$, $P_{Hertz} = 1$ GPa.

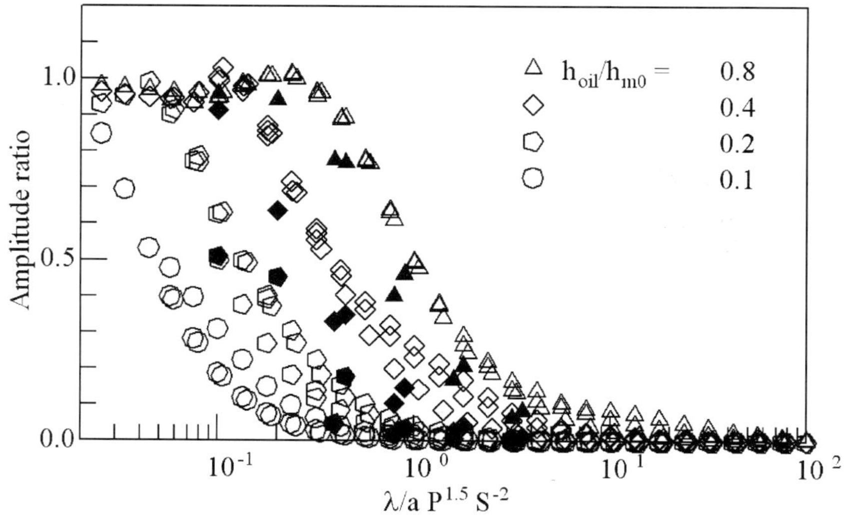

Figure 6. Amplitude ratio against the ratio of wavelength to fully flooded inlet length for different degrees of starvation for three different operating conditions: $P = 22$, $S = 11.4$, $P_{\text{Hertz}} = 1$ GPa; $P = 33$, $S = 10.3$, $P_{\text{Hertz}} = 1.5$ GPa; $P = 44$, $S = 9.6$, $P_{\text{Hertz}} = 2$ GPa.

the results for all the operating conditions were plotted in the form of amplitude ratio against this wavelength ratio and the resulting graph is shown in Figure 6. It may be seen that, although there is some scatter, the results for each degree of starvation fall, broadly, onto a single curve. Thus the amplitude ratio appears to depend solely on the ratio of the wavelength to the length of the fully flooded inlet and on the degree of starvation.

Examining Figure 2 it may be seen that as h_{oil}/h_{m0} decreases the point at which the film forms moves closer to the Hertz contact boundary, reducing the length of the inlet. This suggests that for starved contacts the amplitude ratio might be related, directly, to this reduced inlet length. In order to explore this, the fully flooded inlet length was taken as being equal to $aP^{-1.5}S^2$ while the inlet length under starved conditions was assumed to be a multiple of the length x_{start}. A number of values of this multiple were tried but a multiple of 2.5 was eventually selected as given the best correlation with the fully flooded results. Thus the inlet length L was selected as:

$$\begin{aligned} L &= aP^{-1.5}S^2 \quad \text{– fully flooded,} \\ L &= 2.5x_{\text{start}} \quad \text{– starved.} \end{aligned} \tag{3}$$

Values of $x_{\text{start}}/aP^{1.5}S^{-2}$ are readily obtained from Figure 2 allowing the ratio λ/L to be calculated for each point. Where the degree of starvation is small the inlet length, calculated for starved conditions, may become larger than the fully

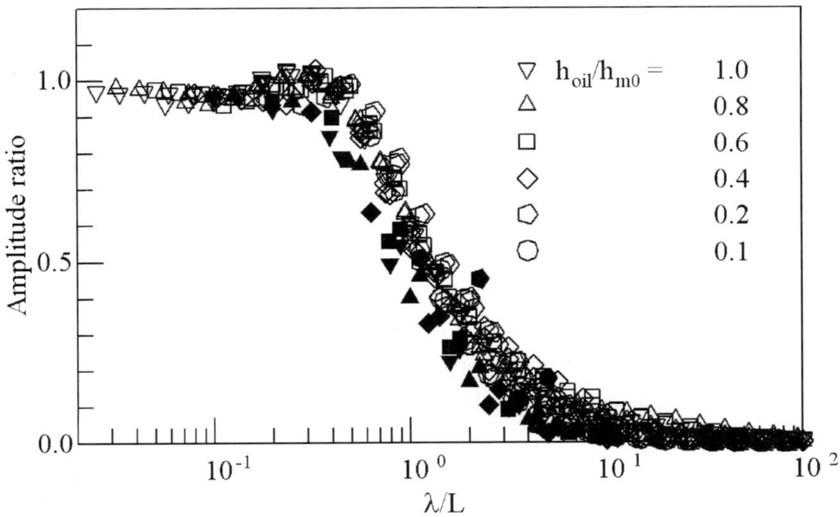

Figure 7. Amplitude ratio vs wavelength ratio.

flooded inlet length. Under these conditions the fully flooded length needs to be used.

Figure 7 shows the result. It is clear that, within a reasonable band, all the results fall onto a single master curve. Thus, it is the only ratio of the wavelength to the inlet length that determines the amplitude ratio. This result applies both to line and point contacts and means that a single curve defines the amplitude ratio under all conditions.

Where only the amplitude ratio is required the equation given in [15] may be used:

$$\frac{\delta h}{A} = \frac{1}{1 + 0.351\nabla + 0.315\nabla^2} \quad \text{where} \quad \nabla = \frac{\lambda}{L}. \tag{4}$$

Where information about the phase of the attenuated wave or the magnitude of the pressure ripple is required the result given in [16] is to be preferred:

$$\frac{\delta h}{A} = 1 + \frac{2}{\pi} wp \quad \text{and} \quad \frac{\delta PL}{E'A} = p, \tag{5}$$

where

$$p = -\frac{\pi}{2} \frac{1}{c_0/w + c_1 + w}, \qquad w = \frac{\lambda}{L}$$

and

$$c_0 = -0.22 + 0.40_i, \quad c_1 = 0.55 + 0.55i.$$

6. CONCLUSIONS

A detailed analysis of roughness effects in line and point EHL contacts under starved conditions has been carried out.

It has been shown, first, that for line contacts, there is a unique relationship between the degree of starvation and the contact's inlet length. This relationship applies, also, to highly starved point contacts.

Second, it has been shown that the roughness attenuation inside the conjunction contacts depends only on the ratio of the wavelength to the inlet length. This applies whatever the operating conditions and whatever the degree of starvation. It also applies whether the contact is a line or a point conjunction and whatever the contact's ellipticity ratio.

A simple equation is given for the relationship between the amplitude ratio and this wavelength ratio with slightly more complex equations for the magnitude of the pressure ripple and for the phases of the clearances and pressures.

With this information it is straightforward to carry out rapid assessments of the likely effects of rough surfaces under any conditions using discrete (fast) Fourier transform methods irrespective of the type of contact, the operating conditions or the degree of starvation.

APPENDIX: STARVATION IN HEAVILY LOADED CONTACTS

It was shown [17] that, under fully flooded conditions, the central clearance in heavily loaded, incompressible contacts can be obtained, in non-dimensional form, from the integral:

$$\bar{h}_m^{4/3} = \frac{2}{3}\int_0^1 y^{1/3}(1-y)^{2/3}dy = \frac{1}{3}\Gamma\left(\frac{4}{3}\right)\Gamma\left(\frac{5}{3}\right),$$

where y is equal to $\bar{h}_m/\bar{h}$ and $\bar{h} = \bar{h}_m + (-\bar{x})^{1.5}$.

The lower bound of the integral corresponds to a point well to the left of the Hertz boundary, the upper limit to Hertz boundary itself. Under starved conditions all that is required is to replace the lower bound by $\bar{h}_m/\bar{h}_{\text{start}}$ and to numerically evaluate the integral. Then, with $\bar{h}_m$ defined, $\bar{h}_{\text{start}}$ and $\bar{x}_{\text{start}}$ may be found. Finally, noting that $\bar{x} = 2.17732x/aP^{1.5}S^{-2}$, values of h_{oil}/h_m and $x_{\text{start}}/aP^{1.5}S^{-2}$ may be obtained. Repetition with a range of values of $\bar{h}_m/\bar{h}_{\text{start}}$ enables the curve in Figure 2 to be plotted.

For most purposes the above calculation is inconvenient and it is preferable to replace the exact relationship between the inlet length ratio and the degree of starvation by an approximate equation. Although a number of forms for this are possible, the relationship:

$$\frac{h_{\text{oil}}}{h_{m0}} = 1 - \exp(-2.9q^{0.75}) \quad \text{where} \quad q = x_{\text{start}}/aP^{1.5}S^{-2}$$

combines simplicity with reasonable accuracy.

REFERENCES

[1] Greenwood, J.A. and Morales Espejel, G.E., "The behaviour of transverse roughness in EHL contacts", *Proc. ImechE, Part J*, **218**, 1994, 121–132.

[2] Venner, C.H., Couhier, F., Lubrecht, A.A. and Greenwood, J.A., "Amplitude reduction of waviness in transient EHL line contacts", in *Proc. 22nd Leeds–Lyon Symposium on Tribology*, Elsevier Tribology Series 31, 1996.

[3] Venner, C.H. and Lubrecht, A.A., "Amplitude reduction of non-isotropic harmonic patterns in EHL circular contacts under pure rolling", in *Proc 24th Leeds–Lyon Symposium on Tribology*, Elsevier, 1998.

[4] Hooke, C.J., "Surface roughness modification in EHL line contacts – The effect of roughness wavelength, orientation and operating conditions", in *Proc. 25th Leeds–Lyon Symposium on Tribology*, Lyon, September 1998.

[5] Hooke, C.J., "The behaviour of heavily loaded line contacts with transverse roughness", *Proc. Instn. Mech. Engrs., Part C*, **213**, 1999, 309–320.

[6] Hooke, C.J., "The behaviour of low amplitude surface roughness under line contacts", *Proc. Instn. Mech. Engrs., J. of Eng. Trib.*, **213**, 1999, 275–286.

[7] Hooke, C.J., "The behaviour of low amplitude surface roughness under line contacts: Non-Newtonian fluids", *Proc. Instn. Mech. Engrs., Part C*, **214**, 2000, 253–265.

[8] Venner, C.H., Kaneta, M. and Lubrecht, A.A., "Surface roughness in elastohydrodynamically lubricated contacts", in *Proc. 26th Leeds–Lyon Symposium on Tribology*, Elsevier Tribology Series 38, 2000.

[9] Hooke, C.J. and Venner, C.H., "Surface roughness attenuation in line and point contacts", *Proc. Instn. Mech. Engrs., J. of Eng. Tribology*, **214**, 2000, 439–444.

[10] Wedeven, L.D., Evans, D. and Cameron, A., "Optical analysis of ball bearing starvation", *Trans ASME, J. of Tribology*, 1971, 349–363.

[11] Hamrock and Dowson, "Isothermal elastohydrodynamic lubrication of point contacts, Part IV – Starvation results", *Trans. ASME, J. of Tribology*, 1977, 15–23.

[12] Chevalier, F., Lubrecht, A.A., Cann, Colin, P.M.E. and Dalmaz, G., "Film thickness in starved EHL point contacts", *Trans. ASME, J. of Tribology*, **120**, 1998, 126–133.

[13] Wijnant, Y.H., "Contact dynamics in elastohydrodynamic lubrication", PhD Thesis, University of Twente, Enschede, The Netherlands, 1998.

[14] Venner, C.H., Berger, G. and Lugt, P.M., "Waviness deformation in starved EHL circular contacts", *Trans. ASME, J. of Tribology*, **126**, 2004, 248–257.

[15] Venner, C.H. and Morales, G.E., "Amplitude reduction of small amplitude waviness in transient elastohydrodynamically lubricated line contacts", *Proc. Inst. Mech. Engrs., Part J*, **213**, 1999, 487–504.

[16] Hooke, C.J., "Roughness attenuation and pressure rippling in EHL contacts", in *IUTAM Symposium on Elastohydrodynamics and Microelastohydrodynamics*, R.W. Snidle and H.P. Evans (eds), Cardiff, September 2004, Springer, Dordrecht, 2006 (this volume).

[17] Hooke, C.J., "The elastohydrodynamic lubrication of heavily loaded contacts", *J. Mech Eng. Sci.*, **19**, 1977, 149–156.

UNSTEADY EHL: 'NODAL' & 'MODAL' FORMULATIONS

J.F. Booker[1] and S. Boedo[2]
[1]*Sibley School of Mechanical & Aerospace Engineering, Cornell University, Ithaca, NY 14853, U.S.A.*
[2]*Department of Mechanical Engineering, Rochester Institute of Technology, Rochester, NY 14623, U.S.A.*

Abstract A novel and rigorous 'inverse hydrodynamic' approach is described in a companion paper for problems of unsteady elastohydrodynamic lubrication. In it all spatially-distributed quantities are represented by nodal values via finite element discretization. Though the formulation accommodates piezoviscous lubricant models as well as planar, cylindrical, and spherical geometries, it is illustrated here by the prototypical planar example of normal approach of a lightly-loaded rigid indenter to a discretized elastic solid separated by a nearly isoviscous lubricant film. Results from the new 'nodal' formulation are compared with those from a previous (and more approximate) 'modal' approach to unsteady EHL computation.

Keywords: unsteady EHL, nodal formulation, modal formulation, inverse hydrodynamic.

1. INTRODUCTION

A companion paper (Booker, 2005) introduces an apparently novel formulation for unsteady EHL computation incorporating the geometric generalization possible with finite element discretization of both viscous fluid and elastic solid components.

The result is presented there in the form of a clearly-stated initial-value problem with a clearly-stated solution sequence involving no form of iteration whatsoever (absent cavitation), even in the presence of piezoviscosity.

Though the formulation is applicable to arbitrary film geometries (planar, cylindrical, spherical, etc.), it is illustrated in the companion paper by a prototypical planar example of normal approach to an elastic semi-infinite half-space. The present paper is illustrated by a very similar problem of normal approach to a finite elastic solid cylinder. Results of the new 'nodal' formulation are compared to those of a previous (and more approximate) 'modal' formu-

R.W. Snidle and H.P. Evans (eds), IUTAM Symposium on Elastohydrodynamics and Microelastohydrodynamics, 71–80.

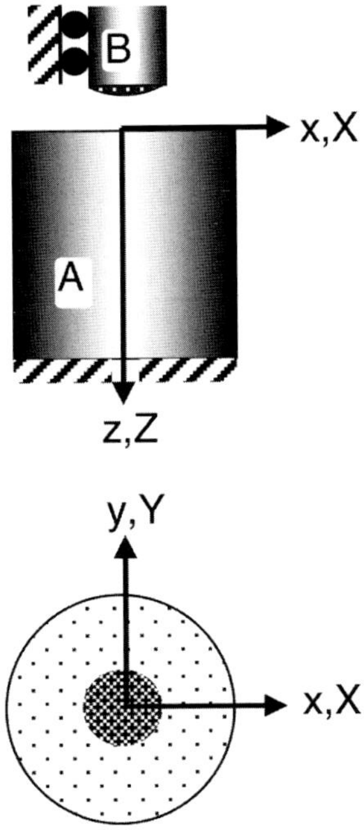

Figure 1. Numerical example: normal approach.

lation described elsewhere in some detail by the authors (Boedo and Booker, 1997).

2. NUMERICAL EXAMPLE

2.1 Test Case

Figure 1 suggests the specific problem of normal approach of a steadily-loaded spherically-crowned rigid circular cylinder ('indenter') B toward an elastic solid circular cylinder ('base') A through a flooded piezoviscous lubricant. (The 'contact zone' of active lubricant film is indicated by the darkened central circle in the lower view.) Figure 1 also shows coincident system coordinates X, Y, Z and film coordinates x, y, z.

Table 1 contains a set of parameters for a numerical study similar (but *not* identical) to physical experiments using mineral oil and a thin base layer of silicone rubber as reported by Gaman et al. (1974) and Higginson (1978). Gaman et al. (1974) call this 'squeeze film' problem the 'Cinderella of EHL' and provide a brief review of its literature, sparse then as now.

Table 1. Test case parameters.

E	$= 1.0$	MPa	Young's modulus
ν	$= 0.45$	–	Poisson's ratio
α	$= 25.0$	GPa^{-1}	Barus coefficient
p_0	$= 0.0$	Pa	reference pressure
μ_0	$= 0.5$	Pa·s	reference viscosity
ρ	$= 850.0$	$\mathrm{kg{\cdot}m}^{-3}$	density
R	$= 22.25$	mm	crown radius
d	$= 3.0$	mm	indenter diameter
D	$= 6.0$	mm	base diameter
L	$= 6.0$	mm	base length
$u^x = u^y$	$= 0.0$	$\mathrm{m{\cdot}s}^{-1}$	tangential entrainment velocity
f^Z	$= 0.13$	N	normal external force

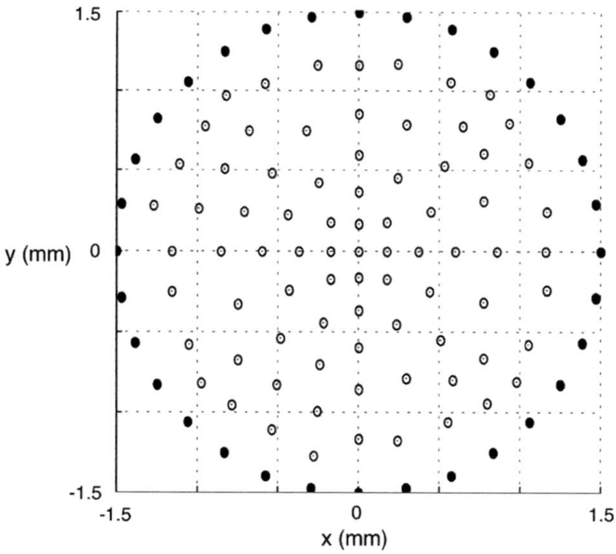

Figure 2. Film mesh.

2.2 Analysis Methods

Figure 2 shows a rather coarse and somewhat irregular discretization of the 'contact zone' as produced with a commercial finite element package (ANSYS). The 111 nodes shown are related by triangular fluid finite elements (described by Boedo and Booker, 1997) and elastic solid stiffness matrices (obtained from tetrahedral elements via ANSYS).

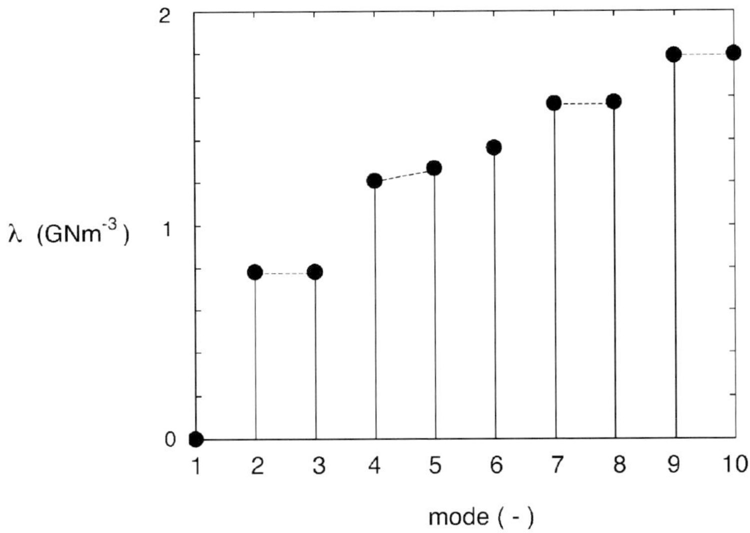

Figure 3. Modal eigenvalues 1–10.

A comparison is made between solutions based on the new 'nodal' approach described by Booker (2005) and an earlier 'modal' approach reviewed by Boedo and Booker (1997).

In the 'nodal' approach the film thicknesses (or equivalently, relative surface displacements) at interior nodes (shown as open circles in Figure 2) serve as independent state variables or generalized coordinates.

In the 'modal' approach the independent generalized coordinates are modal displacements corresponding to a truncated subset of orthogonal generalized eigenmodes based on the elastic structure. Figure 3 shows the first 10 eigenvalues for the present problem, corresponding to successively more complicated eigenmodes, the first 8 of which are shown in Figure 4. The near-pairing of most eigenvalues corresponds to the near-pairing of orthogonal eigenmodes occasioned by the physical axi-symmetry of the problem. (The visible difference in eigenvalues 4 and 5 is presumably caused by irregularity in the nodal mesh of Figure 2.) Isolated eigenmode 1 is a uniform rigid-body displacement, while isolated eigenmode 6 is a rather hat-shaped near-axisymmetric elastic displacement.

The problem chosen is a particularly difficult one for the 'modal' approach, in that deformations are more *local* than *global* (as would be the case in such ring-like structures as connecting rod bearings). Consequently, the sequence of eigenvalues (corresponding to modal energies) in Figure 3 rises relatively slowly, resulting in the need to carry a relatively large number of modes (30 in the present case) throughout the computation to maintain acceptable accuracy.

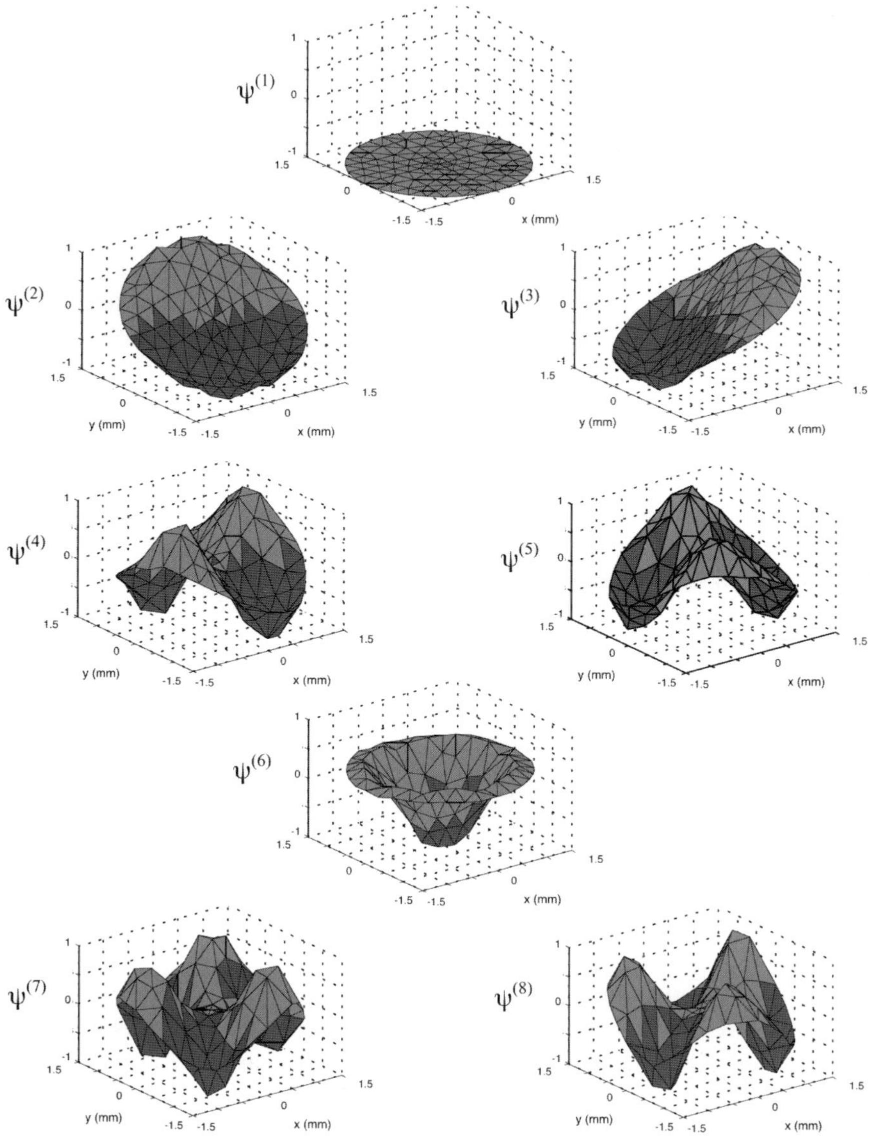

Figure 4. Modal eigenvectors 1–8.

A rigorous comparison of the two methods would require identical initial conditions. For example, the 'nodal' generalized coordinates might be derived from an arbitrary choice of 'modal' generalized coordinates (if this resulted in positive initial film thicknesses at exterior nodes). In the present (somewhat imperfect) comparison, however, initial conditions have been chosen independently for the two methods. Each solution has then been continued until the detailed effects of initial conditions have disappeared; corresponding 'snapshots' of results have then been compared at instants when minimum film thicknesses are essentially identical.

2.3 Results

Figure 5 compares instantaneous film thickness distributions for both 'nodal' and 'modal' solutions. The discrepancies are obvious, as are the similarities. Both distributions show a surprising lack of axisymmetry, perhaps owing to irregularity of the very coarse discretization shown in Figure 2. Decomposition of the results shows that both 'nodal' and 'modal' distributions are dominated by axisymmetric eigenmode 6, an unsurprising finding.

Figure 6 shows the same comparison for a central section (at $y = 0$); the same discrepancies and similarities are visible as in Figure 5.

Figure 7 compares film pressure distributions corresponding to the film thickness distributions of Figure 5 and derivable from it by structural considerations alone. The results are very similar, though not identical.

Figure 8 shows the same comparison for a central section (at $y = 0$); very minor differences are visible as in Figure 7.

The very low pressures of Figures 7 and 8 result in essentially uniform viscosity distributions (not shown).

Figure 9 compares resultant outward nodal flows arising hydrodynamically from the situation of Figures 5–8. The two results are significantly different overall, with a surprising lack of axisymmetry shown by the 'modal' predictions.

3. CONCLUSIONS

The lack of axisymmetry in both 'nodal' and 'modal' results suggests a need for further studies using much finer and more regular meshes. Discrepancies between results suggest further studies with identical initial conditions and/or much longer runs, possibly leading to much thinner films.

Future work might also consider axisymmetric problems of normal approach to a semi-infinite half-space (as a means of eliminating possible effects of irregular discretization).

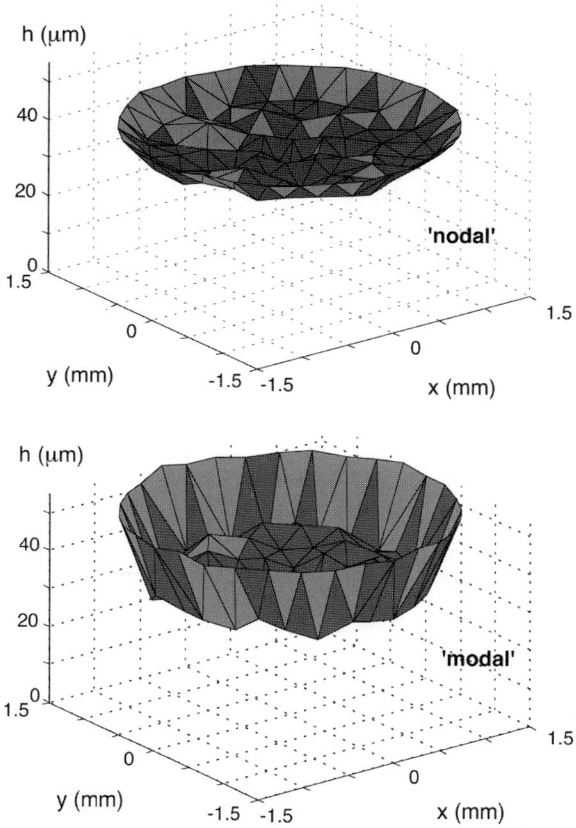

Figure 5. Film thickness; 'nodal': $h_{\min} = 29.92\ \mu$m; 'modal': $h_{\min} = 29.91\ \mu$m.

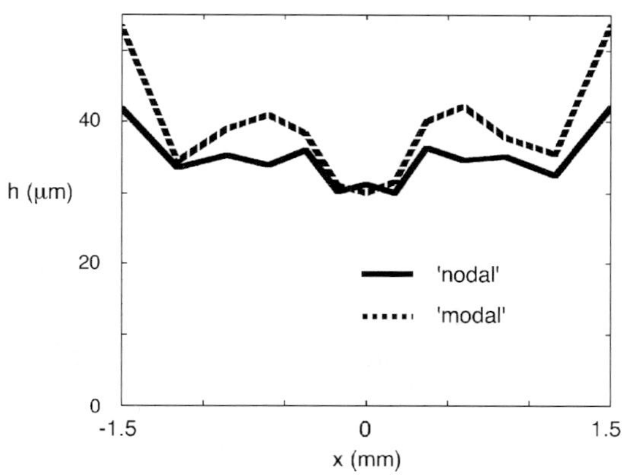

Figure 6. Film thickness (section $y = 0$).

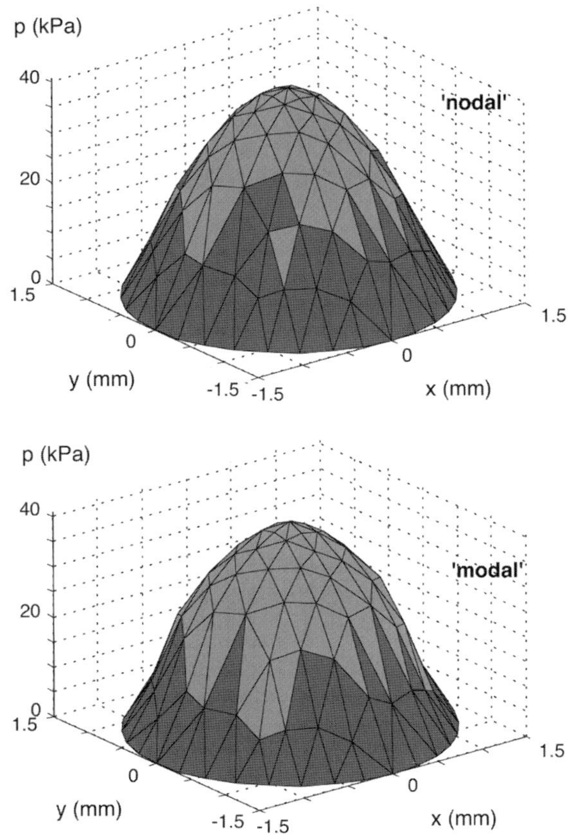

Figure 7. Film pressure; 'nodal': $p_{\max} = 40.78$ kPa; 'modal': $p_{\max} = 41.08$ kPa.

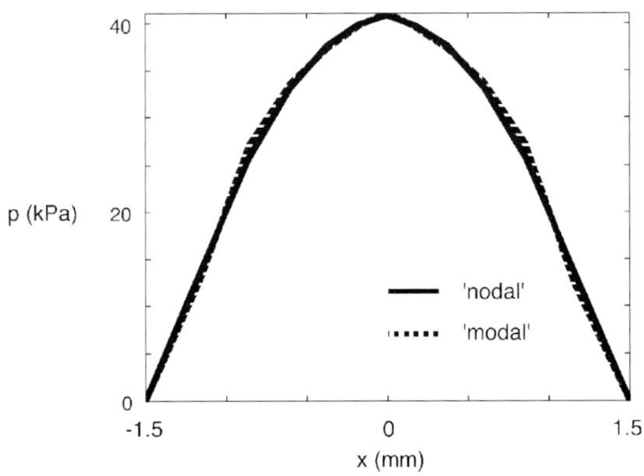

Figure 8. Film pressure (section $y = 0$).

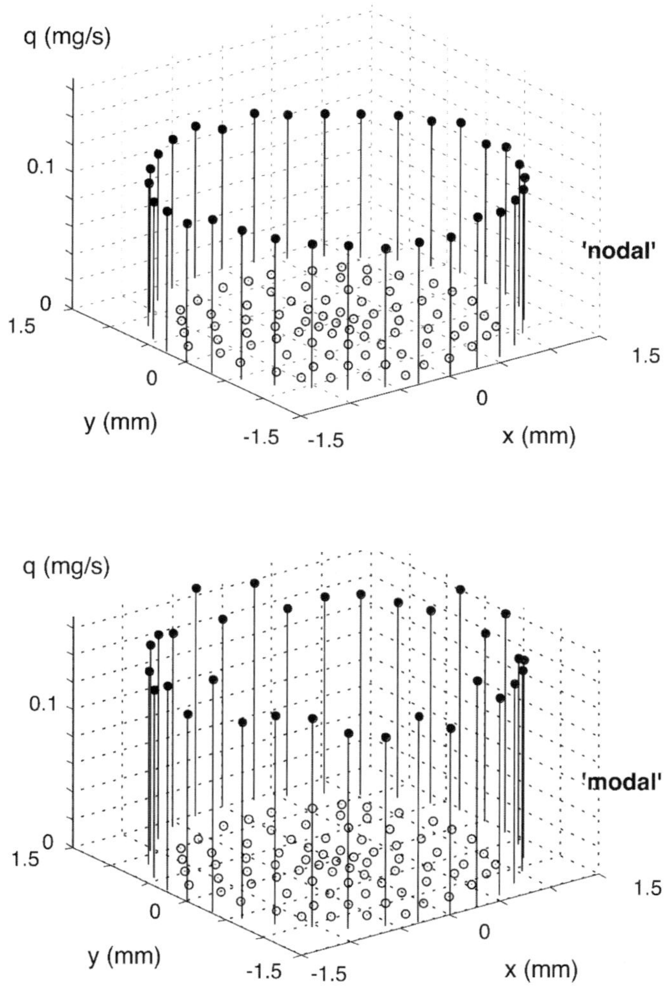

Figure 9. Nodal resultant outward flow;
'nodal': $q_{\max} = 0.1116$ mg/s; $q_{\text{sum}} = 3.384$ mg/s;
'modal': $q_{\max} = 0.1687$ mg/s; $q_{\text{sum}} = 4.695$ mg/s.

ACKNOWLEDGEMENT

The authors greatly appreciate the generous advice given to them by Professor Sir Gordon Higginson.

REFERENCES

Boedo, S. and Booker, J.F., 1997, Surface roughness and structural inertia in a mode-based mass-conserving elastohydrodynamic lubrication model, *Trans. ASME, J. Trib.* **119**(3), 449–455.

Booker, J.F., 2005, Unsteady EHL: An inverse hydrodynamic formulation, in *Life Cycle Tribology, Proc. 31st Leeds–Lyon Symposium on Tribology*, Leeds, 7–10 September 2004, D. Dowson, M. Priest, G. Dalmaz and A.A. Lubrecht (eds), Elsevier, Amsterdam, pp. 65–72 (ISBN 0-444-51687-5).

Gaman, I.D.C., Higginson, G.R. and Norman, R., 1974, Fluid entrapment by a soft surface layer, *Wear* **28**:345–352.

Higginson, G.R., 1978, Squeeze films between compliant solids, *Wear* **46**:387–395.

THE INFLUENCE OF SPINNING ON THE PERFORMANCE OF EHL IN ELLIPTICAL CONTACTS

Peiran Yang and Jinlei Cui
School of Mechanical Engineering, Qingdao Polytechnic University,
11 Fushun Road, Qingdao 266033, China

Abstract Numerical analyses were carried out for the elastohydrodynamic lubrication (EHL) of elliptical contacts with spinning and rolling. The centre of spinning was assumed to be located on the axis passing through the major radius of the contact ellipse. Under steady-state conditions with spinning and rolling it was found that the spinning motion has significant influence on the film profiles along the major axis of the contact ellipse but hardly affects the film profiles along the minor axis. It was proved that pure-spinning can generate a complete lubricant film in an elliptical contact, although at the centre of this contact entrainment is absent. Time-dependent solutions were achieved for the contacts with spinning and rolling with dynamic loading, and it was found that film thicknesses can be increased to some extent by the dynamic loads.

Keywords: elastohydrodynamic lubrication, spinning, elliptical contact, dynamic load.

1. INTRODUCTION

Many machine elements, such as the balls and spherical rollers in some types of roll-element bearings, are working under spinning and rolling conditions. However, because of the limitation of the mathematical model, it is impossible for the effects of spinning to be taken into account in the EHL analyses for line contacts.[1] The effects of spinning on EHL has also been ignored in most EHL analyses for point contacts.[2] Some experimental results for the friction in a pure-spinning elliptical EHL contact have been reported by Allen *et al.*[3] A hydrodynamic analysis for a simple spinning problem has

R.W. Snidle and H.P. Evans (eds), IUTAM Symposium on Elastohydrodynamics and Micro-elastohydrodynamics, 81–92.

been carried out by Snidle and Archard,[4] and numerical EHL results for the film thickness in the spinning-and-rolling elliptical contacts have been given by Dowson *et al.*[5] and Zou *et al.*[6]

In most point contact elastohydrodynamic lubrication analyses in the literature, the lubricated bodies were simplified as an ellipsoid and an infinite plane. Usually the ellipsoid was assumed to rotate with respect to its central axis while the infinite plane was assumed to move in a fixed direction such as along the minor axis of the contact ellipse. In this paper, however, although the ellipsoid was also assumed to rotate with respect to its central axis, the infinite plane was assumed to rotate around a fixed point called the centre of spinning, which was located on the line coincident with the major axis of the contact ellipse. In such a way the influence of spinning on the performance of EHL in elliptical contacts can be investigated theoretically. Because in the domain of numerical solution the nodal velocity components of the infinite plane, no matter their magnitudes or directions, are different with locations, the analysis is more complex than usual. Note that if the distance between the centre of spinning and the centre of the contact ellipse is infinite, the present arrangement is equivalent to the conventional arrangement of EHL point contacts,[2] but, if this distance becomes zero and the ellipsoid is stationary, the present arrangement describes a pure-spinning case.[3, 5]

2. GOVERNING EQUATIONS

As shown in Figure 1, O_a is the rotating centre of the infinite plane, i.e., surface a, O is the centre of the contact ellipse, and r is the distance between O and O_a. When the angular velocity of surface a is denoted by ω, at any point the velocities of surfaces a and b are

$$\begin{cases} u_a = \omega(y+r),\ v_a = -\omega x \\ u_b = \text{constant},\ v_b = 0 \end{cases} \tag{1}$$

Therefore, the entraining velocities in the x and y directions are

$$u = [\omega(y+r) + u_b]/2,\ v = -\,\omega x/2 \tag{2}$$

According to the work of Yang and Wen,[7] under isothermal and Newtonian-fluid conditions the time-dependent Reynolds equation can be written as

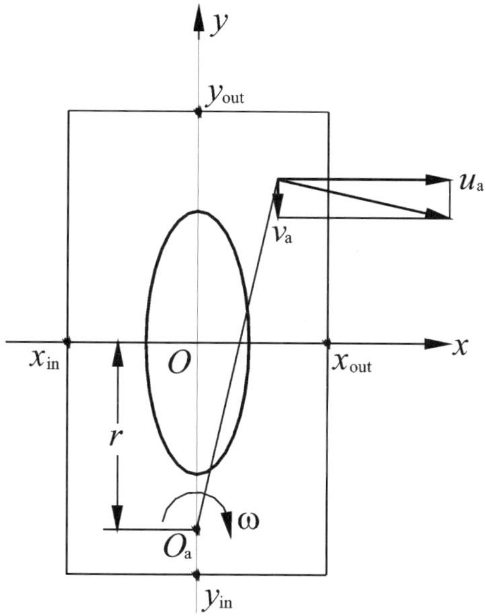

Figure 1. Calculation domain and velocity components at any point on surface a.

$$\frac{\partial}{\partial x}\left(\frac{\rho h^3}{12\eta}\frac{\partial p}{\partial x}\right)+\frac{\partial}{\partial y}\left(\frac{\rho h^3}{12\eta}\frac{\partial p}{\partial y}\right)=\frac{\partial(\rho u h)}{\partial x}+\frac{\partial(\rho v h)}{\partial y}+\frac{\partial(\rho h)}{\partial t} \tag{3}$$

where variable t is time, p and h are pressure and film thickness respectively, and ρ and η are density and viscosity of the lubricant respectively.

The boundary conditions of Eq. (3) are: pressures are equal to zero on all boundaries of the solution domain shown in Figure 1, and over the whole domain, negative pressures are forced to be zero so that the conditions of oil cavitation at unknown boundaries can be satisfied automatically.

The time-dependent film thickness equation can be written as[8]

$$h(x,y,t)=h_{00}(t)+\frac{x^2}{2R_x}+\frac{y^2}{2R_y}+\frac{2}{\pi E'}\iint\frac{p(x',y',t)\mathrm{d}x'\mathrm{d}y'}{\sqrt{(x-x')^2+(y-y')^2}} \tag{4}$$

where E' is the effective modulus of elasticity, R_x and R_y are radii of curvatures in the x and y directions respectively. The unknown function $h_{00}(t)$ represents the rigid distance between contact bodies, and can be determined at any instant from the load balance equation.

In this study it is assumed that the load changes sinusoidally,[9] i.e.,

$$\iint p(x, y, t)\mathrm{d}x\mathrm{d}y = w_0 \cdot C_\mathrm{w}(t) \tag{5}$$

where w_0 is a constant load, and $C_\mathrm{w}(t)$ is a dimensionless load changing function:

$$C_\mathrm{w}(t) = 1 + A_\mathrm{w}\sin(2\pi t/t_\mathrm{p}) \tag{6}$$

where A_w and t_p are the amplitude and period of the dynamic load.

The relationship of the lubricant viscosity and pressure proposed by Roelands[10] was used in this study:

$$\eta = \eta_0 \exp\left\{A_1\left[(1 + A_2 p)^Z - 1\right]\right\} \tag{7}$$

where $A_1 = \ln\eta_0 + 9.67$, $A_2 = 5.1\times10^{-9}$ Pa^{-1}, $Z = \alpha/(A_1 A_2)$, here α is the Barus viscosity-pressure coefficient.

The density-pressure relationship proposed by Dowson and Higginson[1] was adopted:

$$\rho = \rho_0\left[1 + 0.6\times10^{-9} p/\left(1 + 1.7\times10^{-9} p\right)\right] \tag{8}$$

The above is a time-dependent model. Since the load is a periodic function, the solution must also be periodic. When the last term in the right-hand side of Eq. (3) is omitted and we let $C_\mathrm{w}(t)$ in Eq. (5) be equal to unity, the model can also describe a steady-state problem.

3. NUMERICAL APPROACH AND INPUT DATA

The periodic solution of the above problem was achieved time step after time step starting from the steady-state solution at the load w_0. 360 time steps with equal distance were arranged in a load changing period t_p. The periodic errors of pressures and film thicknesses were checked at the end of each period and, if these errors were larger than the required accuracy, another period was started from the results obtained at the last instant of the last period. Usually after five or six periods the relative periodic errors could be reduced to be less than 0.001 and then the results of the last period were accepted as the pursued solution. At each time step the numerical approach used in this study is very similar to that developed by Jin *et al.*[8] That is, the pressure was solved with a multilevel method,[11] and the surface deformation was evaluated by the multilevel multi-integration method.[12] When applying the multilevel techniques, 5 levels of grids were employed. The number of nodes on the finest grid level is 257 in both directions.

It was assumed that both the infinite plane and the ellipsoid are made of steel, and the lubricant is a type of mineral oil. The material properties used in the analyses are: $E' = 226$ GPa, $\eta_0 = 0.08$ Pa·s, and $\alpha = 2.2\times10^{-8}$ Pa^{-1}. For all solutions the elliptical ratio is the same, i.e., $k_e = b/a = 6$, where b and a are the major and minor radius of the contact ellipse, respectively. The elliptical ratio can be determined from the values of R_x and R_y with the highly accurate formulae proposed by Markho[13] and the values used in this study are $R_x = 0.04$ m and $R_y = 0.633$ m, respectively.

4. STEADY-STATE RESULTS

Denote p_H as the maximum Hertzian dry contact pressure at the constant load w_0. Let $p_H = 1.2$ GPa, it can be calculated that $a = 0.829$ mm, and $b = 4.97$ mm. The calculation domain for the solutions of $p_H = 1.2$ GPa is: $x_{in} = -4.4a$, $x_{out} = 2.0a$, $y_{in} = -9.0a$, and $y_{out} = 9.0a$. For the case of $r = 5$ mm, $\omega = 250\pi$ rad/s (7500 rpm), and $u_b = 0.1$ m/s, the three-dimensional distributions of pressure and film thickness are shown in Figure 2. It can be seen that the spinning motion has very strong effect on the film thickness. In this case since r is only 5 mm, the centre of spinning is very close to the lower end of the contact ellipse (the point on the y axis with $y/a = -6$), u_a at the lower end is only 0.02 m/s, and therefore the entraining velocity in the x direction is only 0.06 m/s, so the film thickness in the vicinity of this point is very small. In detail, the minimum film thickness is 0.209 μm, which occurs at the node with $x/a = 0.125$ and $y/a = -5.73$. On the contrary, at the upper end of the contact ellipse (the point on the y axis with $y/a = 6$), since the angular velocity is very high, the surface velocity is $u_a = 7.83$ m/s, so the film thickness is very large (around 2 μm). The shape of the film, therefore, becomes an inclined flat plateau before the exit restriction. Corresponding to this film shape, the pressure at the upper end is a little higher than at the lower end. Note that in Figure 2, in order to show the distribution of pressure more clearly, the directions of both coordinates for the pressure are different from those for the film thickness.

The influence of the spinning motion on the film thicknesses and pressures decreases, drastically in the beginning, with the increase of the distance between the centre of spinning and the centre of the contact ellipse, if the entrainment at the centre keeps constant. This can be seen from Figure 3, where the pressure and film profiles on the y axis are plotted for different values of r while the central velocity of surface a, $r\omega$, is kept to be $u_a(0, 0) =$

$1.25\pi = 3.93$ m/s. It can be noticed from Figure 3 that, when r is larger than 40 mm, the effect of the spinning motion becomes very weak.

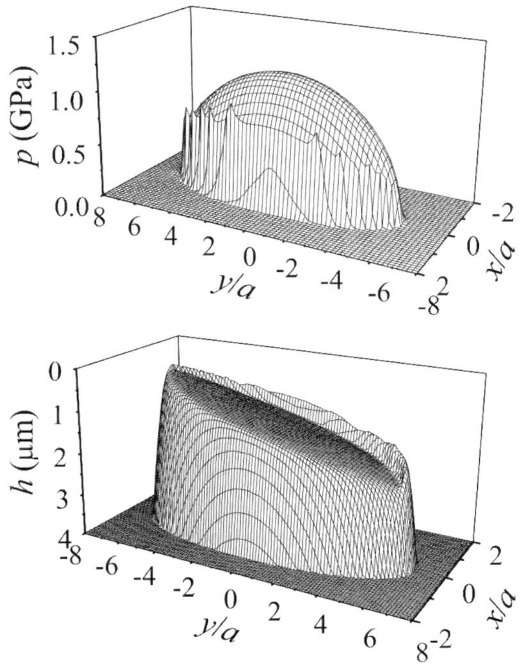

Figure 2. Steady-state pressure and film thickness for the case of $p_H = 1.2$ GPa, $\omega = 250\pi$ rad/s, $r = 5$ mm, and $u_b = 0.1$ m/s.

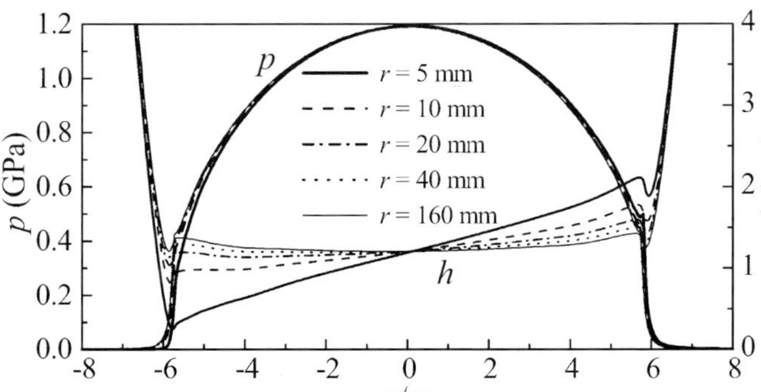

Figure 3. Distributions of pressures and film thicknesses on the y axis for various central distance. $p_H = 1.2$ GPa, $r\omega = 3.93$ m/s , and $u_b = 0.1$ m/s.

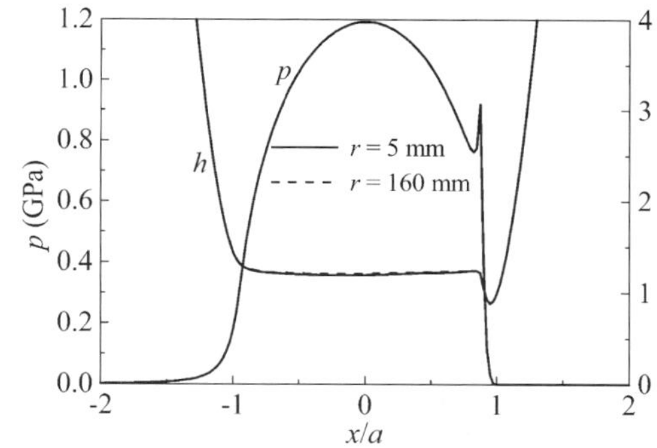

Figure 4. Distributions of pressures and film thicknesses on the x axis for various central distance. $p_H = 1.2$ GPa, $r\omega = 3.93$ m/s , and $u_b = 0.1$ m/s.

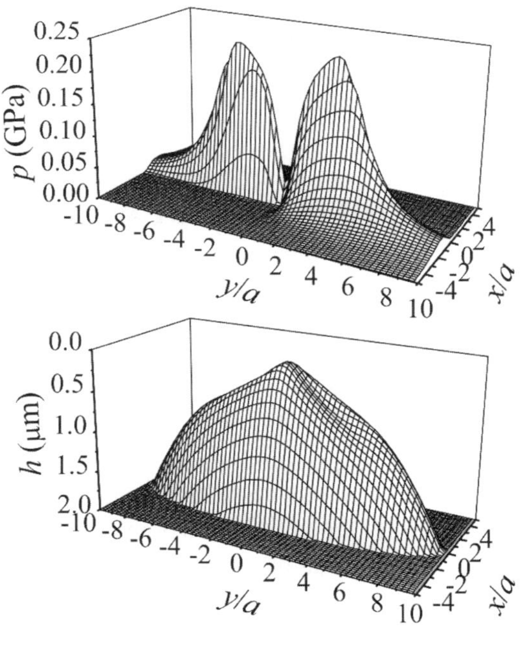

Figure 5. Pressure and film thickness under pure-spinning condition for the case of $p_H = 0.25$ GPa and $\omega = 600\pi$ rad/s.

It is interesting to note that in Figure 3 the five curves for film thicknesses almost pass through the same point, which is near the centre of the contact ellipse. This suggests that the spinning motion has little effect on the behaviour of lubrication along the minor axis of the contact ellipse. In Figure 4 the distributions of pressures and film thicknesses on the x axis are plotted for the two extreme cases in Figure 3, and these results are very similar to the results of the conventional EHL solutions. However, this is not true in the case of pure-spinning.

When both u_b and r are equal to zero, the centre of the contact and the centre of spinning are at the same point and, at this point the velocities are zero, so such a situation is called pure-spinning. If p_H is not large but the angular velocity is very high, hydrodynamic pressure and therefore full fluid film can be established by the pure-spinning motion. For $p_H = 0.25$ GPa and $\omega = 600\pi$ rad/s, the pressure and film thickness produced by the pure-spinning are illustrated in Figure 5. This is a lightly loaded case, the minor radius of the contact ellipse is only 0.173 mm. The calculation domain for this solution is: $x_{in} = -6.4a$, $x_{out} = 6.4a$, $y_{in} = -12.8a$, and $y_{out} = 12.8a$. It can be seen that the behaviour of the pure-spinning film is very different from the conventional EHL film. That is, a pair of pressure peaks,[5] which are symmetrical with respect to the centre of the contact ellipse, is generated, and a pair of small dents corresponding to the pressure peak pair is produced through elastic deformation due to the pressure peaks. Furthermore, in this pure-spinning case the minimum film thickness occurs at the centre of the contact and its value is 0.102 μm.

5. TIME-DEPENDENT RESULTS

For $p_H = 0.8$ GPa (at the load of w_0), $A_w = 0.8$, $r = 5$ mm, $\omega = 250\pi$ rad/s, and $u_b = 0.1$ m/s, periodic solutions were obtained for $t_p = 0.002$, 0.02, and 0.2 s, respectively. Corresponding to $p_H = 0.8$ GPa, the minor radius of the contact ellipse is $a = 0.553$ mm, and the calculation domain is also: $x_{in} = -4.4a$, $x_{out} = 2.0a$, $y_{in} = -9.0a$, and $y_{out} = 9.0a$. About six hours was spent for the periodic solution of each case on a personal computer with a 3.0 GHz central processing unit. Obviously among these cases the strongest effect of the dynamic loads on the behaviour of EHL must occur in the case of $t_p = 0.002$ s, so at four representative instants of this case the pressures and film thicknesses on the x axis are given in Figure 6, and those on the y axis are given in Figure 7. For more information, the contour maps of the film thicknesses at these instants are given in Figure 8. Note in each map in Figure 8 the outmost ellipse is the contour for 4 μm of film thickness.

From Figures 6 to 8 it can be understood that the influence of spinning on lubricating performance of the time-dependent EHL is similar to that under steady-state conditions. The transient effect of the dynamic load is reflected clearly by the curves in Figures 6 and 7. The operating condition of $A_w = 0.8$ means that the instantaneous load at $t = 0.25t_p$ is 9 times as that at $t = 0.75t_p$, however, because of the squeezing or separating action due to the dynamic

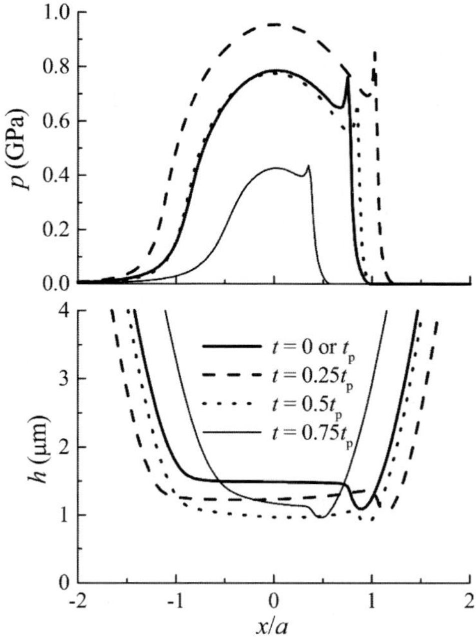

Figure 6. Distributions of pressures and film thicknesses on the x axis at four instants for the case of $p_H = 0.8$ GPa, $A_w = 0.8$, $r = 5$ mm, $\omega = 250\pi$ rad/s, $u_b = 0.1$ m/s, and $t_p = 0.002$ s.

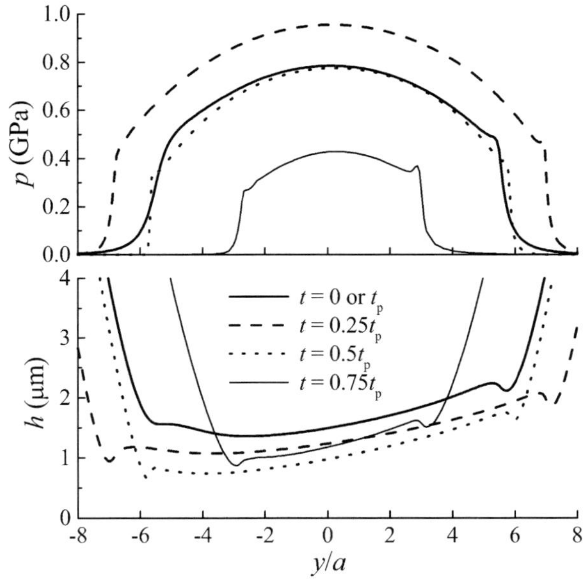

Figure 7. Distributions of pressures and film thicknesses on the y axis at four instants for the same case as in Figure 6.

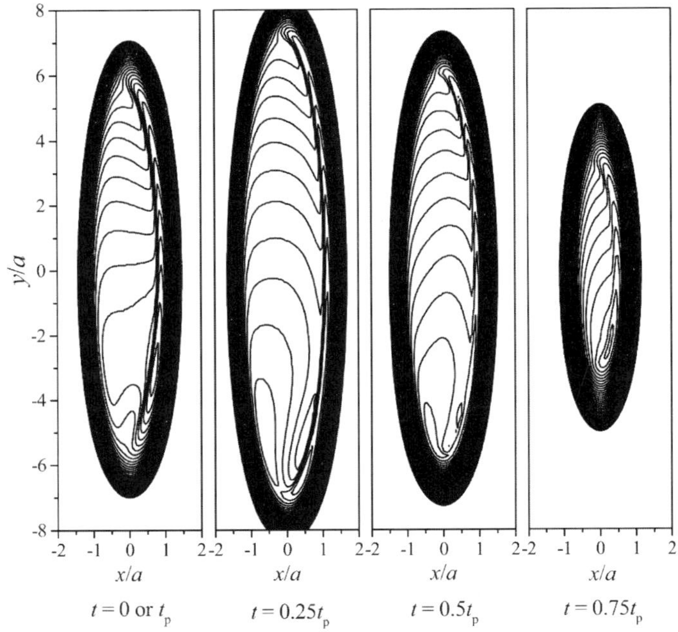

Figure 8. Contour maps of film thicknesses at four instants for the same case as in Figures 6 and 7.

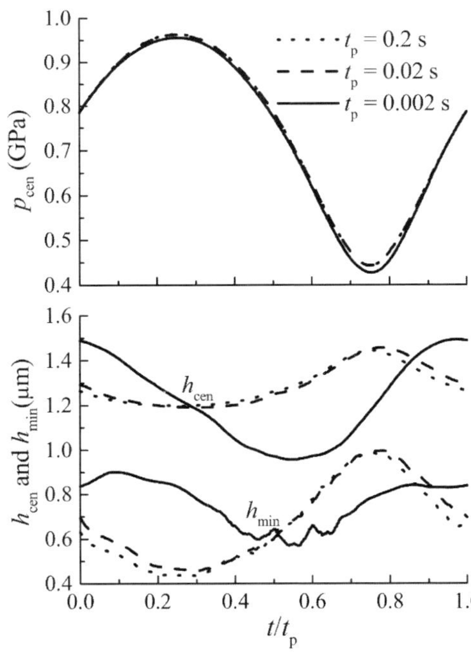

Figure 9. Variations in the central pressures, central and minimum film thicknesses versus time in load periods with different length.

load, the differences between the minimum film thicknesses on the x axis at these instants are very small. The instantaneous loads at $t = 0$ and $t = 0.5t_p$ are of the same size, however, due to the squeezing action at $t = 0$ and the separating action at $t = 0.5t_p$, the film thicknesses at these instants are quite different, and this makes that, in Figure 8 the outmost ellipse at $t = 0.5t_p$ is larger than that at $t = 0$.

The length of the load changing period is the most important factor affecting the transient behaviour of the EHL contacts. For $t_p = 0.002$, 0.02 and 0.2 s, Figure 9 gives the variations in the central pressures p_{cen}, and the central and minimum film thicknesses h_{cen} and h_{min} versus time. In the case of $t_p = 0.2$ s, the time-dependent solution at any instant is very close to the corresponding steady-state solution, suggesting that the dynamic effect is negligibly small. In the case of $t_p = 0.02$ s the dynamic effect is evident but still quite weak, so the curves of h_{cen} and h_{min} are slightly separated from those of $t_p = 0.2$ s. In the case of $t_p = 0.002$ s, however, the dynamic effect is so strong that the curves of h_{cen} and h_{min} are far away from those of $t_p = 0.02$ and 0.2 s. Furthermore, in the case of $t_p = 0.002$ s, the minimum film thickness in a period does not occur at the instant with the maximum instantaneous load. It can be noticed from the curves of p_{cen} in Figure 9 that, since the instantaneous central pressure is mainly determined by the instantaneous load, the length of t_p has little influence on the variation in p_{cen}.

6. CONCLUSIONS

1. Both steady-state and time-dependent solutions to the point contact EHL problems inclusive of the spinning motion can be achieved using the mathematical model and numerical technique described in this paper.
2. The influence of the spinning motion on the EHL performance is important if machine elements are working under conditions similar to those described in this study.
3. Complete fluid films can be created in elliptical contacts with large elliptical ratios by the pure-spinning motion if the load is light enough and the angular velocity of spinning is high enough.
4. The dynamic load can affect the EHL behaviour markedly if the length of the load period is short enough and the amplitude of the load change is large enough.

ACKNOWLEDGEMENTS

The authors are grateful for the financial support of the Natural Science Foundation of Shandong Province, P. R. China through grant No. Z2003 F01, and the Science & Technology Foundation of Qingdao City, P. R. China through grant No. 03-2-JZP-15.

REFERENCES

1. D. Dowson, and G. R. Higginson, *Elastohydrodynamic Lubrication*, Pergamon Press, Oxford (1966).
2. B. J. Hamrock, and D. Dowson, Isothermal elastohydrodynamic lubrication of point contacts, Part III — Fully flooded results, *ASME J. Lubr. Tech.* **99**(2), 264–276 (1977).
3. C. W. Allen, D. P. Townsend, and E. V. Zaretsky, Elastohydrodynamic lubrication of a spinning ball in a nonconforming groove, *ASME J. Lubr. Tech.* **92**(1), 89–95 (1970).
4. R. W. Snidle, and J. F. Archard, Theory of hydrodynamic lubrication for a spinning sphere, *Proc. Instn. Mech. Engrs.,* **184**(L44), 839-848 (1969-1970).
5. D. Dowson, C. M. Taylor, and H. Xu, Elastohydrodynamic lubrication of elliptical contacts with pure spin, *Proc. Instn. Mech. Engrs., Part C: J. Mechanical Engineering Science*, **207**(C2), 83–92 (1993).
6. Q. Zou, C. Huang, and S. Wen, Elastohydrodynamic film thickness in elliptical contacts with spinning and rolling, *ASME J. Tribol.* **121**(4), 686–692 (1999).
7. P. Yang, and S. Wen, A generalized Reynolds equation for non-Newtonian thermal elastohydrodynamic lubrication, *ASME J. Tribol.* **112**(4), 631–636 (1990).
8. Z. M. Jin, P. Yang, J. Cui, and D. Dowson, Transient elastohydrodynamic analysis of elliptical contacts. Part 1: Isothermal and Newtonian lubricant solution, *Proc. Instn. Mech. Engrs., Part J: J. Engineering Tribology*, **218**(J3), 211–224 (2004).
9. P. Yang, and S. Wen, The behaviour of non-Newtonian thermal EHL film in line contacts at dynamic loads, *ASME J. Tribol.* **114**(1), 81–85 (1992).
10. C. J. A. Roelands, Correlation aspects of viscosity-temperature-pressure relationship of lubricating oils, Ph.D. Thesis, Delft University of Technology, the Netherlands (1966).
11. C. H. Venner, Multilevel solution of the EHL line and point contact problems, Ph.D. Thesis, Twente University, Enschede, the Netherlands (1991).
12. A. Brandt, and A. A. Lubrecht, Multilevel matrix multiplication and fast solution of integral equations, *J. Comput. Phys.*, **90,** 348–370 (1990).
13. P. H. Markho, Highly accurate formulas for rapid calculation of the key geometrical parameters of elliptic Hertzian contacts, *ASME J. Tribol.* **109**(4), 640–647 (1987).

SESSION 3

REVIEW AND PROSPECTS FOR THE DEVELOPMENT OF EHL OF FINITE LINE CONTACTS

Xiaoyang Chen, Haoyang Sun and Xuejin Shen
Research Institue of Bearings, Shanghai University, 149 Yanchang Road, Shanghai 200072, P.R. China, E-mail: xychen@mail.shu.edu.cn

Abstract The elastic contacts in finite line contacts (such as those in roller bearings) are non-Hertzian, and there is a wide variety of generatrix used in practice. Optical interferometry studies have shown that the elastohydrodynamic lubrication (EHL) behaviour of such contacts is quite different from that of elliptical or infinite line Hertzian contacts. Few studies of these contacts have been reported during the past four decades. There is therefore a need for research in this area. This paper describes the background, development and recent research on finite line contacts, including that carried out in the authors' laboratory.

Keywords: EHL, finite line contacts, review, film shape and thickness, numerical methods.

1. INTRODUCTION

Line contacts are widely used in engineering components for heavy radial load and high rotational speed conditions, for example, as shown in Figure 1, in roller bearings, cam-followers, traction drives and gear teeth. But all of the elements mentioned above are of finite length, so most of them are normally profiled at their ends to overcome the problems of edge effects caused by their finite length and by misalignment. Extensive mathematical analyses[1] and computer calculations[2~6] have shown that the most favorable effect in respect of edge stress reduction is obtained when the generatrix of finite line contacts corresponds to a certain logarithmic function. But the theoretical basis of these calculation methods for the roller axial profile

R.W. Snidle and H.P. Evans (eds), IUTAM Symposium on Elastohydrodynamics and Micro-elastohydrodynamics, 95–106.

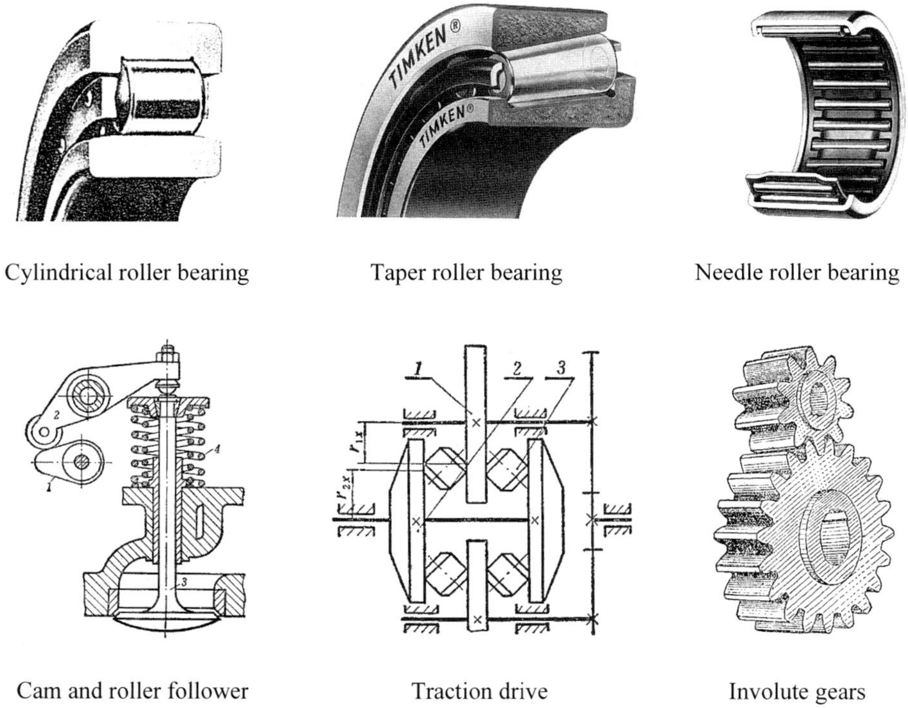

Figure 1. Application of finite line contacts.

design are mainly finite elastostatic line contacts which are based on the classical Hertzian theory and Boussinesq force-displacement relationships for a half-space with an influence coefficient method. This means that the profile of a finite line contact is ideally designed for only one condition of loading. However, it is well known that the finite line contact is separated by a viscous oil film during rolling, which results in the EHL state and the pressure distribution of the oil film is different from that in the elastostatic contact state. The load-bearing capacity, working velocity and service life of roller contacts is deeply influenced by the film thickness, film shape and pressure distribution of the film. Obviously, the optimum crowning value of logarithmic profile designed by elastostatic analysis does not consider the effects of lubricant and rolling speed on the contact stress. It follows that the oil film thickness cannot be estimated by this method. Therefore several new problems arise. What kind of modified profile is the best line contact geometry in the EHL state? Is the crowning value of Lundberg's profile still the optimum value in the EHL state? And how do we design finite line contact profiles for EHL conditions?

Many researchers have studied EHL problems during the past five decades. But most of the researches are restricted to the case of infinite line contacts or elliptical contacts; only a few studies have been carried out on finite line contacts. Existing EHL theory can be used to predict film thickness and pressure in the rolling direction near the central part of a finite line contact, but cannot be used to predict the detailed variation of film thickness and pressure in the longitudinal direction, particularly near the ends of the contact. Infinite line contact theory does not consider sideways flow of lubricant in the inlet region of a contact. Figure 2 shows the behaviour of elliptical contacts of varying aspect ratio. Under increasing load the extent of the contact in the longitudinal direction may exceed the design value in which case the contact may also behave as a finite length contact, presenting the same problem as a finite line contact. But the sideways flow of different profile in finite line contacts is quite different, which can be found clearly in Figure 2 by comparing the interferograms.

Obviously, for the engineering design of finite line contact elements, the oil film thickness and film shape in the whole contact area are the two important working parameters. Further research work is therefore needed on EHL problems of this type even under ideal working conditions of smooth surfaces, isothermal behaviour and a fully flooded contact. Particular attention should be focused upon the effects of the crowning values, loads and the rolling velocity on the film shape, film thickness as well as pressure distribution in the film, especially near the ends of the contact area.

2. REVIEW OF DEVELOPMENTS

2.1 Experimental Techniques

The first experimental investigation of EHL performance for both blended and unblended rollers was made by Gohar and Cameron[9] who measured oil film thickness using optical interferometry in sliding contact, which gave several important images which clearly showed the effect of roller blending. Using the vacuum deposition technique[10], a great improvement in fringe quality was obtained by Wymer and Cameron[8] who developed a very flexible loading system which enabled the optical interferometry to be used to study EHL of finite line contacts under pure rolling and high contact pressure conditions. The test results shown in Figure 2 reveal detailed information on film shape and thickness, and show the effects of end blending. Some other phenomena such as lubricant starvation, deep scratches and static oil entrapments also be found in their experiment. The film thickness near the contact ends is far thinner than that in the central part.

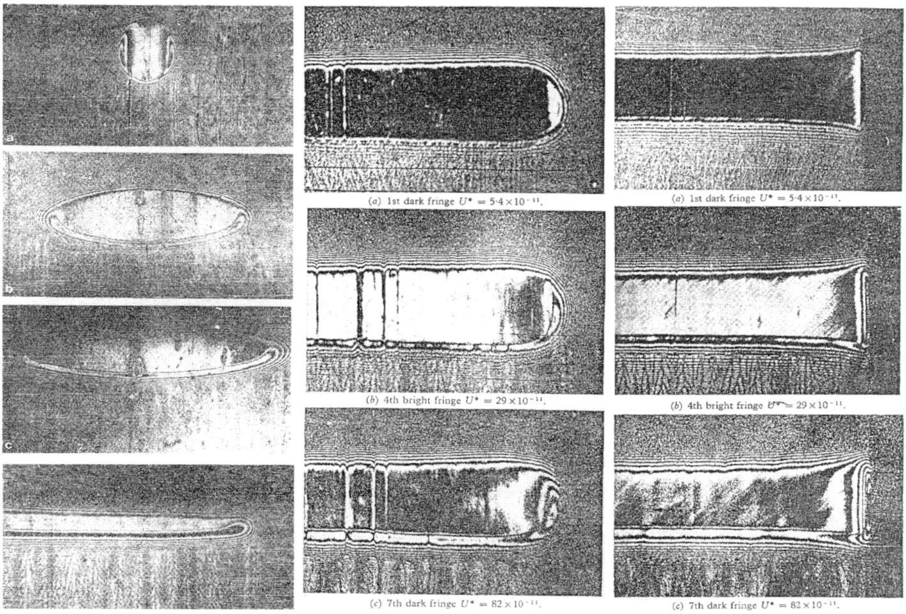

Elliptical contact[7] Blended roller[8] Unblended roller[8]

Figure 2. Difference between interferograms of elliptical and finite line contacts.

The experimental technique developed by Wymer et al. represented a significant breakthrough in optical interferometry of EHL contacts since the study of line contacts requires considerably heavier loads than those needed for circular or elliptical contacts in which the overall contact area is much smaller. The research works of Wymer et al. not only reveal the unique features of finite line contact EHL, but also encourage further research work, and the development of numerical models in particular. It is perhaps unfortunate that they did not investigate a wider range of profiles.

Three types of optical interferometric testing have been used in experimental research. The first type is shown in Figure 3. In this case a point contact optical EHL rig was utilized and the test ball was replaced with a roller[11, 12]. The second type is shown in Figure 4. In this case the real roller bearing working condition was simulated in the test. Because the transparent material (such as glass) cannot withstand high tensile stress, this type of test rig can only be used under lightly loaded working conditions, and the effects of modified roller profile were not easily observed using these methods.

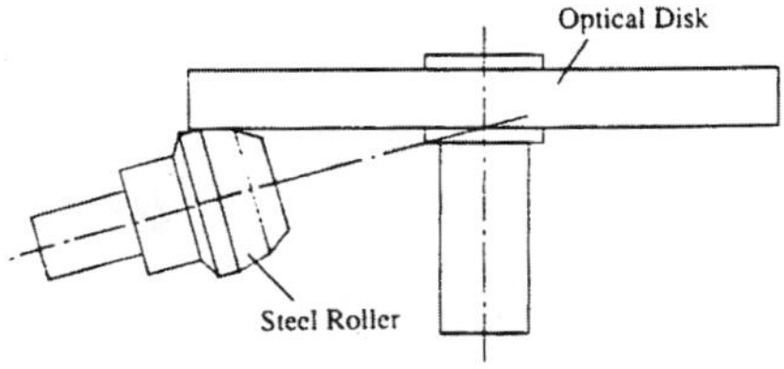

Figure 3. Utilizing the point contact optical EHL rig.

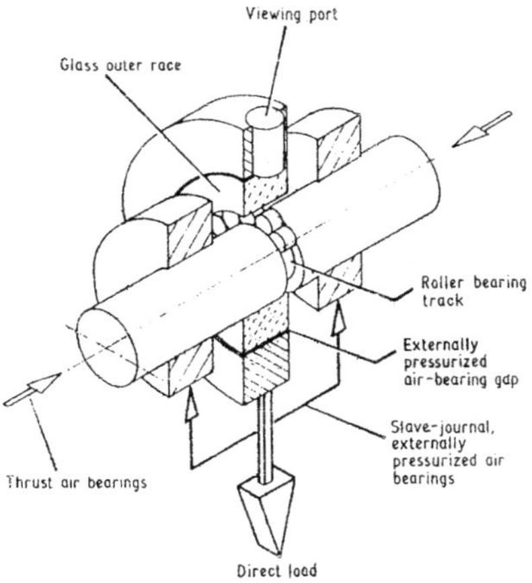

Between outer race and rollers[13]

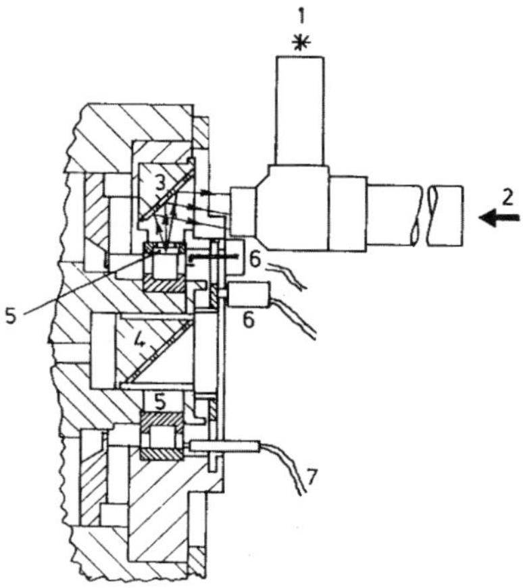

Outer or inner race and rollers[14]

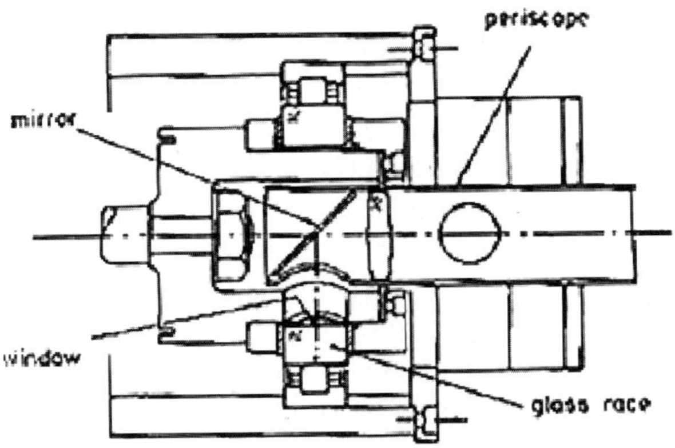

Between inner race and rollers[15]

Figure 4. Simulation of the real roller bearing working state.

The oil film thickness between a tapered roller and a glass disk was measured in a third technique, which is shown in Figure 5. This test method can be used in highly loaded working conditions. The optical interferometric experiments reported by Chen et al.[18] have shown how the oil film thickness and shape is affected by the crowning value of the roller with Lundberg's (logarithmic) profile under heavy load. This indicated that there must be an optimum crowning value that will induce the thickest and most uniform film for given operating conditions. This value is found to be larger than the value that would be obtained from dry contact analysis for the same working conditions.

Parkins and Rudd[19] measured the film thickness in thrust cone bearings using the interference method. The results revealed a very load-sensitive film thinning effect at the edges of such contacts.

2.2 Numerical Methods

Bahadoran and Gohar[20], Hooke[21], and Evans and Snidle[22] have attempted to study the EHL of finite line contacts using modifications to the infinite solution. Mostofi and Gohar[23] gave a numerical solution of the EHL problem of an axially profiled cylindrical roller under flooded, moderate load and various material parameter conditions. Since then, a few attempts have followed to solve this problem theoretically, and some progress has been made. Using the finite element method and the Newton-Raphson technique, Kuroda and Arai[24] investigated numerically the EHL problem between two straight elastic rollers in finite line contact. The results are limited to a few cases. Assuming the same profiled roller contacts as Mostofi and Gohar, Park and Kim[25] obtained improved contour plots of dimensionless pressure and film thickness by using a finite difference method and the Newton-Raphson method. The results showed that the minimum film thickness can be affected by different radii of dub off or crown profiles. It was also pointed out that the maximum pressure and the minimum film thickness occur always near the position where the profiling starts and these are highly dependent on the local geometry there. Barragan de Ling et al.[26, 27] described a theoretical analysis of EHD film forming of perfectly conical thrust rims, the results showed that the presence of crowning greatly reduces the edge thinning effect, particularly at heavy loads.

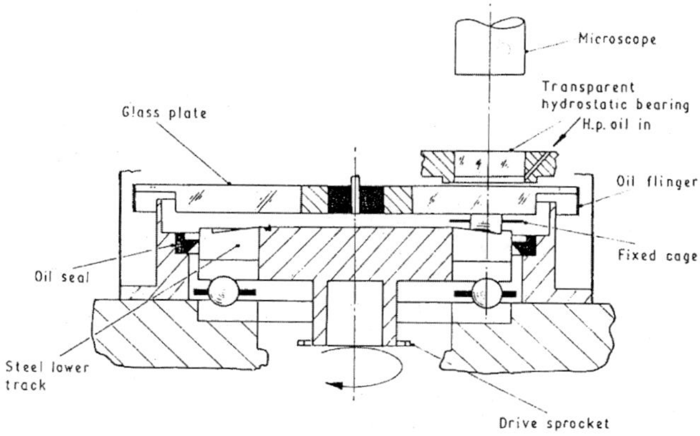

Test rig used in ref. [8] and [16]

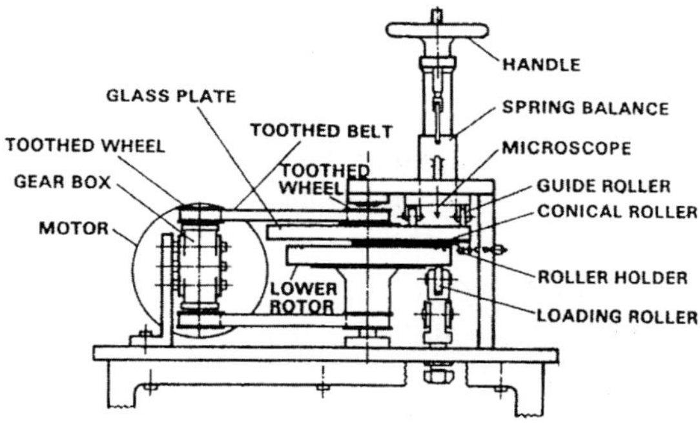

Test rig used in ref. [17]

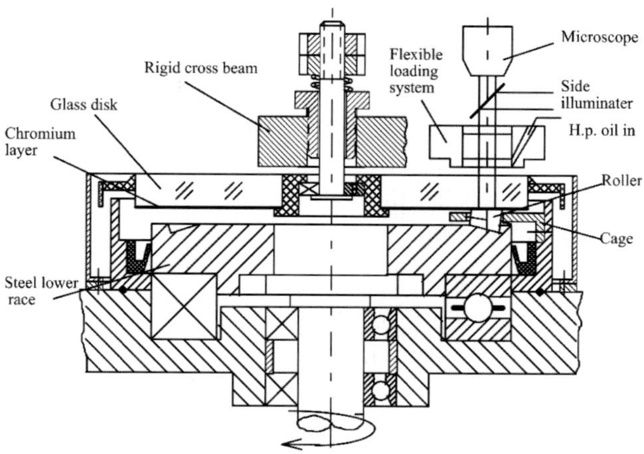

Test rig used in ref. [18]

Figure 5. Measure the film thickness between a taper roller and glass.

For the EHL of finite line contacts with an absolutely straight generatrix, numerical analysis results were given by Xu et al.[28] using an approach based on Jacobi direct methods. It was revealed that the lubricant film in the roller ends could not be established and asperity contact will take place under the condition of low speed and heavy load as shown in Figure 6. Utilizing the above numerical method, the EHL properties of logarithmic profile roller contacts have been studied by Chen et al.[29]. The effect of crown value on finite line contact EHL was investigated. For the same crowning amount, the oil film profile in the lighter loaded condition was more uniform than that in the more heavily loaded one. The experimental phenomenon described in reference [18] was verified, namely that the optimum crowning value required in EHL state is larger than the design value obtained in the elastostatic contact state for the same working conditions. The EHL problems of aligned and misaligned contacts between rollers and races have been researched by Kushwaha et al.[30] using the Newton-Raphson method. They found that the film shape and pressure distribution at the extremities of a finite line contact are not revealed by the traditional line contact solutions usually undertaken. The flow pattern becomes more complex with roller misalignment and the edge effects described are exacerbated.

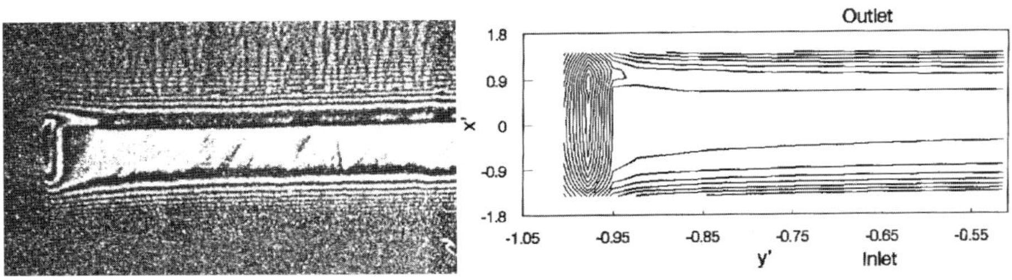

Figure 6. Comparison of film thickness isogram [28] with the optical interferogram of unblended roller [8].

Liu and Yang[31] analyzed the thermal EHL aspects of this problem using multilevel and multi-integration methods. Their results indicated that the thermal solution could not be approximated by the corresponding isothermal solution at large slide/roll ratios. Based on a systematic simulation of most of the test results of reference [8], Figure 7 shows a comparison of numerical solutions of film thickness in the longitudinal direction through the minimum film thickness with the experimental results. Sun and Chen[32] presented a set of results for severe working conditions and investigated the EHL performance of the Lundberg profile. It was found that its crowning

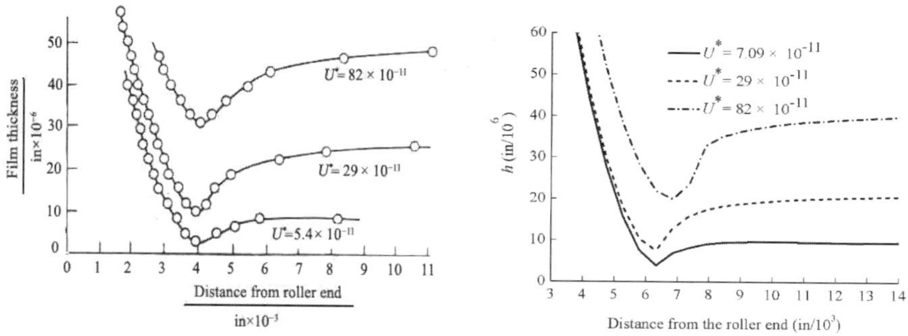

Figure 7. Comparisons of numerical solutions[32] of film thickness through the enclosure in the axial direction with experimental results.[8]

value was insufficient in the EHL state. Larger values are required than those needed in the dry contact condition. This work also shows the effect of velocity and load on the required crowning value.

As mentioned above, most of the numerical solutions assumed ideal working conditions, such as smooth surfaces, steady loading and velocity, flooded lubrication and moderate material parameters. Even under these conditions the problem presents a significant modeling challenge. But thirty years of effort have led to a much better understanding of the behaviour of finite line contacts.

3. ANALYSIS OF EXISTING PROBLEMS

Finite line contacts are non-Hertzian, and the elastostatic contact area can be obtained only by numerical calculation. The accuracy of contact area parameters, such as the half width a and half-length b of the elastostatic footprint, are determined using a fine grid. The dependence on load of the contact dimensions in the x and y directions are different. When the load applied is larger than the design load for which the crowning geometry is calculated the shape of the elastostatic footprint becomes “dog bone” shaped. Because the crowning geometry (logarithmic) is usually specified in terms of two parameters the description of the contact area in a mathematic model and dimensionless variables is difficult.

There is a need to thoroughly verify the predictions of numerical models using the interferometry technique and much remains to be done in this respect before numerical predictions can be reliably applied in machine design.

The key issue in optical interferometry experiments is to achieve high quality interferograms. The quality of the images obtained depends upon the smoothness of the surfaces and the stability of the loading and drive arrangements. Because the contact area is very narrow and the *b/a* ratio is very large, interferograms tend to show only a short length of the contact measured in the longitudinal direction. When glass is used as the transparent member contact pressures are limited to about 1GPa; higher stresses can be tolerated if sapphire is used.

4. SUGGESTIONS FOR FUTURE STUDIES

It has been noted that major progress has recently been achieved in the numerical treatment of the EHL of finite line contacts. Some issues that remain to be resolved are as follows.

The effect of various modified axial profiles and crowning values should be compared over a greater load range.

A summary of validated theoretical results that can be easily used by the practicing engineer is needed.

Extension of the calculated working condition to engineering situations is required.

Studies of the effect of starvation of the lubricated contact are needed.

The effect of rough surfaces should be considered.

An optimum crowning geometry design method based on EHL behaviour is required.

Non-steady-state working should be considered.

Non-Newtonian rheological behaviour of the lubricant should be included in theoretical EHL models.

In real roller bearings there is a need to include consideration of features such as chamfered roller ends and grinding undercuts and their effect on film thickness and pressure.

5. CONCLUSION

It is clear that the EHL behaviour of finite line contacts represents a considerable challenge, both experimentally and theoretically, but considerable progress has been made in understanding the problem.

EHL analyses of finite line contacts can help to optimize the crowning geometry of modified roller ends by introducing a consideration of not only load, but also the lubrication behaviour which depends on lubricant parameters and rolling speed. Pressure distribution and oil film shape are the quantities to be optimized in the practical design process.

This paper has concentrated attention on roller bearings but these considerations of lubricated finite line contacts apply equally to other important machine elements such as cams and followers, traction drive contacts and gears.

ACKNOWLEDGEMENTS

This work was carried out with financial support from the Technical Development Foundation of Shanghai Automobile Industry (No. 0320) and the 'Shuguang' project (No. 2000SG39) of Shanghai Municipality Education Commission and Shanghai Education Development Foundation. The authors also gratefully acknowledge the support of the G.B. Cao Education Foundation for helping with attendance at the IUTAM Symposium.

REFERENCES

1. G. Lundberg, Elastische Beruhrung Zweier Halbraume, *Forschung auf dem Gebiete des Ingenieuwesens*, **10** (5), 201–211 (1939).
2. K. Nikpur and R. Gohar, Profiled Taper Rollers, *Tribology International*, **12**, 203–208 (1975).
3. L. Nayak and K. L. Johnson, Pressure between Elastic Bodies Having a Slender Area of Contact and Arbitrary Profiles, *Int. J. Mech. Sct.*, **21**, 237–347 (1979).
4. M. J. Hartnett and J. W. Kannel, Contact Stresses between Elastic Cylinders: A Comprehensive Theoretical and Experimental Approach, *ASME Trans, J. Lubr. Technol.*, **103** (1), 40–45 (1981).
5. H. Reusner, The Logarithmic Roller Profile – The Key to Superior Performance of Cylindrical and Taper Roller Bearings, *Ball Bearing Journal*, (230), 2–10 (1987).
6. Chen Xiaoyang and Ma Jiaju, A New Numerical Evaluation of Crown Drop for a Needle Roller with Lundberg's Profile, *Mocaxue Xuebao*, **13** (1), 25–31 (1993) (in Chinese).
7. H. Bahadoran and R. Gohar, The Oil Film in Elastohydrodynamic Elliptical Contacts, *Wear*, **29**, 264–270 (1974).
8. D. G. Wymer and A. Cameron, Elastohydrodynamic Lubrication of a Line Contact, *Proc. Instn. Mech. Engrs.*, **188**, 221–238 (1974).
9. R. Gohar and A. Cameron, The Mapping of Elastohydrodynamic Contacts, *ASLE Trans.*, **10**, 215–225 (1967).
10. C. A. Foord, L. D. Wedeven, F. J. Westlake and A. Cameron, Optical Elasto-hydrodynamics, *Proc. Instn mech. Engrs*, **184**, 487–503 (1969–70).
11. N. Ren, D. Zhu and S. Z. Wen, Experimental Method for Quantitative Analysis of Transient EHL, *Tribology International*, **24**, 225–230 (1991).
12. D. Zhu, G. Biresaw, S. J. Clark and T. J. Kasun, Elastohydrodynamic Lubrication with O/W Emulsion, *Trib. Trans. of ASME*, **116**, 310–320 (1994).
13. H. Bahadoran and R. Gohar, Oil Film Thickness in Lightly-Loaded Roller Bearings, *J. Mech. Eng. Sci.*, **16**, 386–390 (1974).

14. J. C. Pemberton and A. Cameron, An Optical Study of the Lubrication of a 65 mm Cylindrical Roller Bearing, *ASME Trans, J. Lubr. Technol.*, **101**, 327–337 (1979).
15. R. A. J. Ford and C. A. Foord, Studies on the Separating Oil Film (EHD Oil Film Thickness) between the Inner Race and Rollers of a Roller Bearing, *Mech. Engng. Trans., The Institution of Engineers, Australia*, 140–144 (1981).
16. N. Dmytrychenko, A. Aksyonov, R. Gohar and G. Wan, Elastohydrodynamic Lubrication of Line Contacts, *Wear*, **151**, 303–313 (1991).
17. M. Hirata and A. Cameron, The Use of Optical Elastohydrodynamics to Investigate Viscosity Loss in Polymer-Thickened Oils, *ASLE Trans.*, **27**, 114–121 (1984).
18. X.-Y. Chen, S.-Q. Zhou and J.-J. Ma, Oil Film Thickness and Shape in Lundberg's Profile Roller Contacts, *Proc. 23rd Leeds-Lyon Sym. on Trib.*, 415–422 (1996).
19. D. W. Parkins and L. Rudd, Thrust Cone Lubrication, Part 3: A Test Facility and Preliminary Measured Data, *Proc. Instn. Mech. Engrs., Part J*, **210**, 107–112 (1996).
20. H. Bahadoran and R. Gohar, End Closure in Elastohydrodynamic Line Contact, *J. Mech. Eng. Sci.*, **16** (4), 276–278 (1974).
21. C. J. Hooke, The Elastohydrodymamic Lubrication of Heavily Loaded Contacts, *J. Mech. Eng. Sci*, **19** (4), 149–156 (1977).
22. H. P. Evans and R. W. Snidle, Film Thinning Factor for Rollers of Finite Width under Elastohydrodynamics Conditions, *Wear*, **175**, 17–23 (1994).
23. A. Mostofi and R. Gohar, Elastohydrodynamic Lubrication of Finite Line Contacts, *ASME Trans, J. Lubr. Technol.*, **105**, 598–604 (1983).
24. S. Kuroda and K. Arai, Elastohydrodynamic Lubrication between Two Rollers, *J. JSME*, **28**, 1367–1372 (1985).
25. T.-J. Park and K.-W. Kim, Elastohydrodynamic Lubrication of a Finite Line Contact, *Wear*, **223**, 102–109 (1998).
26. F. dM. Barragan de Ling, H. P. Evans and R. W. Snidle, Thrust Cone Lubrication, Part 1: Elastohydrodynamic Analysis of Conical Rims, *Proc. Instn. Mech. Engrs.*, **210**, 85–96 (1996).
27. F. dM. Barragan de Ling, H. P. Evans and R. W. Snidle, Thrust Cone Lubrication, Part 2: Elastohydrodynamic Analysis of Crowned Rims, *Proc. Instn. Mech. Engrs.*, **210**, 97–105 (1996).
28. Wen Xu, Jiaju Ma and Xiaoyang Chen, Elastohydrodynamic Lubrication of Finite Line Contacts, Part I: A Numerical Analysis of EHL for Finite Cylindrical Roller Contacts, *Mocaxue Xuebao*, **18** (1), 39–44 (1993) (in Chinese).
29. Xiaoyang Chen, Xuejin Shen, Wen Xu and Jiaju Ma, Elastohydrodynamic Lubrication of Logarithmic Profile Roller Contacts, *Chinese J. Mech. Engrs.* **14** (4), 347–352 (2001).
30. M. Kushwaha, H. Rahnejat and R. Gohar, Aligned and Misaligned Contacts of Rollers to Races in Elastohydrodynamic Finite Line Conjunctions, *Proc. Instn. Mech. Engrs.*, **216**, 1051–1069 (2002).
31. Xiaoling Liu and Peiran Yang, Analysis of the Thermal Elastohydrodynamic Lubrication of a Finite Line Contact, *Tribology International*, **35**, 137–144 (2002).
32. Haoyang Sun and Xiaoyang Chen, Thermal EHL Analysis of Cylinderical Roller under Heavy Load, in *IUTAM Symposium on Elastohydrodynamics and Microelasto-hydrodynamics*, R.W. Snidle and H.P. Evans, Cardiff, September 2004, Springer, Dordrecht, 2006 (this volume).

THERMAL EHL ANALYSIS OF CYLINDRICAL ROLLER UNDER HEAVY LOAD

Haoyang Sun and Xiaoyang Chen

Research Institue of Bearings, Shanghai University, 149 Yanchang Road, Shanghai 200072, P.R. China, E-mail: hysun@mail.shu.edu.cn, xychen@mail.shu.edu.cn

Abstract Full numerical solutions to finite line contact elastohydrodynamic lubrication (EHL) problems are presented. Using the multi-level multi-integration (MLMI) method and some dedicated numerical techniques, convergence has been obtained for severe working conditions (contact pressure can reach 1.0 GPa and the length-width ratio of contact area are over 100 for a roller contact). The well known experiments of Wymer and Cameron have been quantitatively simulated and satisfactory agreement has been obtained. A modification coefficient of the Lundberg profile, δ, has been defined and it has been found that in EHL conditions the crowning value may be several times larger than that in the dry contact state. Both pressure and film thickness distributions should be considered in the EHL state and there is an optimum range of crowning value, expressed by δ as $[\delta_1, \delta_2]$, for particular operating conditions. As load increases δ_1 increases and δ_2 is almost unchanged, and a smaller crowning value is needed for higher velocities. However, the effect of velocity on the crowning value is small compared with that of the load.

Keywords: EHL, finite line contact, crowning design, Lundberg profile.

1. INTRODUCTION

EHL contacts of finite length occur in many machine elements such as rolling element bearings, and involute gear teeth. In order to relieve the high edge stress concentrations caused by their finite length, the rollers are usually crowned slightly toward their ends. Although notable progress in EHL studies have been gained,[1,2] the development of finite line contact EHL theory is very slow. Crowning design is still mainly based on elastostatic theory and optimum crowning geometry under lubricated conditions is difficult to obtain. Cameron's group were the first to study this problem

R.W. Snidle and H.P. Evans (eds), IUTAM Symposium on Elastohydrodynamics and Micro-elastohydrodynamics, 107–120.

experimentally[3,4]. Subsequently, several numerical isothermal steady solutions were obtained.[5–10] Recently, complex factors such as roller misalignment, transient and thermal effects have been considered.[11–13] As the numerical convergence problem is quite severe, the above solutions were usually limited to several cases of moderate working conditions with an assumption that the rollers have a theoretical dub-off profile. Although comparisons between the numerical[5,6,9,11] and experimental[3] results showed some similarities, the value of such comparisons may be doubted as the assumed input data for calculations were quite different from the experimental ones.

Recently, we provided complete EHL solutions for cylindrical rollers with crowned end profile under severe working conditions (the contact pressures exceed 1.0 GPa and the length-width ratio of contact area is over 100).[14] Based on that work, a better simulation of the experiments[3] has been carried out in this paper, and the EHL performance of the Lundberg profile is discussed.

2. PROBLEM FORMULATION

The friction pair is shown in Figure 1. The xy coordinate system is employed for the oil film and the two solids. The z coordinate system is used for both lubricant and solid parts. In the oil film it is named as z, and in the two solids it is written as z_a and z_b, respectively. The generalized Reynolds equation[5] is:

$$\frac{\partial}{\partial x}\left[\left(\frac{\rho}{\eta}\right)_{\mathrm{e}} h^3 \frac{\partial p}{\partial x}\right]+\frac{\partial}{\partial y}\left[\left(\frac{\rho}{\eta}\right)_{\mathrm{e}} h^3 \frac{\partial p}{\partial y}\right]=12u_{\mathrm{e}} \frac{\partial}{\partial x}\left(\rho^* h\right) \tag{1}$$

where $\left(\dfrac{\rho}{\eta}\right)_{\mathrm{e}} = 12(\eta_{\mathrm{e}}\, \rho'_{\mathrm{e}} / \eta'_{\mathrm{e}} - \rho''_{\mathrm{e}})$, $\rho^* = \dfrac{\rho'_{\mathrm{e}}\eta_{\mathrm{e}}(u_{\mathrm{b}}-u_{\mathrm{a}})+\rho_{\mathrm{e}}u_{\mathrm{a}}}{u_{\mathrm{e}}}$,

$\rho_{\mathrm{e}} = \dfrac{1}{h}\displaystyle\int_0^h \rho \mathrm{d}z$, $\rho'_{\mathrm{e}} = \dfrac{1}{h^2}\displaystyle\int_0^h \rho \int_0^z \frac{\mathrm{d}z'}{\eta} \mathrm{d}z$, $\rho''_{\mathrm{e}} = \dfrac{1}{h^3}\displaystyle\int_0^h \rho \int_0^z \frac{z'}{\eta} \mathrm{d}z'\mathrm{d}z$,

$\dfrac{1}{\eta_{\mathrm{e}}} = \dfrac{1}{h}\displaystyle\int_0^h \frac{1}{\eta}\mathrm{d}z$, $\dfrac{1}{\eta'_{\mathrm{e}}} = \dfrac{1}{h^2}\displaystyle\int_0^h \frac{z}{\eta}\mathrm{d}z$

The boundary conditions are:

$$\begin{cases} p(x_{\text{in}}, y) = p(x_{\text{out}}, y) = p(x, -y_{\text{out}}) = p(x, y_{\text{out}}) = 0 \\ p(x, y) \geq 0,\ x_{\text{in}} < x < x_{\text{out}},\ -y_{\text{out}} < y < y_{\text{out}} \end{cases}$$

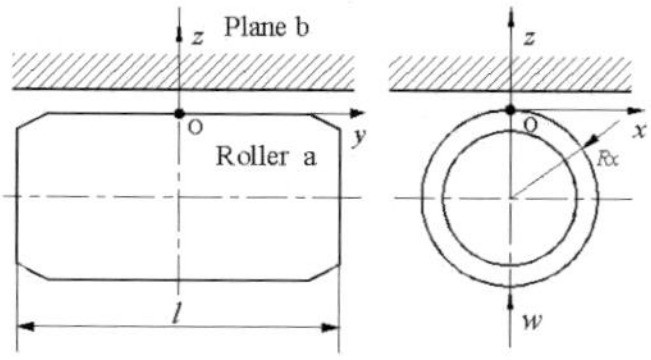

Figure 1. Friction pair and the coordinate system.

The film thickness equation can be written as:

$$h(x) = h_{00} + s(x, y) + \frac{2}{\pi E'} \iint \frac{p(x', y')}{\sqrt{(x - x')^2 + (y - y')^2}} \mathrm{d}x' \mathrm{d}y' \tag{2}$$

where $s(x, y)$ corresponds to the axial profile of the rollers.

The Roelands viscosity equation[16] is:

$$\eta = \eta_0 \exp\left\{A_1\left[-1 + (1 + A_2 p)^{Z_0} (A_3 t - A_4)^{-S_0}\right]\right\} \tag{3}$$

where $Z_0 = \alpha/(A_1 A_2)$, $S_0 = \beta/(A_1 A_3)$, $A_1 = \ln \eta_0 + 9.67$, $A_2 = 5.1 \times 10^{-9}$ Pa^{-1}, $A_3 = 1/(t_0 - 138)$ K^{-1}, $A_4 = 138/(t_0 - 138)$.

The Dowson and Higginson density relationship[17] is adopted:

$$\rho = \rho_0 \left[1 + C_1 p/(1 + C_2 p) - C_3 (t - t_0)\right] \tag{4}$$

Neglecting the heat conduction in the x and y directions, the energy equation for the oil film is:

$$c\left[\rho u \frac{\partial t}{\partial x} + \rho v \frac{\partial t}{\partial y} - \left(\frac{\partial}{\partial x}\int_0^z \rho u \mathrm{d}z' + \frac{\partial}{\partial y}\int_0^z \rho v \mathrm{d}z'\right)\frac{\partial t}{\partial z}\right]$$
$$= k\frac{\partial^2 t}{\partial z^2} - \frac{t}{\rho}\frac{\partial \rho}{\partial t}\left(u\frac{\partial p}{\partial x} + v\frac{\partial p}{\partial y}\right) + \eta\left[\left(\frac{\partial u}{\partial z}\right)^2 + \left(\frac{\partial v}{\partial z}\right)^2\right] \tag{5}$$

The boundary condition are required only at the inlet region if $u(x_{in}, y, z) \geq 0$, it is $t(x_{in}, y, z) = t_0$.

The energy equations of the contact solids are:

$$c_a \rho_a u_a \frac{\partial t}{\partial x} = k_a \frac{\partial^2 t}{\partial z_a^2}, c_b \rho_b u_b \frac{\partial t}{\partial x} = k_b \frac{\partial^2 t}{\partial z_b^2} \quad (6)$$

The temperature domains of the two solids are assumed to be infinite layers with finite thickness d, a fixed ambient temperature is assumed at depth $d = 3.15b$ in the contact solids. The boundary conditions can therefore be written as $t(x_{in}, y, z_a) = t(x, y, -d) = t(x_{in}, y, z_b) = t(x, y, d) = t_0$

At the interface of the solids and oil film, the heat flux is:

$$k_a \cdot \left.\frac{\partial t}{\partial z_a}\right|_{z_a=0} = k \cdot \left.\frac{\partial t}{\partial z}\right|_{z=0}, \quad k_b \cdot \left.\frac{\partial t}{\partial z_b}\right|_{z_b=0} = k \cdot \left.\frac{\partial t}{\partial z}\right|_{z=h}$$

The following load balance equation must be satisfied:

$$\iint p \mathrm{d}x\mathrm{d}y = w \quad (7)$$

3. NUMERICAL APPROACH

In the x direction, the computation domain is [-3.6b, 1.8b]; in the y direction only half the domain is required due to symmetry; in the z direction it is the film thickness and extends to a depth of 3.15b into the two solids. In the multigrid solver for the Reynolds equations six levels are employed. The finest level, consisting of 257 nodes in the x direction and 513 nodes in the y direction, is used for the solution of the energy equations. Across the film, 11 equidistant nodes are used. Within the bounding solids, 6 non-equidistant nodes are used, taking account of the temperature gradient in the solids. The mesh size is small at the interface and then increases in a geometrical series. The numerical model was non-dimensionalisation before numerical analysis. The non-dimensionalisation and solution technique are basically the same as that developed by Yang's group[18]. The MLMI method was used to obtain the pressure, and a sequential column sweeping technique was used to obtain the temperature field. However, convergence could be obtained only for moderate working conditions.[13] After a different dedicated technique was employed, results for quite severe working conditions could be provided: the pressure and temperature are restricted in range during the solution process;

and the coefficient matrix of the discrete energy equation must be constructed as a dominant one. The convergence criteria for pressure, temperature and load are 10^{-5}, 10^{-6}, 10^{-5}, respectively.

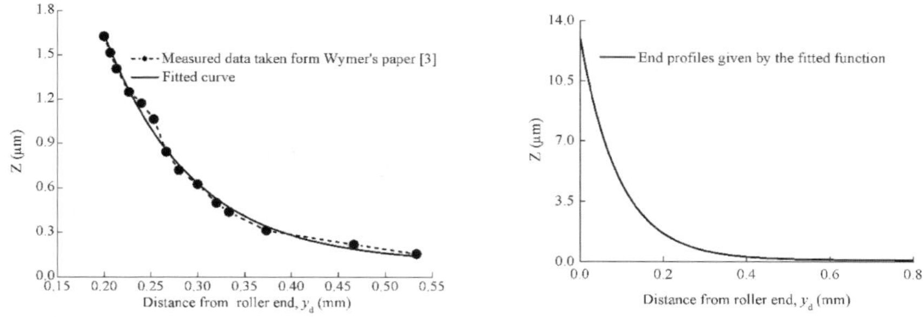

Figure 2. Roller profile fitting.

4. SIMULATION OF EXPERIMENTS

A tapered roller 8.2 mm in the mean diameter and 13.7 mm long, with a total cone angle of 7.9° was used in the experiments,[3] the roller profile was simplified to be symmetrical to its center in the simulations and R_x = 4.1 mm. The solid dots on the left graph of Figure 2 were taken from reference [3] with the assumption that the location of first measured dots was $y_d = 0.2$ mm, where y_d is the distance from the roller end. The relation of $Z(\mu m)$ to y_d (mm) is described as:

$$Z(y_d) = Z_0 + a_1 \exp\left(\frac{y_0 - y_d}{t_1}\right) + a_2 \exp\left(\frac{y_0 - y_d}{t_2}\right) \quad (0 \le y_d \le 6.85) \qquad (8)$$

where Z_0 = 0.09705, y_0 = 0.06293, a_1= 2.9814, t_1 = 0.0942, a_2= 3.6143, t_2 = 0.09421. The right graph of figure 2 shows the simulated roller edge profile, it shows that Z is very close to zero when $y_d > 0.6$ mm, and $Z \geq 0$ is always satisfied. Therefore, $s(x,y)$ in Eq. (2) is:

$$s(x, y) = R_x - \sqrt{\left(R_x - Z \times 10^{-6}\right)^2 - x^2} \qquad (9)$$

The following lubricant properties at an ambient temperature t_0 = 313 K were assumed.[3,4,19] η_0 = 0.507 Pa·s, α = 48 GPa^{-1}, β = 0.046 K^{-1}, ρ_0 = 900 kg/m^3, c = 2000 J/(kg·K), k = 0.14 W/(m·K). The properties of the steel and glass are taken as: $E_a = 206$ GPa, $E_b = 75$ GPa, $k_a = 46$ W/(m·k), $k_b = 0.78$ W/(m·k), $\rho_a = 7850$ kg/m^3, $\rho_b = 2500$ kg/m^3, $c_a = 470$ J/(kg·K), $c_b = 840$ J/(kg·K). As the definition of dimensionless parameters is different in reference[3] and the current paper (see nomenclature), a conversion is needed. The applied load is $w = 349$ lb $= 1560$ N ($p_H = 0.7193$ GPa). For the first case ($U^* = 5.4 \times 10^{-11}$), convergence could not be obtained, so results are shown for a higher value of the non-dimensional velocity of $U^* = 7.09 \times 10^{-11}$.

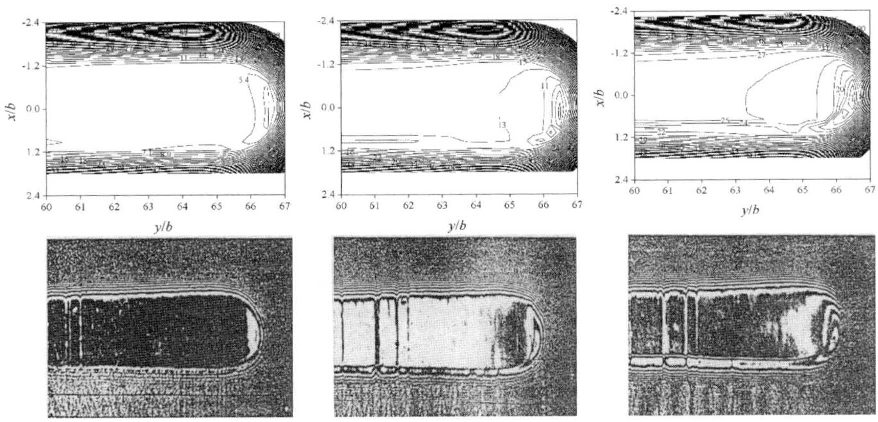

a) $U^* = 7.09 \times 10^{-11}$ numerically; b) $U^* = 29 \times 10^{-11}$ c) $U^* = 82 \times 10^{-11}$
$U^* = 5.4 \times 10^{-11}$ experimentally

Figure 3. Film shape of current solutions and interferograms obtained in reference [3] ($W = 2.3737 \times 10^{-4}$ ($p_H = 0.7193$ GPa), $E' = 117$ GPa, $G = 5616$, $S = 0$).

In Figure 3, the upper row of contour maps shows the numerical film thickness results expressed as $10^5 h/R_x$, and corresponding interferograms[3] are shown in the lower part of the figure. Simulations and experimental results are seen to agree well. In Figure 4, the results are expressed in inches so as to be consistent with the published experimental results. Again, there is good agreement between the predicted and measured values. On the centerline section the discrepancies are less than 5%. On the sections through the end closures (where the minimum film thickness occurs), the numerical values are about 67% to 77% of the experimental ones. The discrepancies may be attributed to three factors: small differences in roller end profiles between the simulated and experimental (Figure 4(b)); the assumption that the roller is symmetrical about its center; and the effect of thin film lubrication. As the calculated film thickness is thinner than the measured, a crowning design based on the current algorithm could be expected to provide "safe" results.

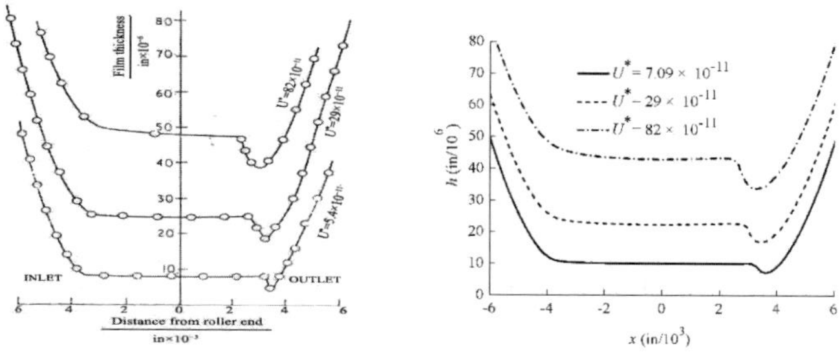

a) Film thickness on the centerline section along the rolling direction

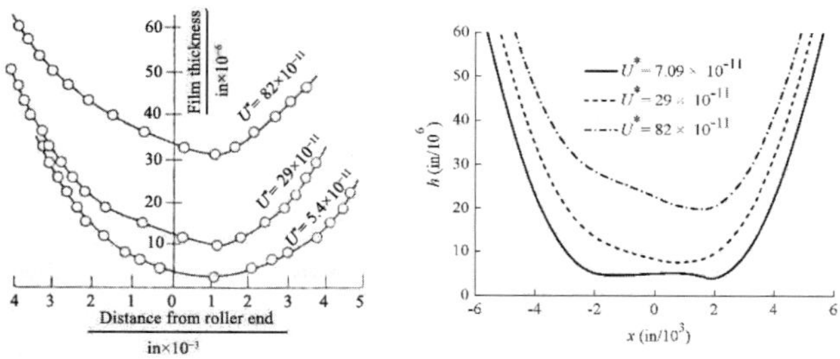

b) Film thickness through the enclosures in the rolling direction

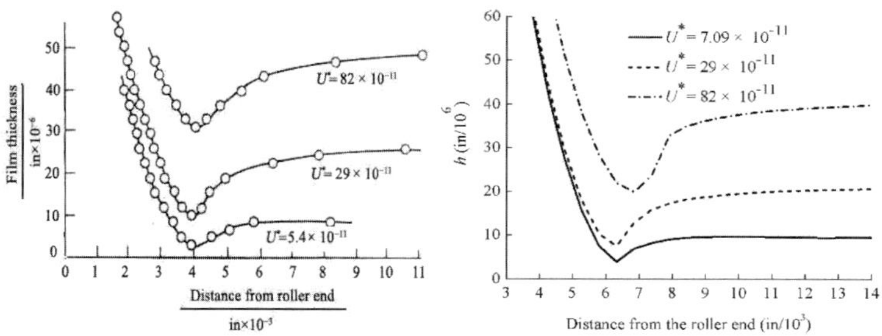

c) Film thickness through the enclosure in the axial direction

Figure 4. Comparisons of current solutions and experimental results obtained in reference [3] ($W = 2.3737 \times 10^{-4}$ ($p_H = 0.7193$ GPa), $E' = 117$ GPa, $G = 5616$, $S = 0$).

5. STUDY OF LOGARITHMIC PROFILE ROLLERS

It is generally accepted that the best profile for a finite line contact in the elasto-static (dry contact) state is the Lundberg profile [20]:

$$\begin{cases} s_{\mathrm{lu}}(y) = s_{\mathrm{lu}}(-y) = -\dfrac{2w}{\pi E'l}\ln\left(1-\left(\dfrac{2y}{l}\right)^2\right) & (0 \le y < 0.5l) \\ s_{\mathrm{lu}}(\pm 0.5l) = \dfrac{2w}{\pi E'l}\left(1.1932 + \ln\left(\dfrac{l}{2b}\right)\right) & \end{cases} \tag{10}$$

A modification coefficient δ is introduced in this paper to enable investigation of the effect of profile changes under EHL conditions.

$$s = \delta \cdot s_{\mathrm{lu}} \tag{11}$$

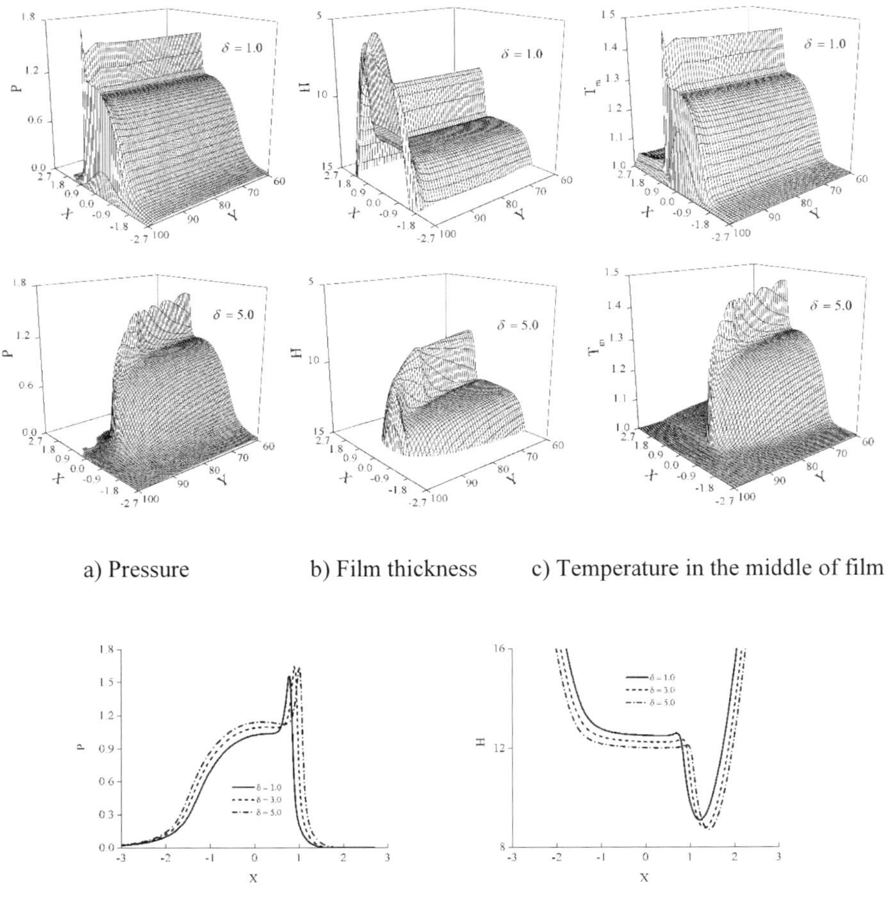

a) Pressure b) Film thickness c) Temperature in the middle of film

d) Pressure and film thickness in the center of the roller

Figure 5. Effect of the modification coefficient δ($W = 0.885\times10^{-4}$ ($p_{\mathrm{H}} = 0.848$ GPa), $E' = 226$ GPa, $G = 4949$, $U = 1.42\times10^{-10}$).

The following lubricant properties were assumed in the simulations to be described. Steel properties are as given above. η_0 = 0.08 Pa·s, α = 21.9 GPa^{-1}, β = 0.0476 K^{-1}, ρ_0 = 870 kg/m^3, c = 2000 J/(kg·K), k = 0.14 W/(m·K). The roller length is l = 0.01 m, roller radius R_x = 0.005 m. The basic operating parameters are: applied load w = 1000 N, entrainment velocity u_e = 2.0 m/s, slide-roll ratio S = 0.5.

Several numerical trials were carried out with different values of δ. Figure 5 shows that the edge pressure and temperature both increase and the film is very thin in the case of δ=1.0. Therefore, the crowning value of the Lundberg profile is insufficient for the EHL state. If δ is too large, such as δ=5.0, the pressure shifts to the center though the film thickness remains thick enough at the edge. The temperature in the middle of the oil film is quite similar in form to the pressure distribution. Figure 5(d) also shows that in the center of the roller, the pressure and the contact width increase and the minimum film thickness decreases with the increase in δ. Current solutions show that if δ=3.0 (see figure 6), with the above input dada, the distributions of pressure, film thickness, as well as the temperature are quite satisfactory.

δ is kept at 3.0 and the cases of u_e = 2.0 m/s and 10 m/s are calculated to show the effect of velocity. Figure 6 shows that, with the increase of velocity, the pressure spike decreses in magnitude, and the film thickness and temperature in the middle of the oil film increase considerably. Figure 6(d) shows that at this higher velocity the pressure distribution tends to shift to the roller's center. Therefore, a smaller crowning value is needed at a higher velocity. However, since such shift is very slow, the velocity effect on the crowning value is not very pronounced if a fundamentally suitable crowning value is provided.

Two groups of solutions representing respectively heavy and low loads are presented to show the effect of load. The pressure and film thickness on section X = 0 are shown in Figure 7. In each group of solutions, δ is selected as 1.5, 2.0 and 3.0 respectively. It should be noted that the crowning value for the heavy and low loads are different even in the case of the same δ. As Eq. (10) and Eq. (11) show, the crowning values are decided not only by δ but also by the applied load. Figure 7 shows that with an increase of δ, the minimum film thickness at the edge also increases, at the same time the pressure distribution shifts to the roller center. This leads to an unsatisfactory pressure distribution in the axial direction. Therefore, the pressure and film thickness distributions should both be considered under EHL conditions, and there is an optimum range of the crowing value, which may be expressed by δ (defined in Eq. (11)) as [δ_1, δ_2], where δ_1 is mainly decided by the film thickness, and δ_2 mainly by the pressure distribution. δ_1 is the lower bound and δ_2 is the upper limit of the crowning value for a certain operating condition. It can be seen that in the case of $\delta = 1.5$ the edge effect of the film thickness at heavy loads is more severe than that at low loads. Therefore, δ_1 should be increased with the increase of load. It can also be seen that with the increase of δ, the pressure distributions in the axial direction at low and heavy loads are quite similar; therefore, δ_2 is almost independent of the load.

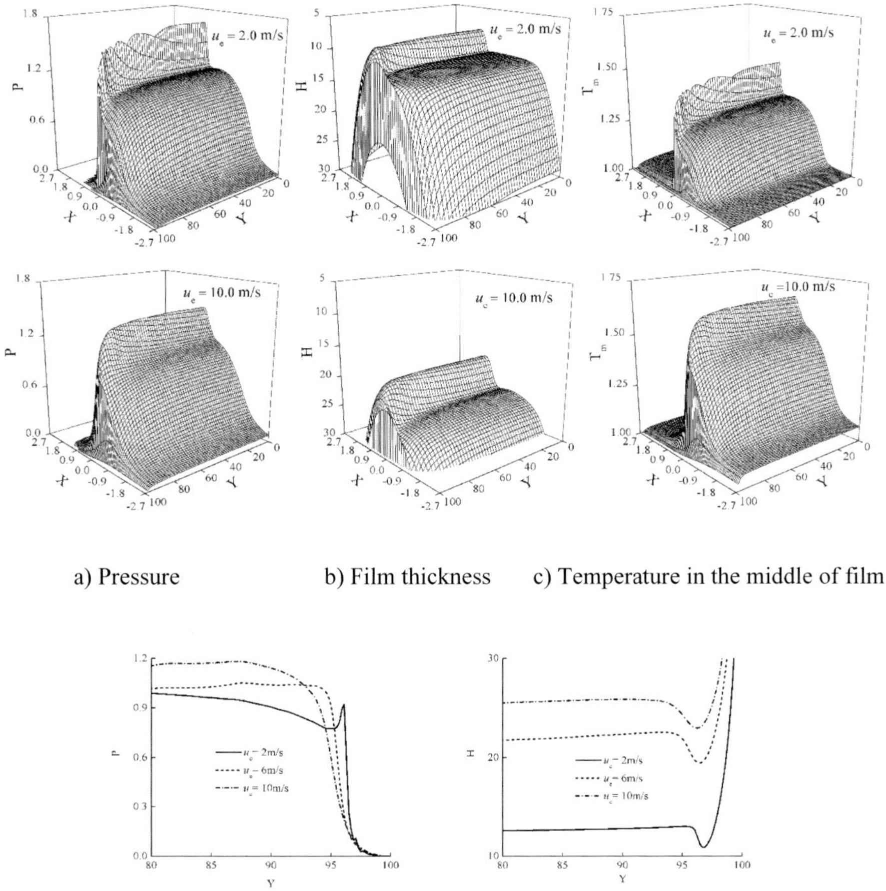

a) Pressure b) Film thickness c) Temperature in the middle of film

d) Pressure and film thickness on section X = 0

Figure 6. Effect of the velocity ($W = 0.885 \times 10^{-4}$ ($p_H = 0.848$ GPa), $E' = 226$ GPa, $G = 4949$, $\delta = 3.0$).

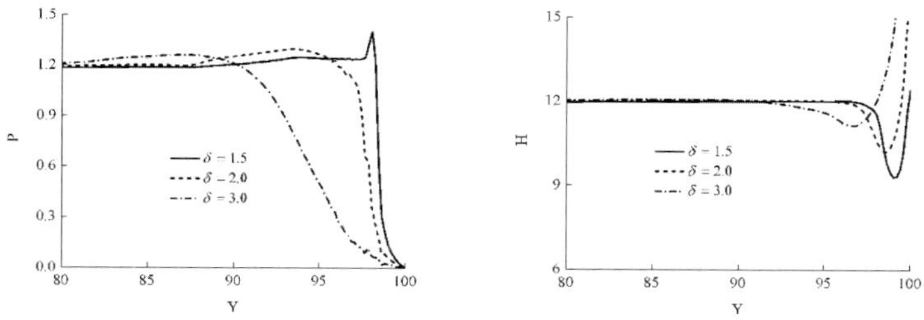

a) Solution for low load ($W = 2.65\times10^{-5}$ ($p_H = 0.465$ GPa))

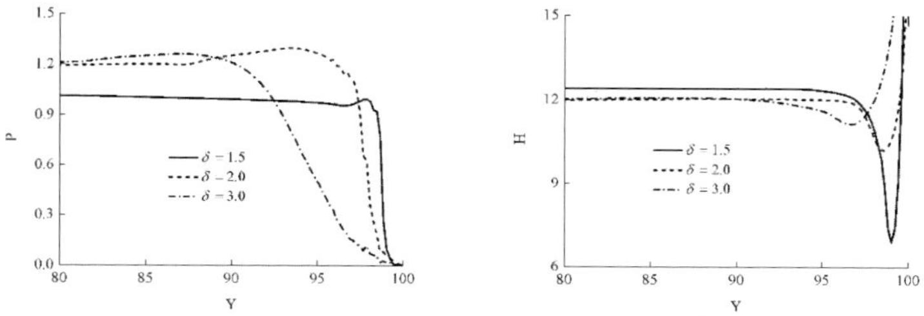

b) Solution of heavy load($W = 1.24\times10^{-4}$ ($p_H = 1.004$ GPa))

Figure 7. Eeffect of the load on δ, pressure and film thickness on section X = 0 ($E' = 226$ GPa, $G = 4949$, $U = 1.42\times10^{-10}$, $S = 0.5$)

6. CONCLUSIONS AND DISCUSION

1. Under EHL conditions the crowning value of the Lundberg profile is insufficient; in order to provide optimum film and pressure behaviour under EHL conditions the crowning may need to be several times that required under dry contact conditions. A modification coefficient of the Lundberg profile, δ, is defined.
2. The pressure and film thickness distributions must both be considered in designing suitable crowning. There is an optimum range of the crowning value expressed by δ as [δ_1, δ_2] for a particular operating condition. With the increase of the load, δ_1 increase whereas δ_2 remains almost unchanged.
3. A smaller crowning value is needed for a higher velocity, but the effect of the velocity on the crowning value is not pronounced compared with that of the load.

NOMENCLATURE

b	half-length of Hertzian contact, $\sqrt{8wR_x/(\pi E'l)}$, m
c, c_a, c_b	specific heat of lubricant and solids, J/(kg·K)
E_a, E_b	elastic modulus of two solids, Pa
E'	reduced elastic modulus, $E' = \left[\left(1-\mu_a^2\right)/\left(2E_a\right)+\left(1-\mu_b^2\right)/\left(2E_b\right)\right]^{-1}$,Pa
E^*	reduced elastic modulus defined in reference [3], $E^* = \left(1/E_a + 1/E_b\right)^{-1}$,Pa
G	material parameter, $\alpha E'$
h	film thickness, m
h_{00}	rigid central film thickness, m
k, k_a, k_b	thermal conductivity of lubricant and solids, W/(m·k)
l	roller length, m
p	film pressure, Pa
p_H	maximum Hertzian contact pressure, $2w/(\pi bl)$, Pa
P	dimensionless pressure, p/p_H
R_x	roller middle radius, m
S	Slide-roll ratio, $(u_a - u_b)/u_e$
t	temperature, K
t_0	ambient temperature, K
T	dimensionless temperature, t/t_0
u, v	film velocity in x and y directions, m/s
u_a, u_b	velocity of surface a and b, m/s
u_e	entrainment velocity, $(u_a + u_b)/2$, m/s
U	velocity parameter, $\eta_0 u_e/(E'R_x)$
U^*	velocity parameter defined in reference [3], $\eta_0 u_e/\left(E^* R_x\right)$
w	applied load, N
W	load parameter, $w/(E'R_x l)$
x, y	coordinates (see fig. 1), m
X, Y	dimensionless coordinates, $100x/R_x$, $100y/R_x$
y_d	distance from roller end in axial directions, mm
x_{in}, x_{out}, y_{out}	domain boundaries, m
z	coordinates across the film, m
z_a, z_b	coordinates in solid a and b, m
α	Barus viscosity-pressure coefficient, Pa^{-1}
β	Reynolds viscosity-temperature coefficient, K^{-1}
δ	modification coefficient of Lundberg profile
μ_a, μ_b	Poisson ratio of solids
η	viscosity of lubricant, Pa·s
η_0	ambient viscosity, Pa·s

ρ, ρ_a, ρ_b density of lubricant and solids, kg/m^3
ρ_0 ambient density of lubricant, kg/m^3

ACKNOWLEDGEMENTS

This work was sponsored by the Technical Development Foundations of Shanghai Automobile Industry (No. 0320). The authors wish to thank Prof. Peiran Yang of Qingdao institute of architecture and Engineering for his valuable research work on the MLMI method, and Prof. D. C. Sun of the State University of New York at Binghamton for his kind help.

REFERENCES

1. A. Lubrecht, W. E. Ten Napel and R. Bosma, Multigrid, an alternative method for calculation film thickness and pressure profiles on the elastohydrodynamic lubricated line contacts, *ASME J. Tribol.* **108**: 551–556 (1986).
2. H. Venner, Multilevel solution of the EHL line and point contact problems, Ph.D. Thesis, Twente University, Enschede, the Netherlands (1991).
3. D. G. Wymer and A. Cameron, Elastohydrodynamic lubrication of a line contact, *Proc. Inst. Mech. Eng.* **188**: 221–238 (1974).
4. E. O. Stejskal and A. Cameron, Optical interferometry study of film formation in lubrication of sliding and/or rolling contacts, *NASA Report*, Lewis Research Center (1972).
5. H. Bahadoran and R. Gohar, End closure in elastohydrodynamic line contact, *J. Mech. Eng. Sci.* **16**: 276–278 (1974).
6. A. Mostofi and R. Gohar, Elastohydrodynamic lubrication of a finite line contact, *J. Lubric. Technol.* **105**: 598–604 (1983).
7. F. D. M. Barragan de Ling, H. P. Evans and R. W. Snidle, Thrust cones lubrication. Part 1: Elastohydrodynamic analysis of conical rims, *Proc. Inst. Mech. Eng., Part J* **210**: 85–96 (1996).
8. F. D. M. Barragan de Ling, H. P. Evans and R. W. Snidle, Thrust cones lubrication. Part 2: Elastohydrodynamic analysis of conical rims, *Proc. Inst. Mech. Eng., Part J* **210**: 97–105 (1996).
9. Park Tae-Jo and Kim Kyung-Woong, Elastohydrodynamic lubrication of a finite line contact, *Wear* **223**: 102–109 (1998).
10. Xiaoyang Chen, Xuejin Shen, Wen Xu and Jiaju Ma, Elastohydrodynamic lubrication of logarithmic profile roller contacts, *Chinese Journal of Mechanical Engineering*, **14**: 347–352 (2001) (in Chinese).
11. M. Kushwaha, H. Rahnejat and R. Gohar, Aligned and misaligned contacts of rollers to races in elastohydrodynamic finite line conjunctions, *Proc. Inst. Mech. Eng., Part C* **216**: 1051–1069 (2002).
12. M. Kushwaha and H. Rahnejat, Transient Elastohydrodynamic lubrication of finite line conjunction of cam to follower concentrated contact, *J. Phys. D: Appl. Phys.* **35**: 2872–2890 (2002).

13. Xiaoling Liu and Peiran Yang, Analysis of the thermal elastohydrodynamic lubrication of a finite line contact, *Tribol. Int.* **35**: 137–144 (2002).
14. Haoyang Sun, Xiaoyang Chen and Peiran Yang, Thermal Elastohydrodynamic lubrication of cylindrical rollers with crown profiled ends at heavy load, *Chinese Journal of Mechanical Engineering* **40**: 99–104 (2004) (in Chinese).
15. Peiran Yang and Shizhu Wen, A generalized Reynolds equation for non-Newtonian thermal elastohydrodynamic lubrication, *ASME J. Tribol.* **112**: 631–636 (1990).
16. J. A. Roelands, Correlation aspects of viscosity-temperature-pressure relationship of lubricating oils, Ph.D. Thesis, Delft University of Technology, the Netherlands (1966).
17. Dowson, D. and G. R. Higginson, *Elastohydrodynamic Lubrication,* Pergamon Press, Oxford (1977).
18. Feng Guo and Peiran Yang, Influence of a ring flat zone in the point contact surface on thermal elastohydrodynamic lubrication, *Tribol. Int.* **32**: 167–175 (1999).
19. A. P. Ranger, C. M. M. Ettles and A. Cameron, The solution of the point contact elastohydrodynamic problem, *Proc. R. Soc.* **346**: 227–244 (1975).
20. G. Lundberg, Elastische Beruhrung zweier Halbraume, *Forschung auf dem Gebiete des Ingenieurwesen* **10**: 201–221 (1939).

EFFECT OF ELASTIC DEFORMATION OF THE JOURNAL BEARING SHELL ON ITS DYNAMIC STABILITY

E. Saber and H.A. El-Gamal
Mechanical and Marine Engineering Department, Arab Academy for Sciences and Technology, Alexandria, Egypt

Abstract The stability of the plain cylindrical journal bearing having an elastic shell (liner) is investigated in this work. A perturbed Reynolds equation is used to determine the unsteady components resulting from given small excitations of the bearing journal. The coupled non-linear perturbation dynamic equations for angular and linear motions of the journal carrying a single mass are solved for different values of the steady state eccentricity ratio and the elastic coefficient of the liner. Values of the stability numbers which give critically stable conditions of motion are determined for each case. It was found in this work that increasing the elasticity of the bearing liner resulted in an increase in bearing stability for eccentricity ratios over approximately 0.45. The opposite was observed for eccentricity ratios below this value. It was also found that shorter bearings are less stable than longer bearings for any value of the elastic coefficient.

Keywords: elastohydrodynamics, elastohydrodynamic lubrication, elastic deformation of bearing shell (liner), bearings dynamic stability, stability of hydrodynamic bearings.

1. INTRODUCTION

The analysis of stability of journal bearings running under different loading conditions is required in various engineering fields. The dynamic behavior of journal bearings and its relation to rotor dynamics in general has been investigated by many researchers [1–8]. Hori [1] examined the stability of an infinitely long bearing theoretically using the linearization technique. Funagawa and Tatara [2] examined the short bearing case both theoretically and experimentally. Gotoda [3] and Nakagawa and Aoki [4] studied the stiffness and damping characteristics of plain journal bearings. Some experiments were also carried out on stiffness and damping coefficients in [4]. Oscutt and Arwas [5] used the linearization technique to study the stiffness and damping behavior of journal

R.W. Snidle and H.P. Evans (eds), IUTAM Symposium on Elastohydrodynamics and Microelastohydrodynamics, 121–131.

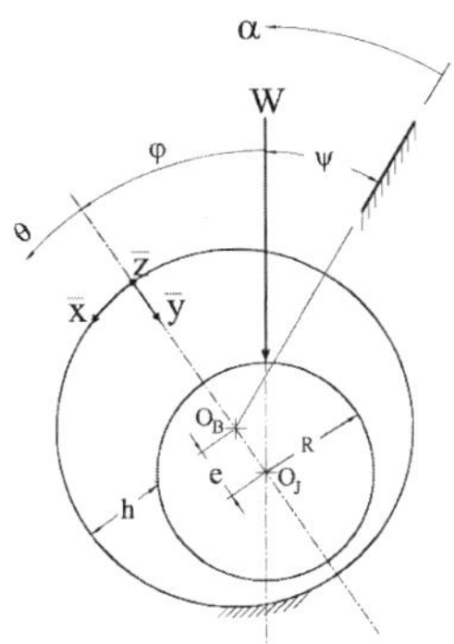

Figure 1. Journal bearing geometry and nomenclature.

bearings running under conditions of laminar and turbulent regimes. On the other hand Wilcock and Pinkus [6] found that the effect of the turbulence on the dynamic bearing stiffness and damping coefficients was small. Lund and Thomsen [7] attempted to improve the accuracy of the linearized techniques used and they also considered the case of non-circular journal bearing. However the thermal effects and the journal angular misalignment were disregarded in his analysis. Nikolakopoulos et al. [8] obtained the dynamic characteristics of the bearing using the finite element analysis for the determination of the oil film pressure. They used Lyapunov's stability criterion to obtain the critical stability of misaligned journal bearings. It is worth noting in this respect that the success of the method depends on the proper selection of Lyapunov function. Fujita et al. [9] examined the stability of infinitely long bearings. They obtained the fluid forces in radial and tangential directions and the stability analysis was made from the solution of the equations of translational and rotational vibrations of the journal. They found that the differences between the results obtained using the exact non-linear equations and those based on linearized equations of motion is small. The variety of bearing geometries, working loading conditions, additives used with the lubricating oil as well as bearing shell flexibility are very important factors that should be taken into consideration. To the authors' knowledge the effect of the elasticity of the bearing shell on the stability of the journal has not been considered in the literature. It is the purpose of this work to fill this gap.

2. ANALYSIS

Figure 1 shows the journal bearing geometry and the system of coordinates used in the analysis. The flow of the oil film is considered laminar, incompressible, Newtonian isoviscous. The dimensionless Reynolds equation governing the dimensionless pressure inside the oil film is

$$\frac{\partial}{\partial\theta}\left(\bar{h}^3\frac{\partial\bar{p}}{\partial\theta}\right)+\left(\frac{R}{L}\right)^2\frac{\partial}{\partial\bar{z}}\left(\bar{h}^3\frac{\partial\bar{p}}{\partial\bar{z}}\right)=6\frac{\partial\bar{h}}{\partial\theta}+12\frac{\partial\bar{h}}{\partial\bar{t}} \tag{1}$$

where $\bar{h}$ is the dimensionless oil film thickness taking into account the elastic deformation of the bearing liner. A perturbation technique is adopted here which assumes that the rotor bearing system to be disturbed by disturbances of small amplitude. The variables are assumed to be composed of steady components on which unsteady small oscillations are superimposed. We may then write the following equation:

$$\left.\begin{aligned}\bar{p}&=p^*(\theta,\bar{z})+p'(\theta,\bar{z},\bar{t})\\ \varepsilon&=\varepsilon_0+\varepsilon'(\bar{t})\\ \bar{h}&=h_0^*+h'(\bar{t})\\ \phi&=\phi_0+\phi'(\bar{t})\\ \bar{W}&=W^*+W'(\bar{t})\end{aligned}\right\} \tag{2}$$

Substituting from expressions (2) into Equation (1) neglecting higher order terms we get:

$$\left.\begin{aligned}&\frac{\partial}{\partial\theta}\left[h_0^{*3}\frac{\partial p^*}{\partial\theta}+3h_0^{*2}h'\frac{\partial p^*}{\partial\theta}+h_0^{*3}\frac{\partial p'}{\partial\theta}\right]\\ &\quad+\left(\frac{R}{L}\right)^2\frac{\partial}{\partial\bar{z}}\left[h_0^{*3}\frac{\partial p^*}{\partial\bar{z}}+3h_0^{*2}h'\frac{\partial p^*}{\partial\bar{z}}+h_0^{*3}\frac{\partial p'}{\partial\bar{z}}\right]\\ &\quad=6\frac{\partial}{\partial\theta}(h_0^*+h')+12\frac{\partial h'}{\partial\bar{t}}\end{aligned}\right\} \tag{3}$$

The equation governing the steady component of pressure p^* becomes

$$\frac{\partial}{\partial\theta}\left(h_0^{*3}\frac{\partial p^*}{\partial\theta}\right)+\left(\frac{R}{L}\right)^2\frac{\partial}{\partial\bar{z}}\left(h_0^{*3}\frac{\partial p^*}{\partial\bar{z}}\right)=6\frac{\partial h_0^*}{\partial\theta} \tag{4}$$

with

$$h_0^*=1+\varepsilon_0\cos\theta+h_e^*$$

where h_e^* is the term that accounts for the elastic deformation of the shell. In general, the determination of h_e^* requires the solution of the three-dimensional elasticity equations of equilibrium and compatibility. However, when the ratio of the thickness of the shell to the radius of the bearing is small ($t/R_b \le 0.1$) and Poisson's ratio ranging between 0.3 and 0.4 it is shown by Hook et al. [10] and Jain and Sinhasan [11] that a simple model of the form

$$h_e^*=C_0p^*=\left[\frac{\mu\omega RR_bt(1-\upsilon^2)}{c^3E}\right]p^*$$

satisfactorily approximates the complex three-dimensional model. This simple model is based on the assumption that the plane stress condition takes place and neglects the tangential strain. It may also result from the assumption of the plane strain condition and neglecting the tangential stress. From Equation (3) the equation governing the unsteady component of the pressure p' is

$$\frac{\partial}{\partial\theta}\left(h_0^{*3}\frac{\partial p'}{\partial\theta}\right)+\left(\frac{R}{L}\right)^2\frac{\partial}{\partial\bar{z}}\left(h_0^{*3}\frac{\partial p'}{\partial\bar{z}}\right)=-6\left[Q(\theta,\bar{z})-\frac{\partial h'}{\partial\theta}-2\frac{\partial h'}{\partial\bar{t}}\right] \quad (5)$$

where

$$Q(\theta,\bar{z})=\frac{1}{2}\left[\frac{\partial}{\partial\theta}\left(h_0^{*2}h'\frac{\partial p^*}{\partial\theta}\right)+\left(\frac{R}{L}\right)^2\frac{\partial}{\partial\bar{z}}\left(h_0^{*2}h'\frac{\partial p^*}{\partial\bar{z}}\right)\right]$$

$$\bar{h}=1+\varepsilon\cos(\alpha-\phi)=1+(\varepsilon_0+\varepsilon')\cos(\alpha-\phi_0-\phi')$$

It can easily be shown from the expression for $\bar{h}$ that

$$h'=\varepsilon'\cos(\alpha-\phi_0)=\varepsilon'\cos\theta$$

Therefore, Equation (5) can be written as

$$\left.\begin{aligned}&\frac{\partial}{\partial\theta}\left(h_0^{*3}\frac{\partial p'}{\partial\theta}\right)+\left(\frac{R}{L}\right)^2\frac{\partial}{\partial\bar{z}}\left(h_0^{*3}\frac{\partial p'}{\partial\bar{z}}\right)\\&\quad=-6\left[\varepsilon'(\sin\theta+P(\theta,\bar{z}))-2\dot{\phi}'\sin\theta-2\dot{\varepsilon}'\cos\theta\right]\end{aligned}\right\} \quad (6)$$

The dots denote differentiation with respect to $\bar{t}$ where

$$P(\theta,\bar{z})=\frac{1}{2}\left[\frac{\partial}{\partial\theta}\left(h_0^{*2}\cos\theta\frac{\partial p^*}{\partial\theta}\right)+\left(\frac{R}{L}\right)^2\frac{\partial}{\partial\bar{z}}\left(h_0^{*2}\cos\theta\frac{\partial p^*}{\partial\bar{z}}\right)\right]$$

The boundary conditions for Equation (4) are

$$p^*(0,\bar{z})=0,\quad p^*(\theta_{\text{eff}},\bar{z})=\left.\frac{\partial p^*}{\partial\theta}\right|_{\theta=\theta_{\text{eff}}}=0$$

$$p^*(\theta,0)=0,\quad p^*(\theta,1)=0$$

The boundary conditions for Equation (6) are

$$p'(0,\bar{z})=0,\quad p'(\theta_{\text{eff}},\bar{z})=\left.\frac{\partial p'}{\partial\theta}\right|_{\theta=\theta_{\text{eff}}}=0$$

$$p'(\theta,0)=0,\quad p'(\theta,1)=0$$

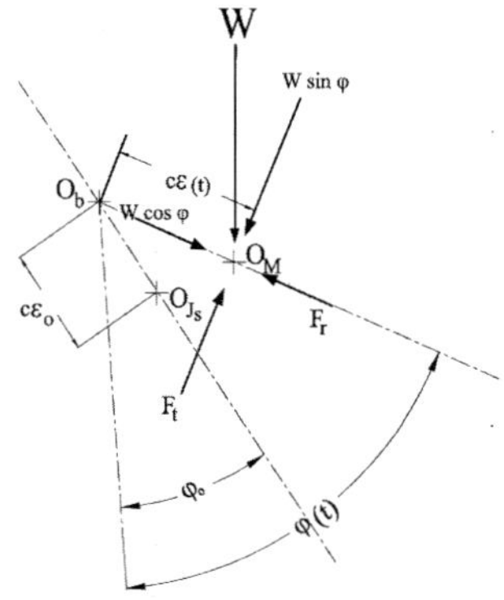

Figure 2. Force component acting on the journal.

In dimensionless form the radial and tangential load components are

$$\bar{F}_r = \frac{F_r}{\mu\omega RL\left(\frac{R}{c}\right)^2} = \int_0^1 \int_0^{\theta_{\text{eff}}} \bar{p}\cos\theta d\theta d\bar{z} \tag{7}$$

$$\bar{F}_t = \frac{F_t}{\mu\omega RL\left(\frac{R}{c}\right)^2} = \int_0^1 \int_0^{\theta_{\text{eff}}} \bar{p}\sin\theta d\theta d\bar{z} \tag{8}$$

The dimensionless resultant load and the attitude angle are given by

$$\bar{W} = \sqrt{\bar{F}_r^2 + \bar{F}_t^2} \quad \text{and} \quad \tan\phi = -\frac{\bar{F}_t}{\bar{F}_r}$$

The equations governing the motion of the journal are required for the study of journal dynamic stability about equilibrium position. For the analysis a weightless rigid shaft supporting a disc of mass $2M$ at its mid-span is selected. The shaft is assumed to be carried by two identical plain journal bearings at its ends. The equations of motion in the radial and transverse directions are (see Figure 2):

$$Mc(\ddot{\varepsilon} - \varepsilon\dot{\phi}^2) = -F_r + W\cos\phi \tag{9}$$

$$Mc(\varepsilon\ddot{\phi} + 2\dot{\varepsilon}\dot{\phi}) = F_t - W\sin\phi \tag{10}$$

In dimensionless form the equations of motion become:

$$\ddot{\varepsilon} - \varepsilon\dot{\phi}^2 + \left(\frac{1}{\Lambda}\right)[\bar{F}_r - \bar{W}\cos\phi] = 0 \tag{11}$$

$$\varepsilon\ddot{\phi} + 2\dot{\varepsilon}\dot{\phi} - \left(\frac{1}{\Lambda}\right)[\bar{F}_t - \bar{W}\sin\phi] = 0 \tag{12}$$

where

$$\Lambda = \frac{\omega M}{\mu (R/c)^3 L}$$

is called here the stability number.

The perturbation load components are obtained from (7) and (8) as

$$F_r' = \int_0^1 \int_0^{\theta_{\text{eff}}} p' \cos\theta d\theta d\bar{z} \tag{13}$$

$$F_t' = \int_0^1 \int_0^{\theta_{\text{eff}}} p' \sin\theta d\theta d\bar{z} \tag{14}$$

The loading on the bearing is assumed steady and consequently $W' = 0$, the perturbation equations governing the radial and tangential motion of the journal are:

$$\ddot{\varepsilon}' + (\varepsilon_0 + \varepsilon')\dot{\phi}'^2 + \frac{1}{\Lambda}(F_r' + W^*\phi' \sin\phi_0) = 0 \tag{15}$$

$$(\varepsilon_0 + \varepsilon')\ddot{\phi} + 2\dot{\varepsilon}'\dot{\phi}' - \frac{1}{\Lambda}(F_t' - W^*\phi' \cos\phi_0) = 0 \tag{16}$$

The dimensionless loading (steady) W^* and the attitude angle ϕ_0 are:

$$W^* = \sqrt{F_r^{*2} + F_t^{*2}} \quad \text{and} \quad \tan\phi_0 = -\frac{F_t^*}{F_r^*}$$

where

$$F_r^* = \int_0^1 \int_0^{\theta_{\text{eff}}} p^* \cos\theta d\theta d\bar{z}$$

and

$$F_t^* = \int_0^1 \int_0^{\theta_{\text{eff}}} p^* \sin\theta d\theta d\bar{z}$$

The initial conditions for Equations (15) and (16) become

$$\left.\begin{array}{l} \varepsilon'(0) = \varepsilon_i \\ \dot{\varepsilon}'(0) = 0 \\ \phi'(0) = 0 \\ \dot{\phi}'(0) = 0 \end{array}\right\} \tag{17}$$

Equations (15) and (16) together with their initial conditions (17) can be integrated using the trapezoidal stiff-technique for solving non-linear simultaneous ordinary differential equations. A step size of 0.01 in $\bar{t}$ is used in the analysis.

3. RESULTS AND DISCUSSION

Figure 3 shows the critical stability number Λ_{cr} versus the steady state eccentricity ratio ε_0 for different values of elastic coefficient C_0. It can be seen that at lower eccentricity ratios below approximately 0.45 the larger the value of C_0 the larger the stability number Λ_{cr}, meaning that the bearing becomes more stable. The opposite holds for eccentricity ratios greater than 0.45. This can be clearly demonstrated by reproducing Figure 3 in terms of percentage difference in Λ_{cr} with respect to the value of Λ_{cr} corresponding to the rigid liner case as shown in Figure 4. At lower eccentricity ratios of order 0.2 the percentage increase in Λ_{cr} reaches about 50% of the rigid value. At higher eccentricity ratios the reduction in Λ_{cr} is well over 40% of the rigid value. It is to be noted that for eccentricity ratios around 0.45, despite the bearing with elastic liner is more stable than the rigid one, the degree of stability, that is the critical value of Λ_{cr} stays unaffected by the value of the elastic coefficient in this range. On the other hand for eccentricity ratios in the range of approximately 0.55 the value of the stability number becomes practically close to the value corresponding to the rigid liner case (less than 10%) irrespective of the value of the elastic coefficient. The critical stability is also examined here for an example of short bearing ($L/D = 0.5$). Figure 5 shows that the critical stability number Λ_{cr} decreased in general meaning that shorter bearings are less stable than longer bearings. Also the reproduction of the figure is shown in Figure 6 where it can be seen that for eccentricity ratio below approximately 0.4 the stability number experiences no appreciable change in its value for a specific elastic coefficient C_0. An increase in the value of ε_0 ($\varepsilon_0 \sim 0.55$) at which Λ_{cr} is independent of C_0 is noticed. Figure 7 shows the results for the case of a relatively long bearing ($L/D = 2$). As expected, the critical stability number Λ_{cr} increased in general indicating that long bearings are prone to be more stable than the others. The reproduced Figure 8 shows an interesting linear dependence of the critical stability number Λ_{cr} on the eccentricity ratio ε_0 for higher values of the elastic coefficient C_0. Again the value of ε_0 at which Λ_{cr} is unaffected by the elastic coefficient C_0 in case of long bearing is approximately 0.43. A value very close to the value corresponding to $L/D = 1$.

4. CONCLUSIONS

In this work it can be concluded that the more elastic the bearing liner the less stable the bearing journal for eccentricity ratios over approximately 0.45. The opposite holds for eccentricity ratios below this value. It is also concluded that shorter bearings are less stable than longer bearings for any value of the elastic coefficient.

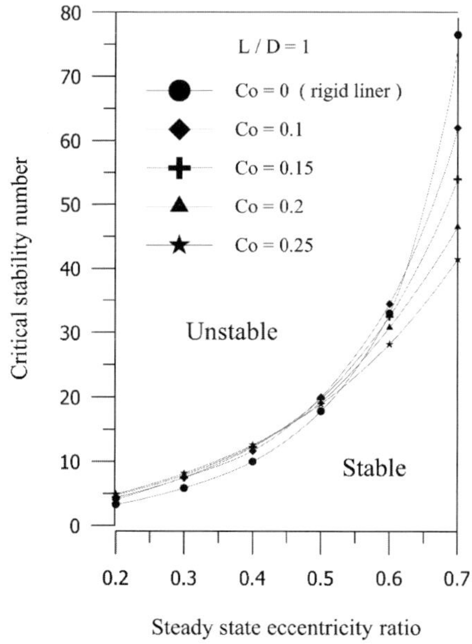

Figure 3. The critical stability number Λ_{cr} versus the steady state eccentricity ratio ε_0 for different liner elastic coefficients C_0.

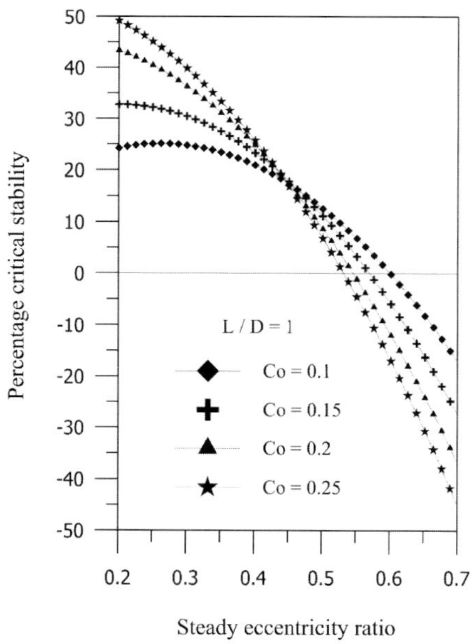

Figure 4. The percentage critical stability versus the steady state eccentricity ratio.

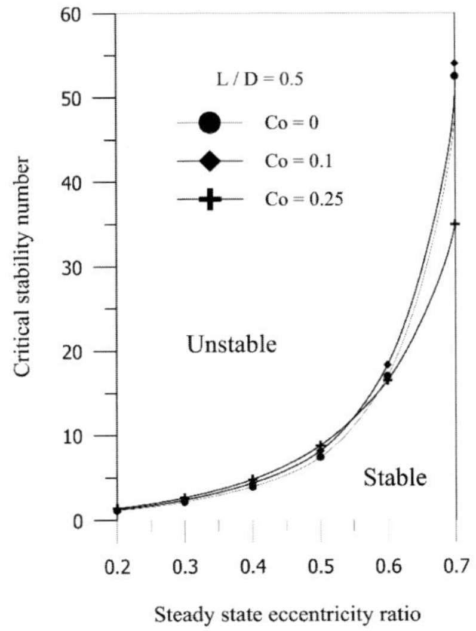

Figure 5. The critical stability number Λ_{cr} versus the steady state eccentricity ratio ε_0 for different liner elastic coefficients C_0 in the case of short bearing ($L/D = 0.5$).

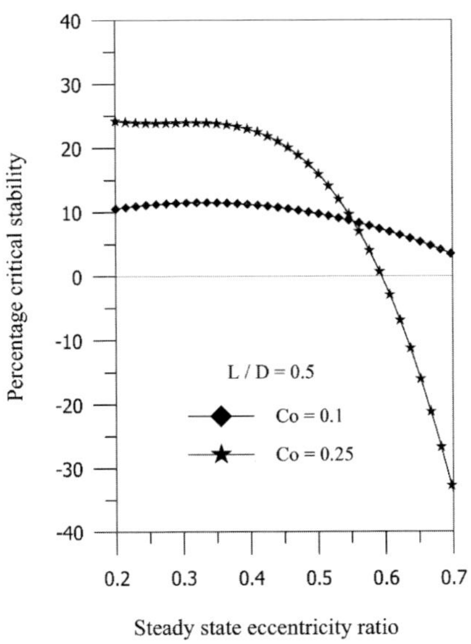

Figure 6. The percentage critical stability versus the steady state eccentricity ratio ε_0 in case of short bearing ($L/D = 0.5$).

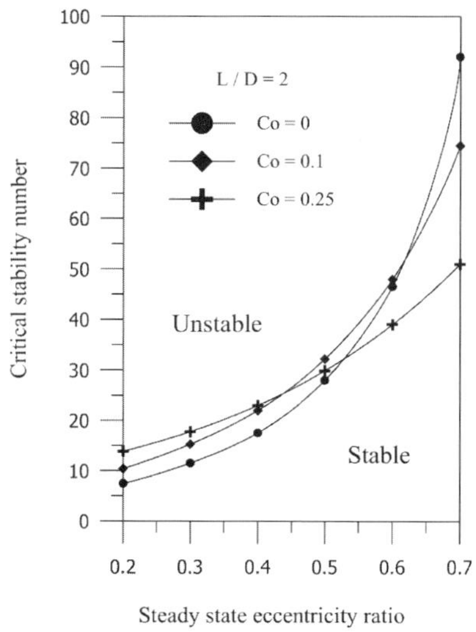

Figure 7. The critical stability number Λ_{cr} versus the steady state eccentricity ratio ε_0 for different liner elastic coefficients C_0 in the case of long bearing ($L/D = 2$).

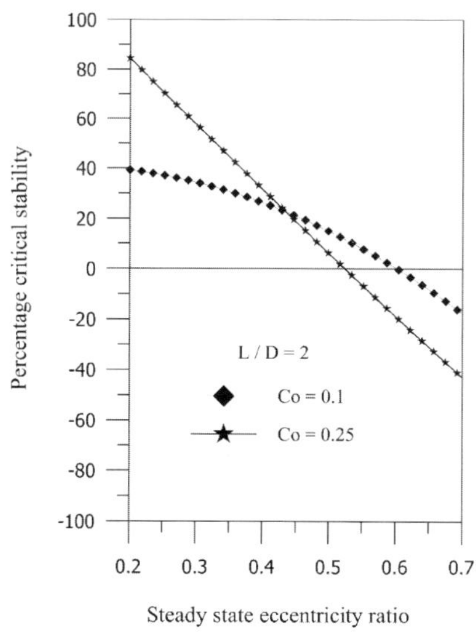

Figure 8. The percentage critical stability versus the steady state eccentricity ratio ε_0 in the case of long bearing ($L/D = 2$).

REFERENCES

[1] Hori, Y., "A theory of oil whip", *Trans. ASME, Journal of Applied Mechanics*, 189–198, 1959.

[2] Funagawa, M. and Tatara, A., "Stability criterion of an elastic rotor in journal bearings", *Trans. of JSME*, Part 1, 30-218:1238–1244, 1964.

[3] Gotoda, H., "On the vibration of a rotating shaft supported by journal bearings (report, unbalance vibration)", *Trans. of JSME*, Part 1, 30-215:887–900, 1964.

[4] Nakagawa, E. and Aoki, O., "Approximate theory of elasticity and damping characteristics of oil film in journal bearings", *Trans. of JSME*, Part 3, 31-229:1398–1408, 1965.

[5] Oscutt, F.K. and Arawas, E.B., "The steady state and dynamic characteristics of a full circular bearing and a partial arc bearing in the laminar and turbulent flow regimes", *Trans. ASME, Journal of Lubrication Technology*, 143–153, 1967.

[6] Wilcock, D.F. and Pinkus, O., "Effect of turbulence and viscosity variation on the dynamic coefficients of fluid film journal bearings", *Trans. ASME, Journal of Tribology*, Vol. 107, 1985.

[7] Lund, J.W. and Thomsen, K.K., "A calculation method and data for the dynamic coefficients of oil lubricated journal bearings", *Topics in Fluid Film Bearing and Rotor Bearing System Design and Optimization*, The Amer. Soc. of Mech. Engrs., New York, 1978.

[8] Nikolakopoulos, P.G. and Papadopoulos, C.A., "Lyapunov's stability of nonlinear misaligned journal bearings", *Trans. ASME, Journal of Engineering for Gas Turbines and Power*, Vol. 117, 576–581, 1985.

[9] Fujita, K., Shintani, A. and Yoshioka, K., "Simulation study on vibration behavior of fluid journal bearing with incompressible fluid", *Proc. of the Intern. Conf. on Flow-Induced Vibration*, FIV 2000, Lucerne, Switzerland, 19–22 June 2000.

[10] Hooke, C.J., Brighton, D.K. and O'Donoghue, J.P., "The effect of elastic deformation on the performance of thin shell bearings", *Proc. I. Mech. Eng.*, Vol. 181 (Part 3D), 63–69, 1966.

[11] Jain, S.C. and Sinhasan, R., "Performance of flexible shell journal bearings with variable viscosity lubricants", *Tribology Int.*, Vol. 16, No. 6, 331–339, 1983.

SESSION 4

A POSSIBLE MOLECULAR DYNAMIC STRUCTURE CONTRIBUTION TO ELASTOHYDRODYNAMIC LUBRICATION

Malcolm F. Fox
Institute of Tribology, School of Mechanical Engineering, University of Leeds, Leeds LS2 9JT, U.K.

Abstract A possible third contribution to the elastohydrodynamic lubrication effect is proposed based upon differential diffusion of higher and lower molecular weight hydrocarbons in the tribological contact. The effects of molecular weight and structure on the physical properties of hydrocarbons show that lower molecular weight, linear hydrocarbons have lower viscosities, and therefore higher self diffusion coefficients, than higher molecular weight, extensively branched structured, hydrocarbons with higher viscosities and lower self diffusion coefficients. In an equilibrium, isotropic, situation separation of lower and higher molecular weight hydrocarbons does not occur. In the elastohydrodynamic contact, under certain rates of relative movement of the metal boundaries, a non-equilibrium, anisotropic, situation will allow the lower molecular weight hydrocarbons to preferentially diffuse forwards in the direction of movement, giving a transient enhancement of higher molecular weight hydrocarbons remaining in the contact. This will give a higher viscosity and higher viscosity-pressure coefficients in the contact generating a fluid pressure above that predicted from the bulk equilibrium physical properties of the mixed fluid. Implications for base oil formulation are discussed.

Keywords: elastohydrodynamic lubrication, low and high molecular weight hydrocarbons, molecular structure, self diffusion coefficients, viscosity, pressure-viscosity coefficients, relative diffusion coefficients.

1. INTRODUCTION

Elastohydrodynamic lubrication theory almost exclusively assumes that the lubricating fluid is a homogenous, structure-less fluid with physical properties of viscosity and its derived temperature and pressure derivatives. The assumption is of an 'ether-like' fluid continuum, in the Victorian scientific sense, which has isotropic and strictly linear physical properties under all applied conditions. As for the majority of papers on elastohydrodynamics, analysis

R.W. Snidle and H.P. Evans (eds), IUTAM Symposium on Elastohydrodynamics and Microelastohydrodynamics, 135–147.

of the papers presented at this Conference shows that the isotropic fluid continuum approach predominates. Sole considerations of molecular structure effects have been concerned with the effect of the elastohydrodynamic contact region on a high molecular weight polymer interspersed into the fluid continuum, such as a Viscosity Index Improver (VII) polymer, passing through it.

The central argument of this paper is that under certain dynamic conditions the molecular structure of the lubricant may contribute to the elastohydrodynamic lubrication effect. This proposal does not challenge the central principles of EHL and these are not proposed to be changed but extended.

2. THE MOLECULAR STRUCTURE, COMPOSITION AND PROPERTIES OF LUBRICANT FLUIDS

If the molecular structure, composition and properties of lubricant fluids are grouped together under the shortened term of 'molecular structure' then there are three possible contributions to elastohydrodynamic lubrication in a moving contact, as:

(i) that lubricant fluids have a range of molecular weights and compound types in terms of:

- molecular size, through molecular weight expressed crudely at this stage as carbon number, with C_{30} as the median for lubricants,
- molecular structure, either as long chain molecules with minimal chain branching as against highly branched, symmetrical molecules or saturated ring/coupled rings structures or similarly arranged aromatic rings. These are the aromatic, napthenic and paraffinic structural divisions of hydrocarbon fluid compounds. Each of these groups has a slightly different physical property, such as density, ρ, and also different responses to externally applied conditions of increased temperature and pressure, such as the coefficients of $d\rho/dt$ and $d\rho/dp$.

(ii) **Viscosity/Pressure Relationships** – a fluid composed of mixed molecular structures of size and structural complexity will nevertheless have an overall, or aggregated, pressure coefficient of viscosity, $d\gamma/dp_{\text{overall}}$. But examining the literature values for the pressure coefficients for fluid viscosity for individual molecules, and it is difficult to find good values for a worthwhile range of relevant compounds (the more recent work of Bair is particularly useful [1] as is also work on viscosity/pressure relationships for higher hydrocarbons [2] and API Report No. 42 [3]) shows that two firm trends are established for the pressure coefficient of viscosity of fluids, such that:

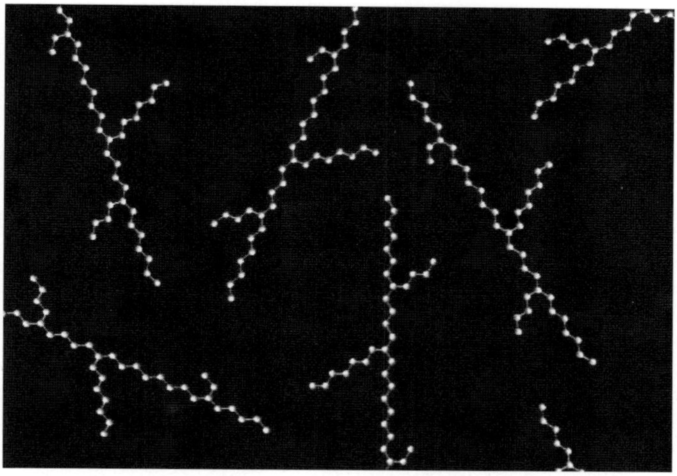

Figure 1. The simplistic approach to the molecular structure of hydrocarbons – the 'Fish Skeleton' model.

- $d\gamma/dp_{\text{individual}}$ increases with molecular size,
- $d\gamma/dp_{\text{individual}}$ increases with structural complexity, where the term 'structural complexity' is defined in terms of increased chain branching and/or the inclusion of ring structures, saturated or unsaturated/aromatic.

The view of hydrocarbon molecular structures can be very simplistic, condensed down to a rigid 'fish backbone' structure, as in Figure 1. However simplistic in terms of structure, it is essentially wrong in depicting hydrocarbons as extended, linear forms.

Reverting to a relatively uncomplicated molecule such as n-butane which is much smaller than the usual lubricant molecule, it consists of a linear arrangement of four carbon atoms. There are several ways in which its structure can be represented (Figures 2a and 2b). Figure 2a represents the spatial arrangement of the carbon and hydrogen atoms, Figure 2b represents the bonding electron density, which holds these molecules together in a very strong manner, with the carbon and hydrogen atoms embedded into it.

From the simplistic starting point of the n-butane structure in Figure 2a, each of the carbon atoms has full valence saturation, denoted as carbon-carbon (principally) and carbon-hydrogen sp3 bond hybridisation by Linus Pauling. There is '*free rotation*' about each carbon-carbon bond and for longer hydrocarbon chains than the n-butane example given,

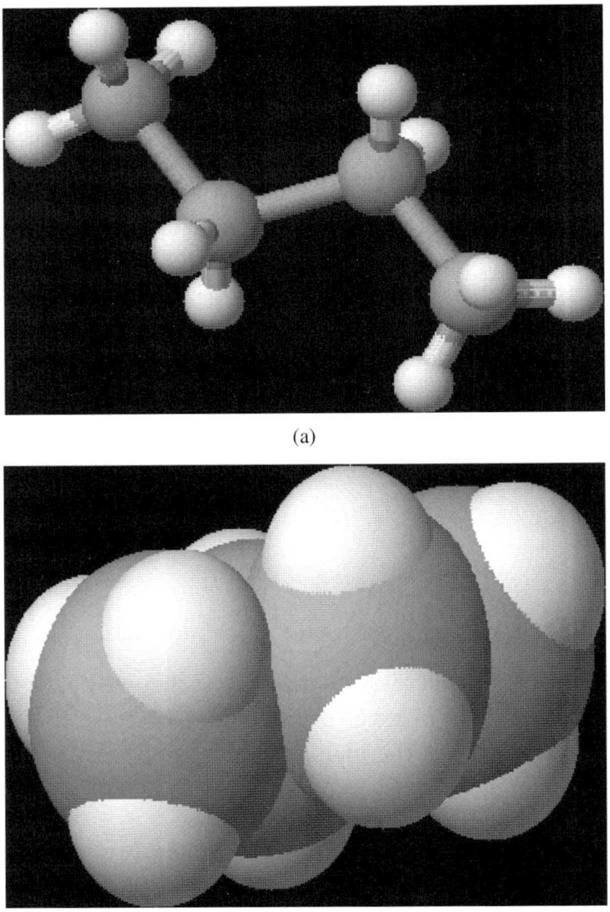

(a)

(b)

Figure 2. (a) A 'ball and stick' representation of the carbon (dark) and hydrogen (light) atoms and their binding in n-butane; (b) hydrogen atoms (light) embedded in the molecular electronic orbital of n-butane.

- longer chain carbon chains will 'fold', by rotation about the assembly of carbon-carbon bonds, to a *minimum energy*,
- configuration which is dependent on temperature, with size measured by the 'radius of gyration',
- longer carbon chains will expand from this folded state with increase in temperature, increasing their 'radius of gyration',
- for higher molecular weight molecules beginning to approach those of interest in lubrication, these conformational effects are seen in Figure 3 for three isomers of twelve-carbon hydrocarbons as C_{12} saturated hydrocarbons, thus $C_{12}H_{26}$,

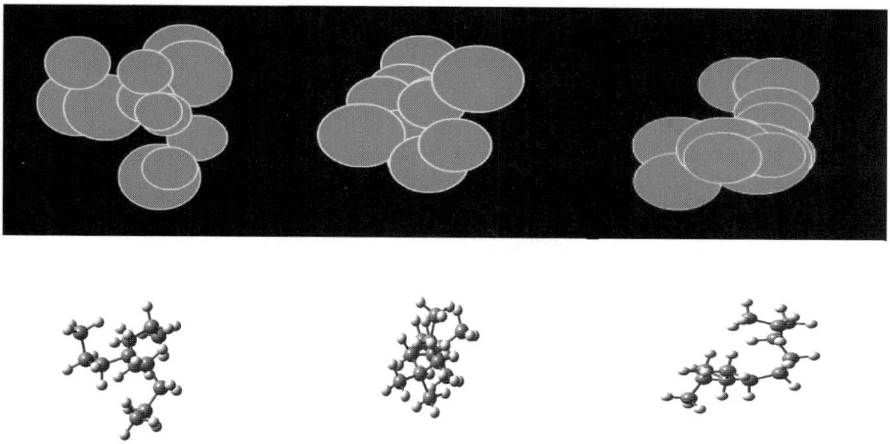

5-Proplylnonane 2,2,4,6,6 **Pentamethylheptane** **Dodecane**

Figure 3. Folded, energy-minimized structures for three types of C_{12} saturated hydrocarbon structures.

- first a 'symmetric star-shaped' $C_{12}H_{26}$ molecule, (or very close to) as 5-propylnonane,
- second, a 'maximum branching' $C_{12}H_{26}$ structure as 2,2,4,6,6 pentamethylheptane, and,
- third, a linear $C_{12}H_{26}$ molecular structure, dodecane, respectively,
- therefore, according to their isomeric conformation, hydrocarbon molecules of the same overall C_xH_y composition can have different forms as either linear or globular, which will affect their physical properties and also their response to temperature and pressure. It is helpful to examine 'structural families' for physical properties which is increasingly significant for the various polyalphaolefins PAO's whose dominant structural features can be varied by the conditions of polymerisation and also future 'Gas to Liquid', or 'GTL' hydrocarbon, whose structures can be varied more precisely by process conditions of catalyst, temperature and pressure, almost 'designer fluids'.

(iii) Structure and Physical Properties

(a) *Isomeric Effects* – The first example of the effect of molecular size on the viscosity of hydrocarbon fluids is given for 'Star Molecules', $CH(C_nH_{2n+1})_3$, for increasing 'n', Figure 4, from $CH(C_3H_7)_3$, or $C_{10}H_{22}$, through to $CH(C_{10}H_{21})_3$ or $C_{31}H_{64}$, with their structures

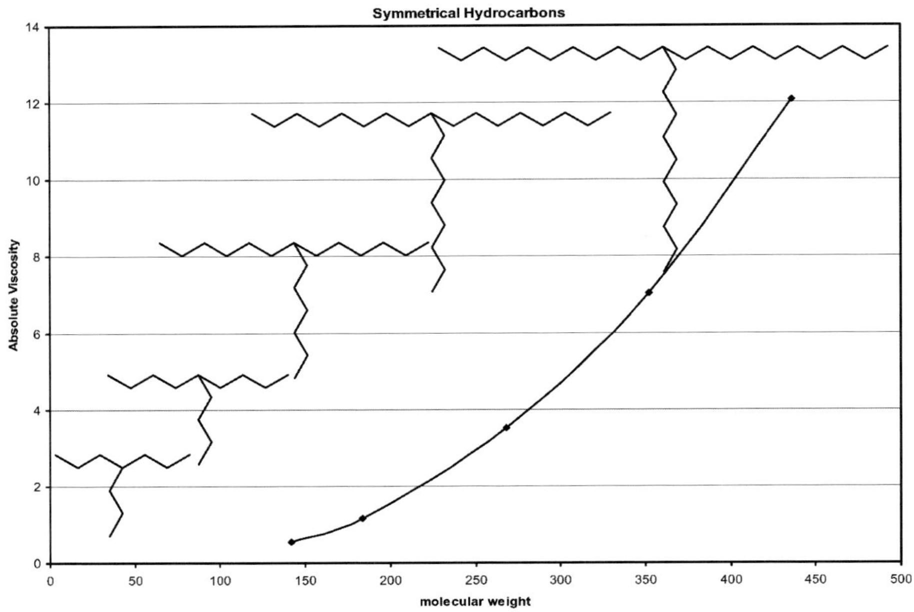

Figure 4. Star hydrocarbon molecule viscosities vs. molecular weight, C_{10}–C_{31}.

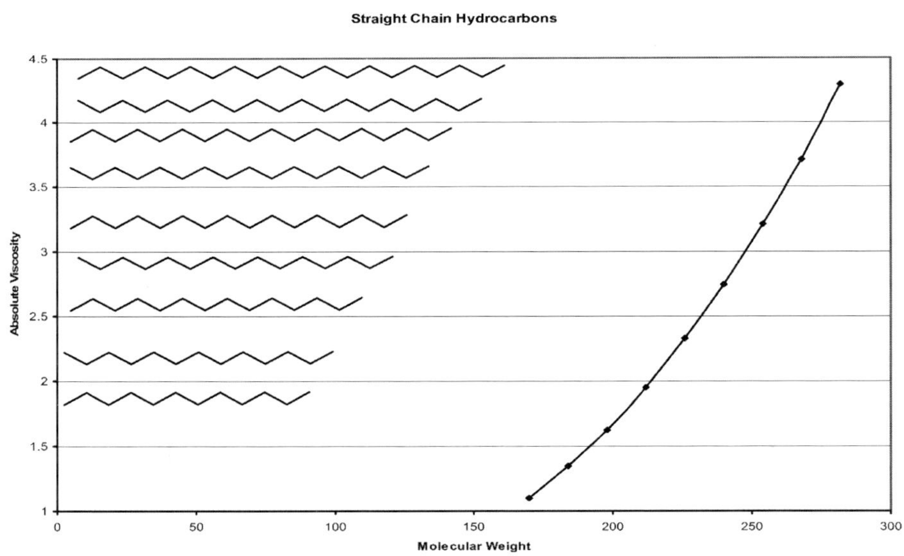

Figure 5. Linear hydrocarbon molecule viscosities vs. molecular weight, C_{12}–C_{20}.

superimposed on the graph. Increase in molecular weight for the same structure shows an initial increasing curvature which tends towards a linear relationship at higher mass (data from [1–3]).

(b) *Molecular Size* – The second example of the effect of molecular size on the viscosity of hydrocarbon fluids is given for linear molecules for increasing '*n*' from C_{12} ($C_{12}H_{26}$), to C_{20} ($C_{20}H_{42}$), with their structures superimposed on the graph. Increase in molecular weight for the same structure shows an initial increasing curvature which tends towards a linear relationship at higher mass. For the same molecular weight ranges where linear molecules and star molecules can be compared, 170 to 280 amu, their viscosities are closely similar but as far as can be seen from the data available, at the limit of the amu ranges given, the star molecular structure viscosities are increasing at a rate double that of the linear molecules.

(c) *Effects of Substituents* – Substitution of groups within a given carbon number molecule containing a cyclohexyl group shows that:

- the close proximity of substituted methyl groups in the carbon chain attached to a cyclic group is not necessarily crucial, as shown for the first, second and fifth structures in Figure 6. Only when the fully substituted α-carbon is combined with the terminal iso-group in the fifth molecular structure in Figure 6 is a substantial viscosity difference seen. This effect is associated with steric hindrance, which is a difficult additional parameter to model.

(d) *Summary of Structure and Properties* – From the structure and physical properties of linear and substituted hydrocarbons, two points are clearly demonstrated, that:

- different sizes and structures of hydrocarbon molecules have individual physical properties,
- lubricant fluids are mixtures of related compounds, therefore there is a range of physical properties.

3. SELF-DIFFUSION COEFFICIENTS OF LIQUIDS

The self-diffusion coefficient of a fluid is one of its fundamental properties in three dimensions or 4π radians and measures the rate at which molecules move relative to reach other. There are several models of molecular movement, the 'isotropic movement' and the 'wait and jump' as limiting examples but for each molecular movement allows fluids to move under the influence of a force.

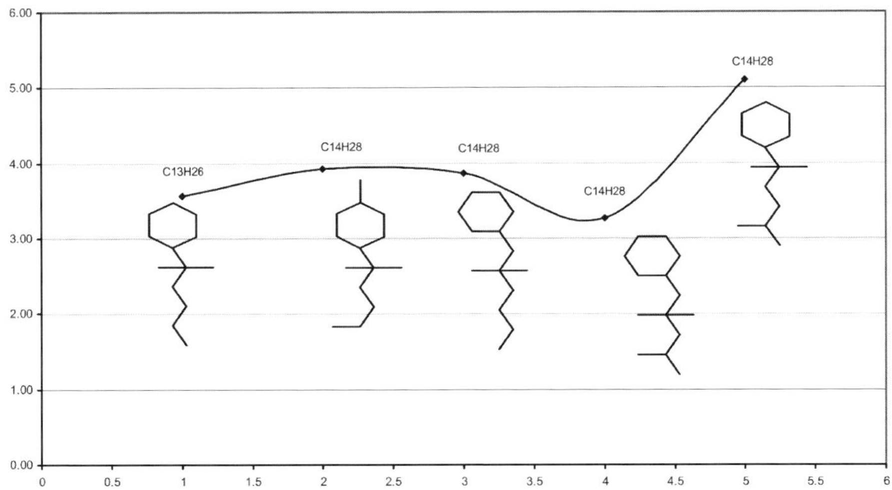

Figure 6. Viscosities of variously substituted AlkylCyclohexanes.

For the unperturbed situation, as contained for a still fluid in an unstirred vessel, molecules of different sizes and structures will remain intimately mixed, provided that they macroscopically miscible, because of thermally driven random diffusion. Self diffusion coefficients measure the movement of individual molecules relative to other molecules. The definition of the self diffusion coefficient, D, for the molecules of a fluid from first principles is:

$$D = kT/6\pi\eta r,$$

where η is the bulk absolute viscosity of the fluid and r is the radius of the diffusing particle.

The temperature dependence is given by $D_T = D_0 \exp(\Delta E_D/rT)$. The essential relationship is that $D \propto 1/\eta$. Therefore, a high viscosity fluid has a relatively low self-diffusion coefficient and conversely, a low viscosity fluid has a relatively high self-diffusion coefficient. This is intuitively reasonable in terms of molecular mass and momentum. This relationship is also very useful in that values for self diffusion coefficients are not readily available for a wide range of fluids whilst their viscosity values are much more known., which is:

- theoretically, inversely related to viscosity,
- in reality, by measurement, to the inverse square root of viscosity.

From this it follows that higher molecular weight compounds have lower linear diffusion velocity of diffusion than lower molecular weight compounds. Thus, if a lubricant fluid is composed of two components, each of which is

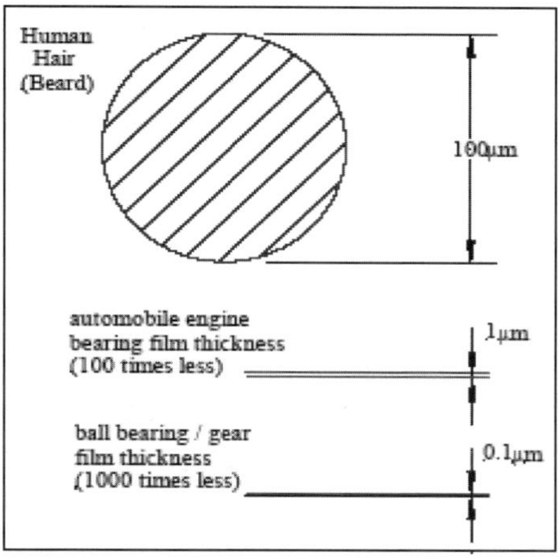

Figure 7. Comparative lubricant film thicknesses for automotive and rolling element bearings, from Dowson [4], reproduced with permission.

centred on two separate, higher and lower, molecular weight distributions and structures, then there are two fluids which will respond in a different quantitative manner to an applied substantial pressure. However, because the two molecular components are intimately mixed and under equilibrium conditions do not separate, the overall physical properties of the mixture are colligative, i.e., where the overall properties are defined solely by the relevant numerical contributions of the individual components.

4. A POSSIBLE MOLECULAR STRUCTURE CONTRIBUTION TO ELASTOHYDRODYNAMIC LUBRICATION

4.1 The Equilibrium, Non-Transport, EHL Model

As described previously in the Introduction, elastohydrodynamic models and analyses assume a uniform, unstructured and isotropic fluid which has physical properties of density, viscosity and pressure. But EHL contacts now have dimensions which are comparable to molecular size, Dowson's [4] illustration of this is very instructive (Figure 7).

Dowson has shown that in rolling element bearings/gear lubricants, the lubricant film thickness is of the order of $\sim 0.1\ \mu$m, which is 100 nm (nanometer). But 100 nm is equivalent to multiple organic molecular dimensions in lubricant fluids (see Figure 8).

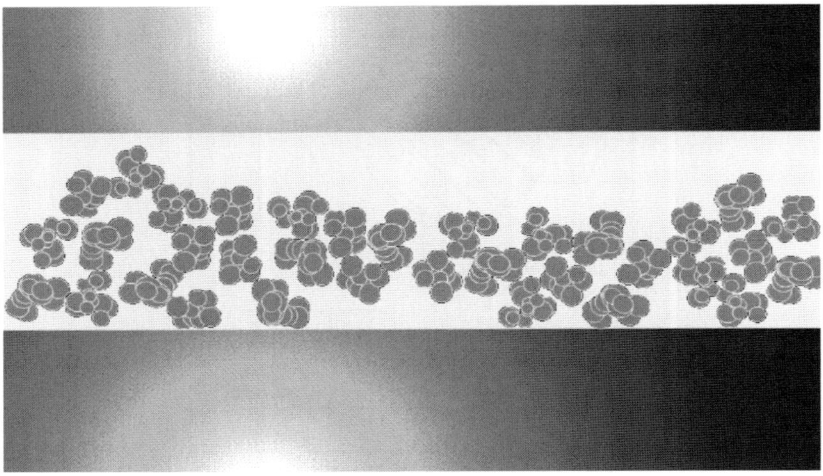

Figure 8. Lubricant molecular size relative to rolling element bearing lubricant film.

4.2 The Non-Equilibrium EHL Contact Model with Differential Fluid Transport

The implications for a mixed hydrocarbon in an EHL contact are that the equilibrium, isotropic, molecular transport properties may no longer apply. The molecular transport properties may become transiently anisotropic within the EHL contact, for:

- in an advancing tribological contact, fluid flow is in the direction of the contact movement and superimposed upon the random thermal diffusive motion such that:
  - diffusive molecular motion against the direction of movement of the contact is restricted,
  - molecular motion with the direction of the contact movement is enhanced,
  - but diffusive molecular motion orthogonal to the contact movement is unchanged.

Two component distributions of significantly different molecular weights and structural types within the lubricant fluid will have different self diffusion coefficients due to the previously described molecular size and structural effects. Thus, hydrocarbons of lower molecular weight with linear structures will diffuse more rapidly out of the EHL contact than higher molecular weight hydrocarbons with highly branched 'star' structures. The net effect within the advancing tribological contact is to generate a transient separation of the higher molecular weight hydrocarbons from the lower molecular weight hydrocarbons in the direction of the contact movement, which is then:

- relatively enriched in the higher molecular weight hydrocarbons which have intrinsically higher viscosities than the smaller molecular weight hydrocarbons,
- where the higher molecular weight, highly branched, hydrocarbons have higher viscosity pressure coefficients than the lower molecular weight, linear hydrocarbons.

The overall effect resulting from the transient separation of the higher and lower molecular hydrocarbons for the pressure developed within the original EHL concept within the contact is to:

- increase the pressure above that predicted from the bulk fluid viscosity pressure coefficient through further enhancement of the fluid viscosity within the contact,
- due first, to the higher viscosities of the transiently enhanced concentrations of larger molecules,
- and second, because the larger molecules have higher viscosity pressure coefficients than the smaller molecules.

It is therefore proposed that under certain conditions of rates of relative mechanical movement within the EHL contact and differential molecular diffusion, the overall effect within the tribolological contact would be to increase the fluid pressure above that predicted from the bulk fluid viscosity pressure coefficients.

4.3 Other Examples of Differential Diffusion

Molecular separation by diffusion is a general phenomenon utilised elsewhere in addition to that proposed here:

- for the separation of gases – ‘Greshams Law’ is used to separate gases of different molecular weights by effusion through a small orifice. The composition of the gaseous mixture after passing through the small orifice is enhanced for the lower molecular weight gas relative to that for the original gaseous mixture,
- for isotopic separation, as used for uranium isotope separation, uses the slight mass, and the related diffusion, difference of the ^{235}U and ^{238}U isotopes in UF_6 by multiple centrifuging in a cascade process.

5. IMPLICATIONS FOR EHL LUBRICANT FORMULATION – 'DUMB-BELLING' AND ITS FURTHER DEVELOPMENT

A well established practice in the formulation of lubricant base oils is to exploit the properties of smaller/larger hydrocarbon molecules as lower/higher viscosity components in an overall formulation, colloquially known as 'dumb-belling'. Suitably optimized by iterative formulation and testing, 'dumb-belling' gives lower wear rates than those obtained from a 'single' component lubricant base oil with a mono-modal hydrocarbon range of molecular size.

The model proposed in this paper is the first to explain the advantages of 'dumb-belling' base oil formulations over otherwise identical, in bulk physical properties, mono-modal hydrocarbon ranges of molecular size.

By proposing an explanation, the model can be used to enhance the 'dumb-belling' effect further. Of the two molecular fluid components, for maximum separation of the two components within a tribological contact:

- the larger hydrocarbon molecules should have lower self diffusion coefficients, thus higher molecular weight, highly branched structures,
- the smaller hydrocarbon molecules should be selected for higher self diffusion coefficients by the opposing arguments of lower molecular weight and linear molecular structures.

The ability to choose between molecular structures at each level of molecular size is now enhanced by the ability (i) to partially choose PAO structures and (ii) the possibility of Gas-to-Liquid', or 'GTL', lubricant fluids designed with specific structures for lubricant molecules.

6. FUTURE ISSUES

The various physical parameter values for individual hydrocarbon molecules needed to move on from the generality of the proposal to more specific experiments are difficult to find, indeed often they do not exist because they have not been measured. Bair has drawn attention to the need for viscosity/pressure measurements of hydrocarbon fluids to be validated [1]. One way forward is to obtain samples of specific molecular structure fluids and undertake physical measurements on them for the purposes of the proposed approach to EHL. Their viscosities, viscosity/pressure coefficients and measurement of their self-diffusion coefficients are needed.

There is then a need to determine the relative self diffusion coefficients of higher and lower molecular hydrocarbons in mixed component systems. This is an issue for relatively better characterised mixtures such as organic alcohol/water systems [5], therefore it is relatively difficult for hydrocarbon mixtures, an area which needs further work.

Finally, comparative EHL effects should be determined for formulated hydrocarbon systems comprising a mixture of hydrocarbon molecular weights and structural, as described above, for a range of relative surface motion velocities in a contact under EHL conditions.

A more detailed theoretical analysis of differential diffusion effects for lower and higher molecular weight molecules in EHL contacts for different rates of relative motion is being developed.

REFERENCES

[1] S Bair, Georgia Institute of Technology, Personal Communications.

[2] D.A. Lowitz, J.W. Spencer, W. Webb and R.W. Schiessler, 'Temperature Pressure-Structure Effects on the Viscosity of Several Higher Hydrocarbons', *J. Chem. Phys.*, **30**(1), 1967, 73–83.

[3] American Petroleum Institute (API) Project Report 42.

[4] D. Dowson, 'Thin Films in Tribology', in *Proceedings of 19th Leeds–Lyon Symposium on Tribology*, Leeds 1992, Elsevier, 1993.

[5] F.Franks, 'The Hydrophobic Interaction', in *Water – A Comprehensive Treative*, Vol. 4, F. Franks (ed.), Plenum Press, New York , 1975, Chapter 1, pp. 1–94 (see also Vol. 2).

MODELLING OF LUBRICANT DEGRADATION AND ELASTOHYDRODYNAMIC LUBRICATION

Ilya I. Kudish[1], Ruben G. Airapetyan[1] and Michael J. Covitch[2]
[1]*Kettering University, 1700 W. Third Ave., Flint, MI 48504, U.S.A.*
[2]*The Lubrizol Corporation, Wickliffe, OH 44092, U.S.A.*

Abstract The paper presents a new approach to modeling elastohydrodynamic contacts with degrading lubricants. Considered lubricants are diluted solutions of polymer molecules with linear structure. The problem is formulated mathematically and analyzed numerically. A parametric analysis of the problem is performed.

Keywords: viscosity index improver, polymers, lubricant degradation, elastohydrodynamic lubrication, non-Newtonian lubricants, viscosity, friction.

1. INTRODUCTION

While in use in machine joints lubricants undergo degradation caused by mechanical stresses, temperature, chemical reactions, contamination etc. The lubricants are formulated by using polymer additives. A fundamental approach to stress-induced lubricant degradation caused by polymer scission is proposed. In the case of a polymer additive of linear structure the degradation phenomenon is described by a single kinetic equation while in the case of star polymer additives it is described by a system of coupled kinetic equations. The equations were solved numerically. The simulation results show that the loss of lubricant viscosity may reach up to 50%. The comparison of experimental and simulated results showed excellent agreement. The above mechanism of lubricant degradation was introduced in models for elastohydrodynamically lubricated contacts. Some new properties of the lubrication phenomenon were uncovered. Lubricants with Newtonian and non-Newtonian rheologies were considered. The simulation results show significant losses of lubricant viscosity (up to 45%) and reduction of frictional stresses (up to 35%). The effect of lubricant

R.W. Snidle and H.P. Evans (eds), IUTAM Symposium on Elastohydrodynamics and Micro-elastohydrodynamics, 149–174.

degradation is milder in the case of non-Newtonian lubricants with limiting stress than in Newtonian lubricants. An approach for lubricant life calculation is proposed.

2. FORMULATION OF THE ELASTOHYDRODYNAMIC PROBLEM FOR DEGRADING LUBRICANT

The recent developments in modelling of polymer degradation in dilute fluid solutions[1-3] allowed for effective description of the kinetics of stress-induced degradation of linear and star polymer additives to lubricants. It was shown that the problems were correctly formulated and the numerical simulation results are in good agreement with independently obtained experimental data[4]. Namely, the simulated and experimentally measured polymer molecular weight distributions and lubricant viscosity behavior are in good qualitative and quantitative agreement. That, in turn, allowed for generalization of the classic elastohydrodynamic problem of lubrication. In the present paper we will limit ourselves to the analysis of a lubricated contact with degrading polymeric additive of linear structure. The same problem for degrading lubricant with a star polymeric additive can be formulated and solved in similar manner. However, the solution of the latter problem is much more computer intensive.

Let us assume that two parallel steadily moving with linear surface speeds u_1 and u_2 infinite cylinders with radii R_1 and R_2 are made of elastic materials with Young's modulus E_1 and E_2 and Poisson ratios ν_1 and ν_2, respectively. The cylinders are separated by a thin lubrication layer and are loaded by the normal to their axes force P. It is clear that the lubricant degrades while flowing through the gap between the deformed elastic cylinders. Therefore, it is necessary to be able to describe each of the above mentioned phenomena: (a) lubricant motion, (b) cylinder deformation, and (c) lubricant degradation.

2.1 Lubricant Motion and Cylinder Deformation

The lubricant is assumed to be an incompressible fluid with a non-Newtonian rheology

$$\frac{\partial u}{\partial z} = \frac{\tau_L}{\mu} G\left(\frac{\tau}{\tau_L}\right), \quad (1)$$

where z is the coordinate of a lubricant particle along the coordinate axis directed across the gap between the cylinders, u is the x-component of the lubricant velocity along the x-axis directed along the cylinders' motion (and the lubrication layer), τ and τ_L are the local and limiting shear stresses in the lubricant, μ is the lubricant viscosity, $\mu=\mu(x,z)$, G is a given function that determines the rheology of the lubricant fluid. According to[5,6] τ_L is usually a linear function of pressure $p(x)$, i.e.

$$\tau_L = \tau_{L0} + \tau_{L1} p, \tag{2}$$

where parameters τ_{L0} and τ_{L1} are certain functions of the lubricant temperature.

Considering steady isothermal lubrication conditions and accepting the classic EHL assumptions for a non-conformal contact[7] the equations governing the isothermal lubrication problem based on the generalized Reynolds and elasticity theory equations can be presented as follows[8,9]

$$\int_{-h/2}^{h/2} z\frac{\tau_L}{\mu}G\left(\frac{1}{\tau_L}(f+z\frac{dp}{dx})\right)dz = \frac{u_1+u_2}{2}\left[\frac{h}{\sqrt{1+\frac{1}{4}\left(\frac{dh}{dx}\right)^2}} - \frac{h_e}{\sqrt{1+\frac{1}{4}\left(\frac{dh(x_e)}{dx}\right)^2}}\right]$$

$$+\int_{-h_e/2}^{h_e/2} z\frac{\tau_L(x_e)}{\mu(x_e,z)}G\left(\frac{f(x_e)}{\tau_L(x_e)}\right)dz, \quad p(x_i) = p(x_e) = \frac{dp(x_e)}{dx} = 0, \tag{3}$$

$$h = h_e + \frac{x^2 - x_e^2}{2R'} + \frac{2}{\pi E'}\int_{x_i}^{x_e} p(x')\ln\frac{x_e - x'}{|x-x'|}dx', \quad \int_{x_i}^{x_e} p(x)dx = P,$$

where x_i and x_e are the coordinates of the lubricant inlet and exit points, respectively, $h(x)$ is the gap between the cylinders, h_e is the exit film thickness, $h_e=h(x_e)$, R' is the effective radius of the cylinders, $1/R'=1/R_1+1/R_2$, E' is the effective elasticity modulus, $1/E'=(1-\nu_1^2)/E_1+(1-\nu_2^2)/E_2$, $f(x)$ is the local sliding frictional stress in the lubricant determined by the equation[9]:

$$\int_{-h/2}^{h/2} \frac{\tau_L}{\mu}G\left(\frac{1}{\tau_L}(f + z\frac{dp}{dx})\right)dz = \frac{u_2 - u_1}{\sqrt{1+\frac{1}{4}\left(\frac{dh}{dx}\right)^2}}. \tag{4}$$

The process of lubricant degradation, i.e. the process of polymer additive scission, occurs while the additive molecules dissolved in the lubricant move along the lubricant flow streamlines. Therefore, it is necessary to formulate the equations for the lubricant flow streamlines $z=z(x)$. Similar to[7] these equations can be represented in differential and integral forms as follows

$$\frac{d}{dx}\int\limits_{-h/2}^{z(x)}u(x,\zeta)d\zeta=0,\ \int\limits_{-h/2}^{z(x)}u(x,\zeta)d\zeta=\frac{h}{2}\frac{u_1}{\sqrt{1+\frac{1}{4}\left(\frac{dh}{dx}\right)^2}}+zu(x,z)-\int\limits_{-h/2}^{z(x)}\zeta\frac{\tau_L}{\mu}G\left(\frac{1}{\tau_L}(f+\zeta\frac{dp}{dx})\right)d\zeta,$$

$$u(x,z)=\frac{u_1}{\sqrt{1+\frac{1}{4}\left(\frac{dh}{dx}\right)^2}}+\int\limits_{-h/2}^{z}\frac{\tau_L}{\mu}G\left(\frac{1}{\tau_L}(f+\zeta\frac{dp}{dx})\right)d\zeta. \quad (5)$$

2.2 Lubricant Degradation

The stress-induced process of lubricant degradation for linear polymer additives is described by the kinetic equation derived and analyzed by Kudish et al.[2]. In particular, for the case of a steady isothermal line contact the kinetic equation of lubricant degradation can be written for the probabilistic density distribution $W(x,z,l)$ of polymer molecular weight, where l is the polymer molecule chain length. The function $W(x,z,l)$ can be introduced in such a way that $W(x,z,l)\Delta l$ is the polymer weight in a unit fluid volume centered at the point with coordinates (x,z) with the polymer molecule chain lengths in the range from l to $l+\Delta l$. In the above mentioned case the kinetic equation can be written as follows[2]

$$u\frac{\partial W}{\partial x}+v\frac{\partial W}{\partial z}=2l\int\limits_{l}^{\infty}R(x,z,L)P_c(l,L)W(x,z,L)\frac{dL}{L}-R(x,z,l)W(x,z,l), \quad (6)$$

where R is the probability of polymer molecule scission over a unit time period and $P_c(l,L)$ is the density of the conditional probability of a polymer molecule with the chain length L to break into two pieces with lengths l and $L-l$. The expressions for the functions R and P_c have been derived by Kudish et al.[2] and have the form

$$R(x,z,l)=0 \text{ if } l\leq L_*,$$

$$R(x,z,l)=1-\left(\frac{l}{L_*}\right)^{2\alpha\frac{U_A}{kT}}\exp\left[-\alpha\frac{U_A}{kT}\left(\frac{l^2}{L_*^2}-1\right)\right] \text{ if } l>L_*, \tag{7}$$

$$P_c(l,L)=\ln(2)\frac{4|L-2l|}{L^2}\exp[-\ln(2)\frac{4l(L-l)}{L^2}],$$

where L_* is calculated according to the formula

$$L_*=\sqrt{\frac{U_A}{Ca_*l_*^2|\tau|}},\quad U_A=\frac{U}{N_A}, \tag{8}$$

and k is the Boltzmann constant, $k=1.381\cdot 10^{-23}$ $J/^{\circ}K$ and N_A is the Avogadro number, $N_A=6.022\cdot 10^{23}$ $mole^{-1}$. In Eqs. (8) parameter C is the shield constant, a_* and l_* are the polymer bead radius and bond length, and U is the polymer bond (C-C bond) dissociation energy. The above functions R and P_c were validated by the comparison of numerically simulated and experimentally obtained data by Kudish et al.[2,3].

We will assume that the distribution of polymer molecules supplied to the contact region is uniform across the lubrication layer, i.e.

$$W(x_i,z,l)=W_a(l) \text{ if } u(x_i,z)>0,\quad W(x_e,z,l)=W_a(l) \text{ if } u(x_e,z)<0, \tag{9}$$

where $W_a(l)$ is the given function of the polymer molecule chain length l. It should be noted that at the inlet to the contact region, $x=x_i$, some of the lubricant enters the contact, some may turn around and exit the contact region through the $x=x_i$ cross section. A similar situation is possible at the exit point $x=x_e$ from the contact if the contact surfaces are moving in the opposite directions. In each case it is assumed that when the lubricant enters the contact region (whether through the inlet $x=x_i$ or the exit $x=x_e$ cross sections) the distribution of the polymer molecules is described by $W_a(l)$ from Eqs. (9).

Some properties of the kinetic equation (6) are established by Kudish et al.[2,3]. Among these properties are the facts that the molecular weight of the polymer additive is conserved along the lubricant flow streamlines while the number of polymer molecules tends to increase due to polymer molecule scission.

As the lubricant contains some linear polymer additive which may degrade while passing the lubricated contact the lubricant viscosity varies as a function of pressure p, the concentration N and the density distribution $W(x,z,l)$ of the polymer additive. In particular, we will assume that the lubricant viscosity depends on the polymer distribution according to the Huggins and Mark Houwink equations[10,11]

$$\mu(p,W) = \mu_a e^{\alpha_p p} \frac{1+N[\eta]+k_H(N[\eta])^2}{1+N[\eta]_a+k_H(N[\eta]_a)^2}, \qquad [\eta]=k' M_W{}^{\beta},$$
$$M_W = \left[\int_0^\infty W(x,z,w) w^\beta dw \Big/ \int_0^\infty W(x,z,w) dw \right]^{1/\beta}, \tag{10}$$

where μ_a is the ambient lubricant viscosity, α_p is the pressure coefficient of viscosity, N and $[\eta]$ are the polymer concentration and the intrinsic viscosity, respectively, $[\eta]_a$ is the ambient (at the inlet point x_i) intrinsic viscosity, k_H is the Huggins constant (k_H=0.25), k' and β are the Mark Houwink constants $k'=2.7\cdot 10^{-4}$ $dL/g(g/mol)^{-\beta}$, $\beta=0.74$[10,11], w is the polymer molecule weight, $w=w_m\, l$, and w_m is the monomer molecular weight.

Therefore, for the given values of x_i, R', E', u_1, u_2, P, μ_a, α_p, α, β, τ_{L0}, τ_{L1}, N, w_m, T_a, U, C, a_*, l_*, k', k_H, N and given ambient distribution of polymer additive $W_a(l)$ the problem is reduced to solution of Eqs. (1)- (10) with respect to constants x_e, h_e and functions $p(x)$, $h(x)$, $f(x)$, $\mu(x,z)$, and $W(x,z,l)$.

After the solution of the problem is complete one can determine the frictional stresses τ_1 and τ_2 applied to the contact surfaces according to the formulas

$$\tau_i = (-1)^{i-1} f - \frac{h}{2}\frac{dp}{dx}, \tag{11}$$

where index $i=1$ is for the lower contact surface and $i=2$ is for the upper contact surfaces, respectively.

In this paper for specific calculations we have chosen the following rheological model of a lubricant

$$G(x) = \tanh^{-1} x \tag{12}$$

proposed by Bair and Winer[5] and based on a series of experimental studies of lubricant rheological behavior. Under the above isothermal conditions the rheological model described by Eq. (12) allows for analytical calculation of integrals in Eqs. (3)-(5) that may simplify and speed up the whole solution process.

2.3 Dimensionless Formulation of the Problem

Let us introduce the following dimensionless variables

$$(x',a,c)=\frac{1}{a_H}(x,x_i,x_e),\ (z',h)'=\frac{1}{h_e}(z,h),\ p'=\frac{p}{p_H},\ \mu'=\frac{\mu}{\mu_a},\ u'=\frac{2u}{u_1+u_2},$$
$$v'=\frac{2a_H v}{(u_1+u_2)h_e},\ \tau_i'=\frac{2h_e\tau}{\mu_a(u_1+u_2)},\ \tau_{L0}'=\frac{\tau_{L0}}{p_H},\ (W',W_a')=\frac{1}{W_0}(W,W_a), \tag{13}$$

where a_H and p_H are the half-width of and the maximum pressure in a dry Hertzian contact, W_0 is the characteristic value of the molecular weight distribution.

The values of parameters x_i, R', E', u_1, u_2, P, μ_a, α_p, α, β, τ_{L0}, τ_{L1}, N, w_m, T_a, U, C, a_*, l_*, k', k_H , and N determine the specific values of a number of dimensionless parameters which uniquely identify the solution of the problem. These dimensionless parameters are τ_{L0}/p_H , τ_{L1} and

$$a=\frac{x_i}{a_H},\ V=\frac{3\pi^2\mu_a(u_1+u_2)R'E'}{P^2},\ Q=\alpha_p p_H,\ s_0=\frac{2(u_2-u_1)}{u_1+u_2},\ \varepsilon=\left(\frac{a_H}{2R'}\right)^2,$$
$$\gamma=\alpha\frac{U}{kT},\ \delta=\frac{\mathrm{U}a_{\mathrm{H}}^2}{\mathrm{C}a_* l_*^2\mu_a(u_1+u_2)R'},\ \nu=k'Nw_m^{\beta},\ \kappa=\frac{\mathrm{u}_1+u_2}{2a_H}\cdot 1s. \tag{14}$$

For simplicity, in the further discussion primes at the dimensionless variables are omitted.

Therefore, the solution of the problem is determined by the values of the dimensionless parameters τ_{L0}, τ_{L1}, and parameters from Eq. (14) and by the function $W_a(l)$. The solution of the problem is represented by the dimensionless functions: pressure $p(x)$, gap $h(x)$, sliding frictional stress $f(x)$, lubricant viscosity $\mu(x,z)$, distribution of molecular weight $W(x,z,l)$ and two dimensionless constants: exit coordinate c and exit film thickness

$$H_0=\frac{2h_e R'}{a_H^2}. \tag{15}$$

2.4 Lubricant Flow Topology

Obviously, it is sufficient to consider just cases of non-positive values of the slide-to-roll ratio s_0. Cases with $s_0>0$ are identical to the latter ones if the upper and lower solid surfaces are interchanged. Before we consider the numerical method employed for solution of the aforementioned problem it is necessary to understand the specifics of the topology of the lubricant flow.

In case of pure rolling, $s_0=0$, the flow is symmetric about the x-axis. It has two sets of flow streamlines running through the whole contact and two sets of flow streamlines that enter the contact region through the inlet gap, turn around, and exit the contact region through the same inlet gap. The flow streamlines running through the whole contact region are adjacent to the contact surfaces while the flow streamlines that turn around are adjacent to the x-axis. It is important to mention that in the flow there is a special point where both the horizontal u and vertical v components of lubricant velocity are equal to zero. This point is the bifurcation point. The bifurcation point is located in the inlet of the lubrication film and outside of the Hertzian region. Due to continuity of the components of the lubricant velocity in the vicinity of the bifurcation point both u and v are small. Moreover, at the bifurcation point which is the point of intersection of the flow separatrices – the curves that separate different zones of the flow, i.e. zones occupied by sets of streamlines running through the contact and the ones that turn around and exit through the inlet gap. In the case of pure rolling there are two symmetric separatrices between the running through and turning around streamlines, one separatrix dividing two sets of turning around streamlines, and one separatrix dividing the running through streamlines. The latter two coincide with the x-axis.

For cases with $-2<s_0<0$ there are still two pairs of flow streamline sets: two sets of running through the whole contact streamlines and two sets of turning around and exiting through the inlet gap ones. However, they are no longer symmetric about the x-axis and in comparison with the pure rolling case the entire flow, including the bifurcation point and the separatrices, are shifted toward the upper surface $z=h/2$. For the case of $s_0=-2$ there is just one set of turning around flow streamlines adjacent to the upper contact surface and one set of flow streamlines running through the contact that is adjacent to the lower contact surface everywhere in the contact. Outside of the inlet region the latter set of running through flow streamlines is also adjacent to the upper contact surface. For cases with $-\infty<s_0<-2$ the lubricant flow is asymmetric. It is represented by two sets of flow streamlines adjacent to the lower and upper contact surfaces and running through the contact in opposite directions as well as by two sets of turning around flow streamlines one of which enters the contact region and exits it through the inlet gap and the other that enters and exits the contact through the exit gap. For $-2<s_0\leq 0$ there is only one bifurcation point located between the contact surfaces in the inlet zone while for $s_0=-2$ the bifurcation point belongs to the upper contact surface in the inlet zone. For $-\infty<s_0<-2$ there are two bifurcation points between the contact surfaces, one in the inlet and the other in the exit zones of the contact.

Under isothermal conditions in the case of no degradation the lubricant viscosity depends only on pressure p which, in turn, is a function of x. Therefore, the lubricant viscosity depends only on the distance of a fluid particle from the inlet point $x=a$, i.e. the lubricant viscosity μ depends on x and is independent from z. That makes the function of lubricant viscosity μ a continuous function of x and z.

In the case of lubricant degradation the situation changes completely. Fluid particles that approach the separatrices between two sets of turning around and running through streamlines from different sides follow different flow streamlines and are subjected to different shear stresses over different time periods. Therefore, these fluid particles degrade differently that, in turn, leads to different viscosities of these fluid particles. Thus, we can conclude that degrading lubricants possess a remarkable property. Namely, for all slide-to-roll ratios s_0 (except for $s_0=0$) the separatrices to which fluid particles in the process of their motion approach from both sides are the curves of the lubricant viscosity $\mu(x,z)$ discontinuity (see Figure 7). In spite of that, the shear stress $\tau(x,z)$ is continuous in the entire lubricated region. This is due to the fact that the sliding frictional stress $f(x)$ and the derivative of pressure dp/dx are continuous functions of x. Taking into account the above indicated properties of $\mu(x,z)$ and $\tau(x,z)$ we can conclude that $\partial u/\partial z$ as a function of x and z is also discontinuous along the aforementioned separatrices.

The above mentioned discontinuities of $\mu(x,z)$ and $\partial u/\partial z$ cause some additional complications in numerical solution of an EHL problem for degrading lubricants with Newtonian and non-Newtonian rheologies.

2.5 Simulation Input Data

For particular simulations the following values were used: $U=347\ kJ/mole$, $T=310^{\circ}K$, $a_*=0.374\ nm$, $l_*=0.154\ nm$, $w_m=35.1\ g/mole$, $\mu_a=0.00125\ Pa\cdot s$, $N=1.204\ g/dL$.

The following dimensionless parameters $a=-2$, $V=0.1$, $Q=5$, $\varepsilon=10^{-4}$, $\tau_{L0}=0.002$, $\tau_{L1}=0.046$, $\gamma=1.0774$, $\delta=0.671\cdot 10^{8}$, $\nu=4.524\cdot 10^{-3}$, $\kappa=1$ used in the simulations correspond to the Hertzian half-length of the contact $a_H=10^{-3}\ m$ and the Hertzian maximum pressure $p_H=1.284\ MPa$.

One of the challenges in the application of the above approach to practical EHL problems is the necessity to calibrate the kinetics part of the model, i.e. to be able to chose the proper values for the constants C and α. The values of the latter parameters were found from the comparison with experimental data[2]. However, in these experimental data the shear stress to which the flow was subjected and the lubricant temperature were unknown. The lack of knowledge of these parameters was compensated by the proper choice of the value of the shielding constant C that provided the right value for the group $C|\tau|/T$ that affects the degradation process. Therefore, the actual values of these parameters must be determined from detailed experimental data with all necessary input and output data available. In the following calculations we took $C=15.5$ and $\alpha=0.008$.

The initial distribution of the polymer molecular weight $W_a(l)$ is taken from Covitch[4].

2.6 Solutions for Lubricants with Newtonian and Non-Newtonian Rheologies without Degradation

In this section we will compare the EHL solutions in cases of lubricants with Newtonian and non-Newtonian rheologies assuming that both lubricants have the same ambient viscosity and do not undergo degradation. That allows determination of the reversible loss of lubricant viscosity.

In the inlet zone of a heavily loaded lubricated contact the pressure is small and the lubricant viscosity is close to one while the gap is large. Therefore, in the inlet zone the local sliding frictional stress $f(x)$ is small. Due to that and the fact that the derivative of pressure dp/dx is relatively small the rheological function G behaves like a linear function. It means that the behavior of G resembles that for a fluid with Newtonian rheology. Therefore, taking into account that in heavily loaded lubricated contacts the film thickness H_0 is primarily determined by the inlet zone[12] we should expect that the difference between the lubrication film thicknesses in the contacts with the Newtonian and non-Newtonian lubricants is relatively small. On the other hand, in the above cases for high slide-to-roll ratio s_0 the sliding frictional stress $f(x)$ for a lubricant with non-Newtonian rheology ($G(x)=tanh^{-1}x$) is noticeably lower than that for a lubricant with Newtonian rheology ($G(x)=x$) given the same ambient lubricant viscosity.

For a lubricant with Newtonian rheology the solution of the isothermal EHL problem is independent of the slide-to-roll ratio s_0. In particular, for the Newtonian lubricant with the above input data we have $H_0=0.1966$ and $c=1.0513$. For lubricants with non-Newtonian rheology the solution of the isothermal EHL problem depends on s_0. For the considered non-Newtonian lubricant, $s_0=0$, and τ_{L0} decreasing from 0.01 to 0.002 the film thickness monotonically decreases from $H_0=0.1967$ to $H_0=0.1963$, respectively, while for $s_0=-0.5$ the film thickness monotonically decreases from $H_0=0.1967$ to $H_0= 0.1961$, respectively. Within the same range of variations of s_0 and τ_{L0} the value of c varies by no more than seven units in the fifth place after the decimal point.

The above data indicate that the film thickness H_0 for a lubricant with the considered non-Newtonian rheology is practically identical to the one for the lubricant with the Newtonian rheology. Also, the value of H_0 is almost independent of the slide-to-roll ratio s_0 and depends significantly on the value of the dimensionless pressure coefficient Q. At the same time, the sliding frictional stress $f(x)$ is noticeably affected by the lubricant rheology, pressure coefficient of viscosity Q, speed-load parameter V, limiting stress τ_L, and the slide-to-roll ratio s_0.

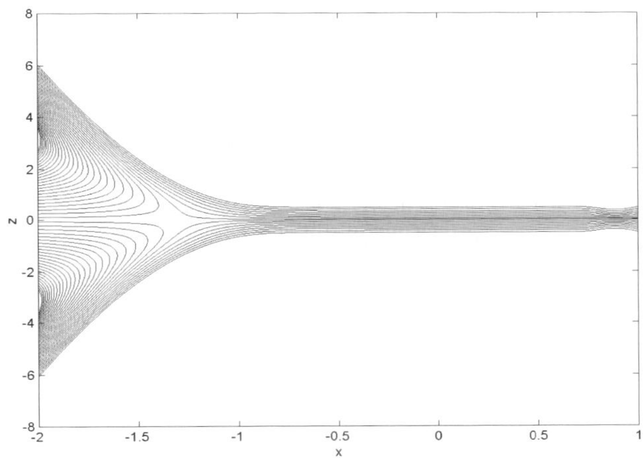

Figure 1. The map of flow streamlines z(x) for the nondegrading lubricant with Newtonian rheology for the slide-to-roll ratio equal to zero.

The behavior of the pressure $p(x)$, gap $h(x)$, film thickness H_0, coordinate of the exit point c, lubricant flow streamlines $z(x)$, and lubricant viscosity $\mu(x,z)$ in the latter cases are very similar. For $s_0=0$ the graphs of the flow streamlines $z(x)$ are given in Figure 1. For the lubricant with the non-Newtonian rheology the graphs of the flow streamlines $z(x)$, pressure $p(x)$ and gap $h(x)$ are very close to the corresponding ones for the Newtonian fluid. In case of $s_0=0$ for any rheological function G the sliding frictional stress $f(x)=0$ because $G(x)$ is an odd function of x. For $s_0=-0.5$ the graphs of $f(x)$ for the non-degrading Newtonian and non-Newtonian lubricants are presented in Figure 2. The behavior of $f(x)$ is distinctly different from the one of pressure $p(x)$: the values of $f(x)$ are greater in the case of the Newtonian lubricant than in the case of the non-Newtonian one. This effect is due to the reversible viscosity loss of the non-Newtonian lubricant in comparison with the Newtonian one. A similar behavior of the surface

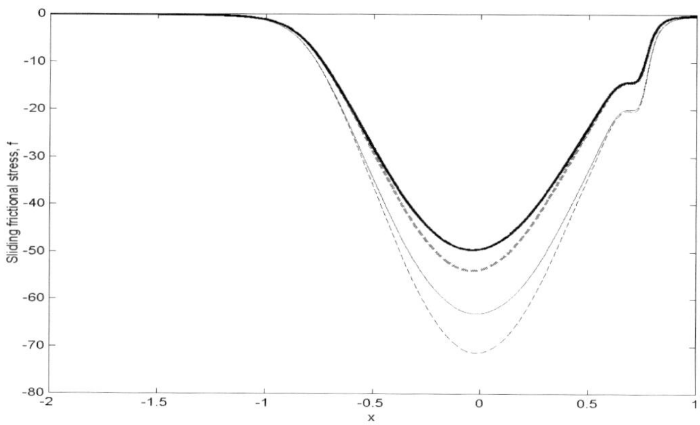

Figure 2. The sliding frictional stress f(x) distributions for the nondegrading lubricants with Newtonian (dashed line) and non-Newtonian (solid line) rheologies and for the degrading lubricants with Newtonian (dashed bold line) and non-Newtonian (solid bold line) rheologies for the slide-to-roll ratio equal to -0.5.

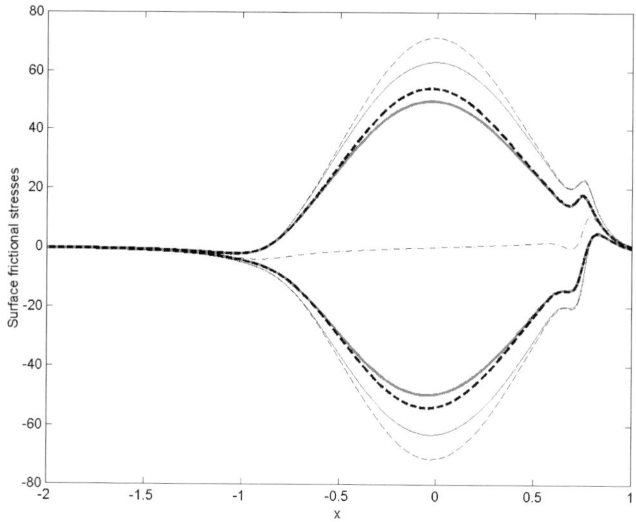

Figure 3. Frictional stresses applied to the upper and lower contact surfaces for nondegrading (dashed line) and degrading (bold dashed line) Newtonian lubricants and for nondegrading (solid line) and degrading (bold solid line) non-Newtonian lubricants for the slide-to-roll ratio equal to -0.5 and the surface frictional stresses for the nondegrading Newtonian lubricant (dashed-dotted line) under pure rolling conditions.

frictional stresses τ_1 and τ_2 can be seen in Figure 3, where τ_1 and τ_2 are given for the lubricants with Newtonian and non-Newtonian rheologies under mixed rolling and sliding conditions ($s_0=-0.5$) as well as the surface shear stress $\tau_1=-\tau_2$ for the lubricants with Newtonian (dashed-dotted line) under pure rolling conditions ($s_0=0$). For the case of the pure rolling ($s_0=0$) for the non-Newtonian lubricant the surface shear stresses $\tau_1=-\tau_2$ are very close to the ones for the case of the Newtonian lubricant.

The reversible loss of the viscosity of the non- Newtonian lubricant can be determined as a ratio of the sliding frictional stresses $f(x)$ for non-Newtonian and Newtonian lubricants in the Hertzian region. It is based on the fact that in the Hertzian region of a heavily loaded lubricated contact $h(x)-1<<1$[12]. Therefore, from Figure 2 it follows that for $s_0= -0.5$ the reversible lubricant viscosity loss of the non-Newtonian lubricant reaches 12% of its original value.

2.7 Solutions for Lubricants with Newtonian and Non-Newtonian Rheologies with Degradation

In this section we will compare the EHL solutions for a lubricant with non-Newtonian rheology experiencing the stress-induced degradation with a solution for a lubricant with a similar rheology but without degradation. That allows determination of the irreversible loss of lubricant viscosity. Moreover, we will compare the EHL solutions for degrading lubricants with Newtonian and non-Newtonian rheologies.

Let us consider some general properties of a solution of the isothermal EHL problem for a degrading lubricant with non-Newtonian rheology. The behavior of the pressure $p(x)$ and gap $h(x)$ distributions as well as of the film thickness H_0 and exit coordinate c of the latter problem with respect to parameters a, V, and Q is similar to the one described for lubricants with Newtonian and non-Newtonian rheologies[9,12,13]. For a lubricant with Newtonian rheology an increase in Q causes a relatively slow increase in H_0 and a rapid increase in $f(x)$ and, therefore, in $\tau(x,z)$. That, in turn, leads to a rapid lubricant degradation. For a lubricant with the considered non-Newtonian rheology the increase of $f(x)$ is moderated by the lubricant rheology. Therefore, the degradation process of such a lubricant runs slower. A usually very small parameter ε almost does not affect the problem solution. For small τ_{L0} and τ_{L1} the shear stress τ in the lubricant is small which, in turn, slows down the process of lubricant degradation (see Eqs. (7) and (8) for R and L_*). As the values of τ_{L0} and τ_{L1} increase the lubricant shear stress τ and the film thickness H_0 increase while as the slide-to-roll ratio s_0 increases the film thickness H_0 slowly decreases. However, variations in H_0 are insignificant while the increase in τ is directly proportional to the increase in τ_L, i.e. in τ_{L0} and τ_{L1}. The latter causes lubricant degradation to run faster (see Eqs. (7) and (8) for R and L_*). Moreover, an increase in the value of the parameter δ leads to a corresponding increase in the characteristic chain length L_* that causes lubricant degradation to slow down (see Eqs. (7) and (8) for R and L_*). For smaller values of the parameter γ the process of lubricant degradation runs slower than for the larger ones (see Eqs. (7) for R). For higher values of ν the lubricant viscosity responds more strongly to changes in the molecular weight distribution, i.e. for higher ν the loss of lubricant viscosity is higher (see Eqs. (10) for μ). The value of the parameter κ in the left-hand side of the kinetic equation controls the rate of lubricant convection and, therefore, the time a lubricant small volume is present in the contact area. For high values of κ the time of lubricant presence in the contact area is small and the rate of lubricant degradation is low and vice versa.

Now, let us examine the case of pure rolling ($s_0=0$). Under these conditions lubricants with Newtonian and non-Newtonian rheologies degrade to a lesser extent than in cases when $s_0\neq 0$. The solutions for Newtonian and non-Newtonian lubricant rheologies are qualitatively and quantitatively very close to each other because $|G^{-1}(x)|\leq|x|$ and $G(x)\to x$ as $x\to 0$. For $s_0=0$ the maps of flow streamlines for the non-Newtonian lubricant are very close to the ones for the Newtonian lubricant (see Figure 1). This map shows that the flow streamlines are symmetric about $z=0$. Due to the symmetry of the problem the lubricant viscosity μ and strain $\partial u/\partial z$ are continuous functions of x and z. All of the flow streamlines enter the contact through the inlet cross section of the lubrication film. Some of them that are adjacent to the contact surfaces run through the whole contact. However, the other streamlines turn around and exit the contact region also through the inlet cross section of the film. The pressure and gap distributions are very close to the ones for the non-degrading Newtonian lubricant. The maximum difference between the gap distributions for Newtonian and non-Newtonian fluids reaches *8%*. In this case the exit film thicknesses and exit coordinates for Newtonian and non-Newtonian degrading lubricants are $H_0=0.1829$, $c=1.0527$ and $H_0=0.1834$, $c=1.0527$, respectively. The comparison of the data for H_0 obtained for degrading and non-degrading lubricants shows a relatively small affect of lubricant degradation on the film thickness H_0 – only about 8%.

The distributions of the lubricant viscosity μ and polymer molecular weight W for lubricants with Newtonian and non-Newtonian rheologies are practically identical. The fact that for $s_0=0$ lubricants with Newtonian and non-Newtonian rheologies degrade relatively slowly can be also seen from the graphs of the reciprocal of the lubricant viscosity (see Figure 4) and the

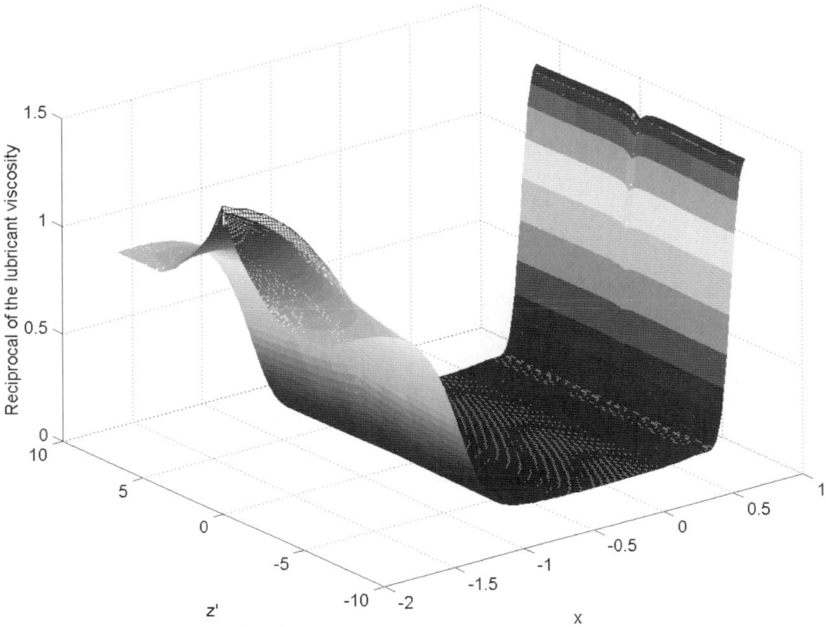

Figure 4. The reciprocal of the lubricant viscosity in the Newtonian and non-Newtonian lubrication film under pure rolling conditions.

distribution of the molecular weight W along the flow streamline $z(x)$ that is closest to $z=0$ and running through the whole contact below $z=0$ (see Figure 5). In Figure 4, the variable z' is an artificially stretched z-coordinate across the film thickness (namely, $z'=zh(a)/h(x)$) to make the relationship more transparent. It is clear from the graphs of the reciprocal of the lubricant viscosity (see Figure 4) that for the Newtonian and non-Newtonian lubricants the maximum irreversible viscosity loss reaches about 31%. Lubricant degradation depends on a number of parameters among which the shear stress τ plays one of the major roles. For $s_0=0$ the sliding frictional stress $f(x)=0$ and, therefore, the shear stress $\tau=(12H_0^2/V)zdp/dx$ reaches its extrema in the inlet and exit zones while in the Hertzian region it is relatively small[12]. That leads to the fact that practically all lubricant degradation occurs in the inlet zone while the lubricant almost does not degrade in the Hertzian region and the exit zone. The latter also can be clearly seen from the distributions of the molecular weight W in Figure 5.

For given lubricant flow parameters the degree of lubricant degradation is higher along longer streamlines. The length of the flow

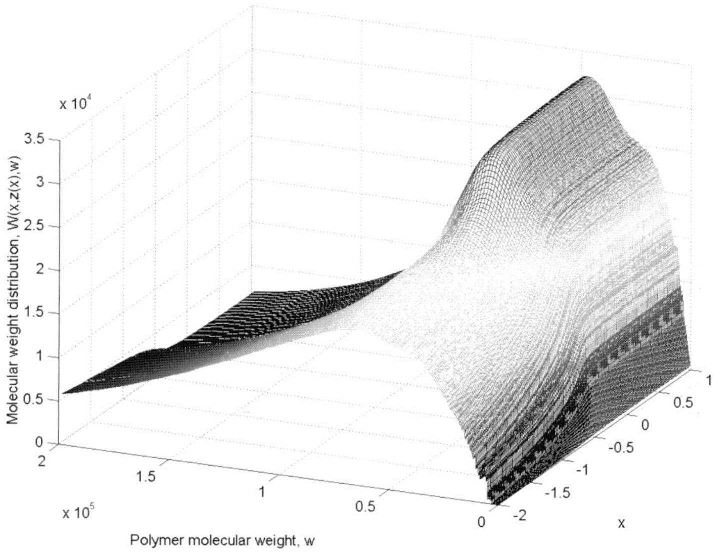

Figure 5. The distribution of the molecular weight W in the degrading lubricants with Newtonian and non-Newtonian rheologies along the flow streamline closest to z=0 and running through the whole contact below z=0 under pure rolling conditions.

streamlines adjacent to the contact surfaces is larger and the absolute value of the shear stress along them is slightly higher than for the ones that turn around and exit through the inlet cross section lubricant. In spite of that fact lubricant degradation along the turning around streamlines is higher because they are completely located in the inlet of the contact – practically the only contact zone where lubricant degradation takes place. It also follows from the expression for the shear stress $\tau=(12H_0^2/V)zdp/dx$ that along the streamlines running through the whole contact the degree of lubricant degradation is higher near the contact surfaces. Among the turning around streamlines the degree of lubricant degradation is higher along the longest streamlines that enter the contact region and are closest to the contact surfaces and exit the contact through the inlet cross section close to $z=0$. Obviously, for higher viscosity μ the film thickness H_0 is higher while dp/dx changes insignificantly. Therefore, the shear stress τ is higher which promotes faster lubricant degradation and higher viscosity loss. Similarly, for higher pressure viscosity coefficients Q one can expect faster degradation and higher viscosity loss.

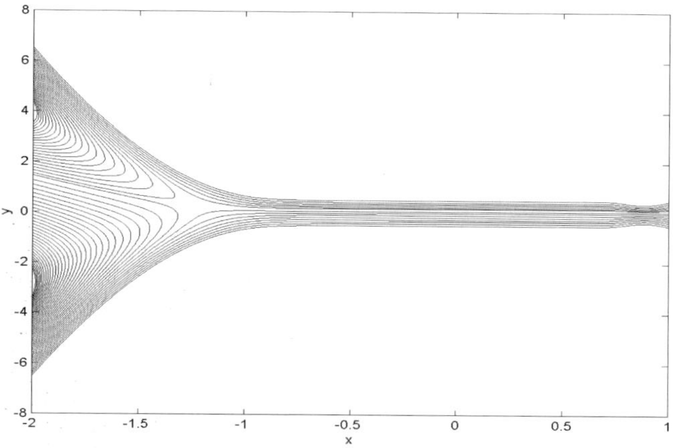

Figure 6. The map of the flow streamlines z(x) for the degrading lubricants with Newtonian and non-Newtonian rheologies for the slide-to-roll ratio equal to -0.5.

In Figures 4 and 7 the variable z' is an artificially stretched z-coordinate across the film thickness (namely, $z' = zh(a)/h(x)$) to make the relationship more transparent.

Now, let us examine a case of mixed rolling and sliding with $s_0 = -0.5$. In this case the solutions of the problem for the lubricants with Newtonian and non-Newtonian rheologies are also very close to each other. The map of the flow streamlines $z(x)$ for the degrading Newtonian and non-Newtonian lubricants is given in Figure 6. The lubricant flows are no longer symmetric about $z=0$ and the maps of their flow streamlines are completely different from the ones for the pure rolling case. There are still two sets of turning around as well as the adjacent to the lower and upper contact surfaces and running through the whole contact flow streamlines. The exit lubrication film thickness H_0 and the exit coordinate c exhibit behavior very similar to the case of pure rolling. For the non-degrading and degrading Newtonian lubricants they are equal to $H_0=0.1961$, $c=1.0513$ and $H_0=0.1819$, $c=1.0513$, respectively. For the non-degrading and degrading non-Newtonian lubricants the film thickness is equal to $H_0=0.1961$ and $H_0=0.1813$, respectively. In all other respects the distributions of pressure p and gap h are very similar to the ones for the case of pure rolling. The numerical results show that in the case of mixed rolling and sliding for both lubricants with Newtonian and non-Newtonian rheologies the sliding frictional stress $f(x)$ (see Figure 2) is smaller than in a similar case without degradation but still large enough to cause lubricant degradation to a much

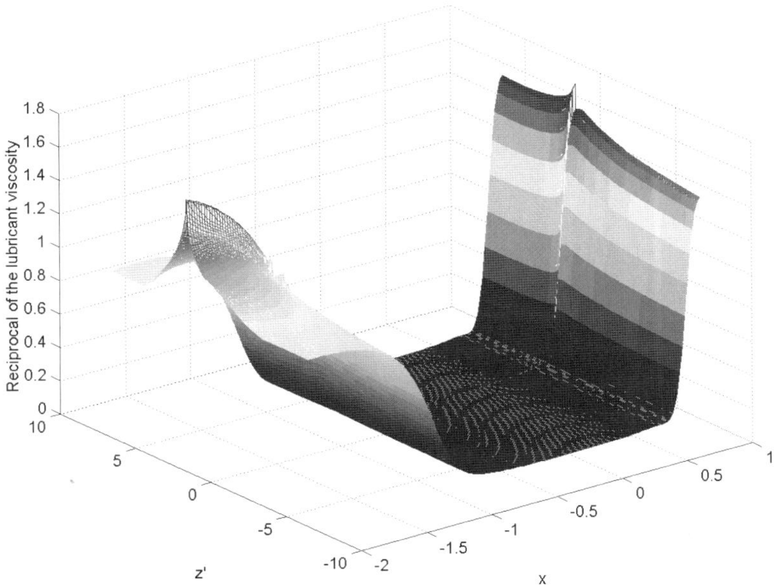

Figure 7. The reciprocal of the viscosity in the lubricants with Newtonian and non-Newtonian rheologies for the slide-to-roll ratio equal to -0.5.

greater extent than in the case of pure rolling. In the case of mixed rolling and sliding the surface frictional stresses τ_1 and τ_2 are much higher (see Figure 3) than the ones for the case of pure rolling ($\tau_2=-\tau_1=(6H_0^2/V)hdp/dx$) and, at the same time, lower than for the case of no degradation. That explains the stronger lubricant degradation in the case of mixed rolling and sliding conditions in comparison with the case of pure rolling conditions (see Figures 4, 7 and 5, 8). For the degrading lubricants the frictional stresses τ_1 and τ_2 applied to the contact surfaces are on average lower than for the non-degrading lubricant by about *26%*. The frictional stress τ in the lubrication film is mostly concentrated near the contact surfaces. Depending on the sign of dp/dx the maximum of the absolute value of the surface frictional stress is reached either on the lower or upper contact surfaces. That leads to faster lubricant degradation near that contact surface and to slower lubricant degradation near the other one. This can be clearly seen from the behavior of the reciprocal of the lubricant viscosity μ in cases of Newtonian and non-Newtonian lubricants (see Figure 7). In the case of $s_0=-0.5$ the maximum loss of the lubricant viscosity reaches about *41%*. That can be seen from the comparison of graphs of the sliding frictional stress $f(x)$ for the non-

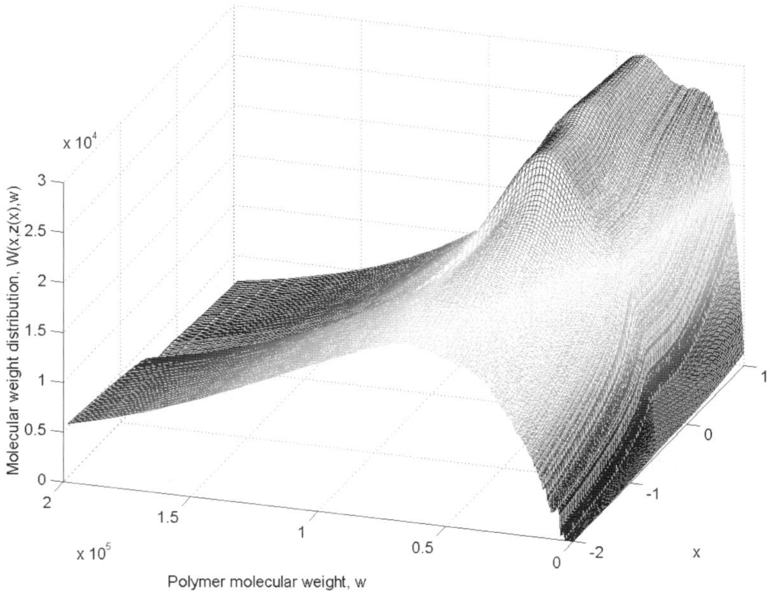

Figure 8. The distribution of the molecular weight W in the degrading lubricants with Newtonian and non-Newtonian rheologies along the flow streamline closest to z=0 and running through the whole contact below z=0 for the slide-to-roll ratio equal to -0.5.

degrading and degrading lubricants (see Figure 2) as well as from the graph of the distribution of the reciprocal of the lubricant viscosity μ (see Figure 7). The behavior of the polymer molecule distribution W for the lubricants with Newtonian and non-Newtonian rheologies along the flow streamline $z_{31}(x)$ running through the whole contact below $z=0$ and located next to the first turning around streamline is given in Figure 8. The comparison of the polymer molecular weight W distributions for the cases of pure rolling and mixed rolling and sliding (see Figures 5 and 8) shows that in the case of pure rolling the lubricant degrades slower than in the case of mixed rolling and sliding. In the latter case a rapid polymer scission occurs throughout the entire contact region while in the former case it is mostly concentrated in the inlet zone of the contact. That can be seen from the values of $W(x,z(x),w)$ for polymer molecules with small molecular weight w (short chain lengths l) at different points along the flow streamlines. Moreover, from Figure 8 it follows that for the considered Newtonian and non-Newtonian lubricants under mixed rolling and sliding conditions the lubricant degradation occurs throughout the contact. For the considered non-Newtonian lubricant for higher values of the slide-to-roll ratio $|s_0|$ the relative impact of the rolling frictional stress $(6H_0^2/V)zdp/dx$ on τ decreases slowly while for the Newtonian lubricant it decreases significantly. Moreover, the described differences in the degradation behavior of the lubricants with Newtonian and non-Newtonian rheologies may be the key to understanding why in practice synthetic lubricants degrade at a much slower pace.

Finally, let us give a qualitative description of the behavior of a contact with degrading lubricant under the conditions of pure sliding, $s_0=-2$. This case corresponds to motionless upper ($z=h/2$) and moving lower ($z=-h/2$) contact surfaces. The lubricant flow is no longer symmetric about $z=0$. There is only one set of turning around as well as adjacent to the lower contact surface and running through the whole contact flow streamlines. The exit film thickness H_0 exhibits behavior very similar to the cases of pure rolling and mixed rolling and sliding. For the case of pure sliding the frictional stresses τ_1 and τ_2 are much higher than for the cases of pure rolling and mixed rolling and sliding. In the lubrication film the frictional stress τ is mostly concentrated near the motionless upper contact surface and it is relatively low near the lower moving contact surface. On average in the case of pure sliding the lubricant undergoes degradation to a much greater extent than in the case of pure rolling. The loss of the lubricant viscosity μ near the moving and motionless contact surfaces may reach about 10-20% and 60-70%, respectively. For the degrading lubricant the frictional stresses applied to the contact surfaces are generally lower than for the non-degrading lubricant by about 20-40%. In case of pure rolling lubricant degradation occurs through out the whole contact region.

It is convenient to analyze the process of polymer degradation based on the behavior of the characteristic polymer chain length L_* (see Eqs. 8) which represents the position of the center of the profile of the probability of polymer scission $R(x,z,l)$. In fact, the probability of polymer scission R is the main key to the control over the lubricant degradation. As the zone of the chain lengths l where $R(x,z,l)$ varies from 0 to 1 is quite narrow one can conclude that the polymer molecules with chain lengths l shorter than L_* practically do not undergo scission while most of the molecules longer that L_* get degraded. It is important to emphasize that the degree of polymer scission varies from point to point together with the shear stress τ which, in turn, is a certain function of x and z. In the Hertzian region the shear stress τ is relatively large due to large lubricant viscosity μ while in the inlet and exit zones the values of the pressure derivative dp/dx are higher than the ones in the Hertzian zone. That explains the above mentioned fact of lubricant degradation throughout the contact.

The effect of lubricant viscosity loss of up to *70%* was observed experimentally and documented by Walker, Sanborn, and Winer[14]. In cases discussed in this paper the loss of lubricant viscosity reached about *41%*.

One has to keep in mind that in reality the value of the pressure coefficient of viscosity Q can be much higher than $Q=5$ which may lead to a much higher viscosity loss.

It is important to realize that the lubricant viscosity loss caused by degradation is controlled by the contact operating conditions as well as by the lubricant rheology, the nature, concentration and molecular weight distribution of the polymer additive in the lubricant supplied to the contact. Moreover, significant changes in the lubricant viscosity μ and film thickness H_0 may also lead to noticeable variations in frictional stresses τ_1 and τ_2 applied to the contact surfaces. That can also significantly change fatigue life of solids involved in a contact with degrading lubricant (see Kudish[15,16]). Therefore, for any given operating conditions it is desirable to select a lubricant with certain rheology and concentration and molecular weight distribution of a polymer additive that will provide sufficiently good lubrication conditions (high lubrication film thickness and low frictional stress) while the viscosity loss is reduced to minimum.

In the case of a non-isothermal EHL problem due to heat generation in the lubrication film and elevated lubricant temperature the lubricant viscosity (in case of no lubricant degradation) is lower than in a comparable isothermal case. That leads to higher values of not only T but also L_* which slows down the degradation process. The extent to which the polymer degradation process slows down depends on the lubricant nature, nature and distribution of the molecular weight of polymer molecules entering the contact, and on the contact operating conditions.

3. LUBRICANT LIFE MODELING

Let us consider a contact of a roller of length L_r with another cylinder with parallel axis of length greater than L_r lubricated by a lubricant supplied from an oil tank with the volume of lubricant V. We will assume that the necessary amount of lubricant is removed from the tank and delivered to the contact and than, after entering and exiting the contact, it is returned back to the oil tank. Let us suppose that the lubricant volume V in the oil tank is constant in time and the oil in the tank is well mixed. Moreover, we will assume that the lubricant moves from left to right while entering the contact which is true if $|s_0|\leq 2$. Based on the assumption that on a time scale equal to the time necessary for lubricant to pass through the contact the flow is quasi-steady we can derive an equation for the density of the polymer molecular weight distribution $W_T(n,l)$ in the oil tank per unit volume at the time moment n for molecules with chain lengths from the interval $[l,l+\Delta l]$

$$\frac{dW_T}{dn} = \frac{L_r}{V}\left\{ -\int_{z_{01}}^{z_{02}} u(x_i,z)W(x_i,z,l)dz + \int_{-h_e/2}^{h_e/2} u(x_e,z)W(x_e,z,l)dz \right.$$
$$\left. -\left[\int_{-h_i/2}^{z_{01}} u(x_i,z)dz + \int_{z_{02}}^{h_i/2} u(x_i,z)dz \right] W_T(n,l) \right\}, \quad W_T(0,l) = W_{T0}(l), \tag{16}$$

where z_{01} and z_{02} are the lower and upper boundaries in the inlet lubricant flow that separate the lubricant exiting through the inlet gap from the lubricant that enters the contact, $h_i=h(x_i)$ and $h_e=h(x_e)$ are the inlet and exit gap, respectively, $W(x_i,z,l)$ and $W(x_e,z,l)$ are the solutions of the EHL problem for the degrading lubricant at the inlet and exit film cross sections, respectively, $W_{T0}(l)$ is the initial polymer molecular weight distribution in the oil tank.

The coupling between the kinetic equation (8) describing lubricant degradation and Eq. (9) occurs naturally through the initial condition for the kinetic equation (8)

$$W(x_i,z,l) = W_T(n,l). \tag{17}$$

It can be shown that in the oil tank the total number of polymer molecules per unit of lubricant volume increases with time n while the total polymer molecular weight per unit of lubricant volume is conserved in time.

The lubricant life can be defined by assuming that the lubricant useful life is determined by the period of time n_{max} during which the lubricant viscosity μ_T in the oil tank is not lower than a certain value μ_{min}, i.e.

$$\mu_T(n) \geq \mu_{\min} \quad \text{for} \quad n \leq n_{\max}; \quad \mu_T(n_{\max}) = \mu_{\min}. \tag{18}$$

Let us introduce new functions

$$M_\beta = \int_0^\infty W(x,z,l)l^\beta dl \Big/ \int_0^\infty W(x,z,l)dl, \quad M_{T\beta} = \int_0^\infty W_T(n,l)l^\beta dl \Big/ \int_0^\infty W_T(n,l)dl.$$

We will assume that in an incompressible lubricant flow there exists a point (x_*,z_*) such that

$$M_{f\beta}(\mathrm{n})=0, \quad M_{\beta}(x_*,z_*)=\alpha M_{T\beta}(n), \quad 0<\alpha<1,$$

$$-\int_{z_{s1}}^{z_{s2}} u(x_i,z)M_{\beta}(x_i,z)dz+\int_{-h_e/2}^{h_e/2} u(x_e,z)M_{\beta}(x_e,z)dz$$

$$\approx\left[-\int_{z_{s1}}^{z_{s2}} u(x_i,z)dz+\int_{-h_e/2}^{h_e/2} u(x_e,z)dz\right]M_{\beta}(x_*,z_*),$$

where α is a certain function that depends on the contact conditions and time n. The function α varies with time n in such a way that as lubricant degradation goes on the value of α monotonically increases and, eventually, approaches *1*. This approximation may serve as a first crude approach to estimation of lubricant life and it leads to an approximate solution

$$M_{T\beta}(n)=M_{T\beta 0}\exp\left[-\int_0^n \lambda(t)dt\right],$$
$$\lambda(n)=\frac{L_r(1-\alpha)}{V}\left[\int_{-h_i/2}^{z_{s1}} u(x_i,z)dz+\int_{z_{s2}}^{h_i/2} u(x_i,z)dz\right], \tag{19}$$

$$M_W(n)=M_{W0}\exp\left[-\frac{1}{\beta}\int_0^n \lambda(t)dt\right], \tag{20}$$

where M_{W0} is the value of M_W at time moment $n=0$. From Eqs. (8) and (11) we obtain

$$\int_0^n \lambda(t)dt=\ln\frac{2k_H N[\eta]_a}{\sqrt{1-4k_H(1-\mu_{\min}/\mu_b)}-1},$$
$$[\eta]_a=k' M_{W0}{}^{\beta}, \; \mu_b=\frac{\mu_a}{1+N[\eta]_a+k_H\left(N[\eta]_a\right)}, \tag{21}$$

where μ_b is the viscosity of the base oil without the polymer additive. It follows from Eqs. (12) that the rough estimate of the time scale necessary for degrading the tank of oil of volume V under given lubricating conditions is

$$n_0 \cong \lambda_0^{-1}, \quad \lambda_0 = \frac{L_r(u_1+u_2)h_e}{2V}\left[1-\frac{M_\beta(x_e,0)}{M_{T\beta}(0)}\right]. \quad (22)$$

A more detailed analysis of lubricant life will be provided in the papers to follow. These results can be used for evaluation of useful lubricant life and for more realistic estimation of the lubrication film thickness and frictional stresses in lubricated contacts. That may provide a solid basis for maintenance scheduling of lubricated joints working under various conditions. Moreover, it may enable researches to create new and more durable lubricant additives for general and specific applications.

4. CONCLUSIONS

A plane isothermal elastohydrodynamic problem for a line contact lubricated by fluids with Newtonian and non-Newtonian rheologies with degrading polymer additives is modeled. The polymer additive to lubricant, viscosity improver (VI), is considered to be of linear structure. The lubricant degradation is caused by stress-induced scission of polymer molecules while they move through the contact along the flow streamlines. The polymer degradation process is controlled by the probabilistic kinetic equation for the polymer molecular weight distribution. The lubricant flow is described by the generalized Reynolds equation. The lubrication and lubricant degradation processes are coupled through the lubricant viscosity that depends on pressure as well as on the concentration and distribution of the polymer molecular weight. The problem is solved numerically. In particular, the lubricant flow streamlines, distribution of the polymer molecular weight, reversible and irreversible lubricant viscosity losses, pressure, gap, and film thickness are determined. The stress-induced scission of polymer molecules leads to irreversible changes in the lubricant viscosity – irreversible viscosity loss. Numerical results show that in the considered cases of pure rolling the viscosity loss reaches about 31% while in the cases of mixed rolling and sliding the reversible and irreversible viscosity losses at the exit from the contact reaches up to *12%* and *41%*, respectively. In cases of pure sliding the viscosity loss may be as high as 70%. Moreover, a new property of a flow of degrading lubricant is discovered – the viscosity of a degrading lubricant is generally a discontinuous function of the position of a fluid particle. The viscosity loss in the inlet zone of the contact causes all parameters of the lubricated contact to change. In particular, in the considered case of mixed rolling and sliding conditions the lubrication film thickness of the degrading lubricants is about

8% thinner than the one for similar non-degrading lubricants. For the considered cases the effect of lubricant degradation on the non-Newtonian lubricant is noticeably lower than on the Newtonian lubricant. The changes in the frictional stresses applied to the contact surfaces may noticeably change their fatigue life. The results of the paper can be used for evaluation of useful lubricant life and for more realistic estimation of the lubrication film thickness, frictional stresses, and fatigue life of solids in lubricated contacts. Such an analysis of the viscosity loss may provide a solid basis for maintenance scheduling of lubricated joints working under various conditions. Moreover, this analysis may enable researches to create new and more durable lubricant additives for general and specific applications.

ACKNOWLEDGEMENT

The support of the National Science Foundation, Surface Engineering and Tribology Program, Grant No. CMS-9912501 is gratefully acknowledged.

REFERENCES

1. I.I. Kudish, R.G. Airapetyan, and M.J. Covitch, Modeling of Kinetics of Strain Induced Degradation of Polymer Additives in Lubricants, J. of Mathematical Models and Methods in Applied Sciences, 12(6), 835–856 (2002).
2. I.I. Kudish, R.G. Airapetyan, and M.J. Covitch, Modeling of Kinetics of Stress Induced Degradation of Polymer Additives in Lubricants and Viscosity Loss, STLE Tribology Trans., 46(1), 1–11 (2003).
3. I.I. Kudish, R.G. Airapetyan, G.R. Hayrapetyan, and M.J. Covitch, Kinetics Approach to Modelling of Stress Induced Degradation of Lubricants Formulated with Star Polymer Additives, STLE Tribology Trans., (in print).
4. M.J. Covitch, How Polymer Architecture Affects Permanent Viscosity Loss of Multigrade Lubricants, SAE Technical Paper Series, Paper No. 982638 (1998).
5. S. Bair and W.O. Winer, Shear Strength Measurements of Lubricants at High Pressure, J. of Lubr. Technol., 101(3), 251–257 (1979).
6. E. Hoglund and B. Jacobson, Experimental Investigations of the Shear Strength of Lubricants Subjected to High Pressure and Temperature, ASME J. of Tribology, 108(4), 571–578 (1986).
7. B.J. Hamrock, *Fundamentals of Fluid Film Lubrication*, NASA Reference Publication 1255 (1991).
8. I.I. Kudish, Asymptotic Methods for Studying Plane Problems of the Elastohydrodynamic Lubrication Theory in Heavily Loaded Regimes. Part 1. Isothermal Problem, Izvestija Akademii Nauk Arm.SSR, Mekhanika, 35(5), 46–64 (1982).
9. I.I. Kudish and R.G. Airapetyan, Modeling of Line Contacts with Degrading Lubricant, ASME J. of Tribology, 125(3), 513–522 (2003).
10. F.W. Billmeyer, Jr., *Textbook of Polymer Science*, John Wiley & Sons, New York (1966).

11. G. Crespi, A. Valvassori, and U. Flisi, in: *The Stereo Rubbers*, W.M. Saltman (ed.), (1977) pp. 365–431.
12. I.I. Kudish, Asymptotic Analysis of a Problem for a Heavily Loaded Lubricated Contact of Elastic Bodies. Pre- and Over-critical Lubrication Regimes for Newtonian Fluids, Dynamic Systems and Applications, Dynamic Publishers, Atlanta, 5(3), September, 451–476 (1996).
13. I.I. Kudish, Some Problems of Elastohydrodynamic Theory of Lubrication for a Lightly Loaded Contact, Mechanics of Solids, 16(3), 81–95 (1981).
14. D.L. Walker, D.M. Sanborn, and W.O. Winer, Molecular Degradation of Lubricants in Sliding Elastohydrodynamic Contacts, ASME J. Lubrication Technology, 97(3), 390–397 (1975).
15. I.I. Kudish, A New Statistical Model of Contact Fatigue, STLE Tribology Transactions, 43(4), 711–721 (2000).
16. I.I. Kudish, Effect of Lubricant Degradation on Contact Fatigue, STLE Tribology Trans. (in print).

THE EFFECT OF VISCOELASTICITY ON THE PERFORMANCE OF JOURNAL BEARINGS

D.Rh. Gwynllyw[1] and T.N. Phillips[2]
[1] *School of Mathematical Sciences, University of the West of England, Bristol BS16 1QY, U.K.,*
E-mail: Rhys.Gwynllyw@uwe.ac.uk
[2] *School of Mathematics, Cardiff University, Cardiff CF24 4AG, U.K.,*
E-mail: PhillipsTN@cardiff.ac.uk

Abstract In this paper we examine the question of whether the viscoelastic properties of multigrade oils can have a measurable effect on the performance of dynamically loaded journal bearings. Our investigation takes the form of a computational study in which the full set of coupled equations (kinematic and constitutive) governing the flow of the lubricant is solved using a moving spectral element method. The problem is formulated in terms of velocity, pressure and extra-stress and the weak formulation of the system of equations is used as the basis of the discretization. The lubricant is modelled using the Phan-Thien Tanner and White–Metzner models, in the latter of which both the viscosity and the relaxation time are functions of shear-rate and pressure. An arbitrary Lagrangian–Eulerian (ALE) method is used to evolve the computational mesh in time.

Keywords: viscoelastic fluid, journal bearing, spectral element method, ALE formulation.

1. INTRODUCTION

The prediction and assessment of friction, wear and load bearing capacity of a lubricated dynamically loaded journal bearing over a wide range of operating conditions is extremely important. The modelling issues involved in dynamically loaded journal bearings are extensive. These include the influences of oil feed, thermal effects and cavitation, changes in viscosity with shear-rate, temperature and pressure, and the effect of viscoelasticity.

This paper presents and discusses some results obtained with a mathematical model that describes the dynamics of journal bearings and the behaviour of viscoelastic lubricants. The formulation of models of this type is extremely important for lubrication engineers since it allows them to predict and understand bearing fatigue. The lubricant model adopted in this work includes viscoelasticity, shear-rate and pressure dependent viscosity and relaxation time

R.W. Snidle and H.P. Evans (eds), IUTAM Symposium on Elastohydrodynamics and Micro-elastohydrodynamics, 175–186.

and a simplified treatment of cavitation. The lubrication approximation is not invoked. For the purpose of this work both the journal and bearing are assumed to be inelastic. This assumption will need to be relaxed for a comprehensive analysis of journal bearing lubrication and deformation.

In some experimental work, Williamson et al. [15] found that the viscoelastic properties of multigrade oils can have a measurable effect on lubrication characteristics. In this study a journal bearing simulator rig was used and extreme eccentricity ratios of around 0.98 were reached. This is in the range where major effects are observed. It may seem rather surprising that multigrade oils, which have relatively low relaxation times, can produce such measurable effects. However, it seems that the first normal stress difference is a stronger function of pressure than the shear stress. If this is the case then this offers an explanation as to why the beneficial effects of viscoelasticity only manifest themselves at higher eccentricity ratios when the pressures are likely to be high enough for the normal stress amplification effect to come into play. The computational studies performed in this paper and in the paper of Phillips et al. [14] would seem to suggest that viscoelasticity can have a beneficial effect on journal bearing performance throughout the range of eccentricity ratios not only at the extreme values.

2. FORMULATION OF THE PROBLEM

Consider the two-dimensional geometry shown schematically in Figure 1. The journal of radius R_J rotates with a constant angular velocity ω in a stationary bearing of radius R_B. Both the journal and the bearing are assumed to be of infinite extent in the axial z-direction. The distance between the axes of the journal and the bearing is given by e. The eccentricity ratio is defined by $\epsilon = e/c$, where $c = R_B - R_J$ is the average gap so that $0 \leq \epsilon \leq 1$. The region between the journal and the bearing, Ω, is occupied by a lubricant. The journal is free to move under the action of an applied load which may be variable, its own weight and the reaction force exerted on it by the lubricant. This means that the centre of the journal traces out a nontrivial path in space and so the flow domain is not fixed in time.

The governing equations for an incompressible viscoelastic fluid comprise the conservation of momentum

$$\rho\left(\frac{\partial \mathbf{v}}{\partial t} + \mathbf{v}.\nabla\mathbf{v}\right) = -\nabla p + \nabla.\mathbf{T}, \tag{1}$$

the conservation of mass

$$\nabla.\mathbf{v} = 0, \tag{2}$$

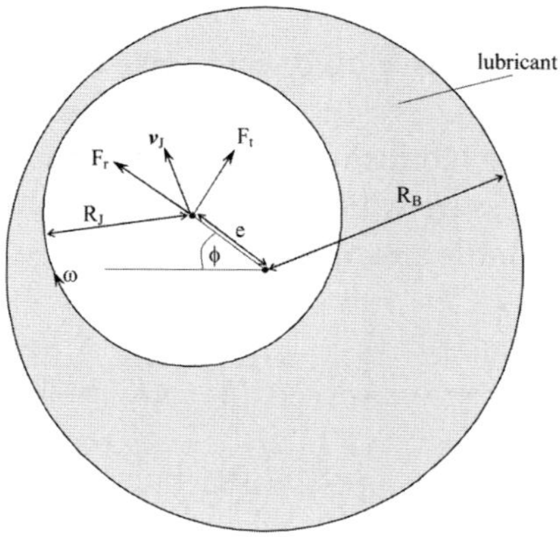

Figure 1. Schematic diagram of the dynamically loaded journal bearing geometry.

and a constitutive equation. In this paper we consider the Phan-Thien Tanner model

$$\left(1 + \frac{tr(\boldsymbol{\tau})}{\eta_p}\right)\mathbf{T} + \lambda \overset{\triangledown}{\mathbf{T}} = \eta\boldsymbol{\gamma} \tag{3}$$

and a form of the White–Metzner model

$$\mathbf{T} + \lambda(p, \|\dot{\boldsymbol{\gamma}}\|) \overset{\triangledown}{\mathbf{T}} = \eta(p, \|\dot{\boldsymbol{\gamma}}\|)\boldsymbol{\gamma}, \tag{4}$$

where $\overset{\triangledown}{\mathbf{T}}$ is the convected time derivative of the extra-stress tensor, $\mathbf{T}$, given by

$$\overset{\triangledown}{\mathbf{T}} = \frac{\partial \mathbf{T}}{\partial t} + \mathbf{v} \cdot \nabla\mathbf{T} - \mathbf{T} \cdot \nabla\mathbf{v} - (\mathbf{T} \cdot \nabla\mathbf{v})^T.$$

The White–Metzner model is one of the simplest coordinate invariant viscoelastic fluid models that allows for a single mode relaxation time λ. In these equations $\mathbf{v}$ is the velocity, p is the pressure, ρ is the density, and η is the viscosity. The rate of strain tensor, $\dot{\boldsymbol{\gamma}}$, is given by

$$\dot{\gamma} = \nabla\mathbf{u} + (\nabla\mathbf{u})^T, \tag{5}$$

and $\|\dot{\boldsymbol{\gamma}}\|$ is the scalar invariant

$$\|\dot{\boldsymbol{\gamma}}\| = \sqrt{(\frac{1}{2}\dot{\boldsymbol{\gamma}} : \dot{\boldsymbol{\gamma}})}.$$

A possible functional form for $\lambda(p, \|\dot{\boldsymbol{\gamma}}\|)$ is to make it proportional to the viscosity [2]

$$\lambda(p, \theta, \|\dot{\gamma}\|) = k_\lambda \eta(p, \theta, \|\dot{\gamma}\|), \tag{6}$$

where k_λ is a constant.

The viscosity law that we have used was proposed by Davies and Li [2] who were interested in modelling pressure effects in the statically loaded journal bearing. The viscosity of lubricating oils is, in general, a function of shear-rate, temperature and pressure which can be modelled using a combination of the well-known Cross, Barus and Vogel laws, viz.

$$\eta(p, \theta, \|\dot{\gamma}\|) = \left\{\eta_\infty + \frac{(\eta_0 - \eta_\infty)}{[1 + (K(p,\theta)\|\dot{\gamma}\|)^m]}\right\} \times \exp\left(\alpha p + \frac{\beta}{\theta}\right), \tag{7}$$

where

$$K(p,\theta) = K_0 \exp\left(\bar{\alpha} p + \frac{\bar{\beta}}{\theta}\right).$$

Here η_∞ is the asymptotic high shear rate viscosity, α is the pressure coefficient of viscosity, β is a reference temperature, and K_0, $\bar{\alpha}$, $\bar{\beta}$ and m are further material constants. The material parameters are estimated by best-fitting experimental data. The parameter α in (7) controls the dependence of the low shear-rate plateau on pressure. It is assumed that the dependence of the high shear-rate plateau on pressure is governed by the same exponential law. The parameter $\bar{\alpha}$ controls the rate of shear-thinning on pressure. It may realistically be assumed that this behaviour is independent of the low and high plateau levels reached. It is important to note that the viscosity law (7) is consistent with experiments [1, 12] which span only limited ranges of the pressures which the lubricants experience under general operating conditions. In this paper an isothermal analysis is performed and therefore changes in viscosity due to temperature are ignored.

In the journal bearing, changes in pressure contribute to the largest variation in viscosity. The variations of viscosity due to shear thinning for lubricants of interest under the conditions investigated so far are no more than 50%, whereas corresponding changes due to pressure can be as much as tenfold.

3. CAVITATION

At low pressures, the likelihood of cavitation must be considered. More importantly, the inclusion of a cavitation model has been shown to be crucial to the stability of journal bearings. A comprehensive treatment of cavitation involves the solution of a two-phase flow problem in which the phases are separated by a permeable free surface (interface). Such a treatment requires the solution of the conservation equations in each of the lubricant and cavitation phases together with appropriate conditions on the interface between the phases. The position of this interface moves in time since the journal is subject to dynamic loading. This presents a formidable computational task.

The inclusion of cavitation modelling in journal bearings has been investigated in numerous papers and a large number of different approaches have been proposed (see Gwynllyw et al. [5, 6] for a discussion). The most popular model is the so-called π-film model, in which it is assumed that the cavitating region occupies the diverging part of the gap. The use of a cavitation model circumvents the problems associated with the solution of a complicated two-phase flow problem in which the fluid/vapour interface is a moving boundary.

In this paper we have followed an approach which defines the cavitating region to be the region where the pressure falls below a prescribed threshold. In this region we modify the viscosity law (7) so that the viscosity is small. This is achieved using a cut-off function. At pressures below some value p^-, we assume that the lubricant cavitates. We make no attempt at modelling rupture, and simply assume that there remains fluid in the cavitated region but that the viscosity quickly but smoothly decreases to an asymptotic value $\eta_{\min}$. Ideally, we would wish to choose $\eta_{\min}$ to be some effective viscosity of the cavitated medium. We have chosen the value $\eta_{\min} = 8.0 \times 10^{-4}$ Pa s. The effect of this is to quickly reduce the viscosity in the cavitating region to a level somewhat above that of air.

4. NUMERICAL APPROXIMATION

The arbitrary Lagrangian–Eulerian (ALE) approach [9, 10, 11] is used to deal with the problem of moving grids in this problem. In the ALE formulation the reference system is considered as a computational reference system, which can be chosen quite arbitrarily. Given a reference frame with velocity $\mathbf{m}$ then the partial time derivative can be evaluated by:

$$\frac{\partial}{\partial t} = \frac{\delta}{\delta t} - (\mathbf{m} \cdot \nabla), \tag{8}$$

where $\delta/\delta t$ is the time derivative with respect to the moving reference frame. When the reference frame is fixed ($\mathbf{m} = \mathbf{0}$) the Eulerian description is recovered. When the reference frame follows the motion of the fluid ($\mathbf{m} = \mathbf{v}$) the Lagrangian description is recovered with $\delta/\delta t$ becoming the material derivative.

In this paper, the reference frame is taken to be the spectral element grid which is determined at each time-step by the position of the journal. With this approach the evaluation of the numerical approximation to $\delta/\delta t$ requires no interpolation. In addition no assumptions are made regarding the grid velocity. The extra convection term introduced in the ALE scheme is absorbed into the iterative temporal scheme with the mesh velocity $\mathbf{m}$ evaluated analytically given the velocity and position of the centre of the journal [8].

The region Ω between the journal and bearing is partitioned into a number of spectral elements, $\Omega_k,\ 1 \le k \le K$, such that $\cup_{k=1}^{K} \bar{\Omega}_k = \bar{\Omega}$ and $\Omega_k \cap \Omega_l = \emptyset$ for all $k \neq l$. We assume that the decomposition is geometrically conforming in the sense that the intersection of two adjacent elements is either a common vertex or an entire edge. Each physical element is mapped onto the parent element $[-1, 1] \times [-1, 1]$ on which a Gauss–Lobatto Legendre (GLL) grid is used. The transfinite mapping technique of Gordon and Hall [4] is used to perform this mapping. The $P_N - P_{N-2}$ spectral element method (Maday et al. [13]) is used in which the velocity and pressure approximations are of degrees N and $N-2$, respectively. The velocity and pressure approximations corresponding to element k are therefore

$$\mathbf{v}_N^k(\xi, \eta) = \sum_{i=0}^{N} \sum_{j=0}^{N} \mathbf{v}_{i,j}^k h_i(\xi) h_j(\zeta), \tag{9}$$

$$p_N^k(\xi, \eta) = \sum_{i=1}^{N-1} \sum_{j=1}^{N-1} p_{i,j}^k \tilde{h}_i(\xi) \tilde{h}_j(\zeta), \tag{10}$$

respectively, where h_i and $\tilde{h}_i$ are Lagrangian interpolants on the full and inner Gauss–Lobatto Legendre grids, respectively. The approximation for stress is similar to that for velocity. With the velocity and pressure approximation spaces thus chosen the Babuška–Brezzi compatibility condition (Maday et al. [13]) is satisfied. There are no spurious pressure modes in the pressure approximation. Care must also be exercised with respect to the choice of the stress space to ensure that a well-posed problem results in which the velocity approximation is not polluted by spurious modes [3].

The discrete variational formulation of the problem results in the following system of algebraic equations

$$A\boldsymbol{\tau}_N - B^*\mathbf{v}_N = \mathbf{f}_N, \tag{11}$$

$$B\boldsymbol{\tau}_N - D^*\mathbf{p}_N - \sigma M\mathbf{v}_N = \mathbf{g}_N, \tag{12}$$

$$D\mathbf{v}_N = \mathbf{h}_N, \tag{13}$$

where the N subscript indicates a vector of numerical values evaluated at the discretisation points. Block Gaussian elimination yields a symmetric semi-positive definite system for the pressure unknowns

$$S\mathbf{p}_N = \mathbf{c}_N, \tag{14}$$

where

$$S = DE^{-1}D^*, \quad E = BA^{-1}B^* - \sigma M \tag{15}$$

and $\mathbf{c}_N = \mathbf{h}_N - DE^{-1}\mathbf{k}_N$ and $\mathbf{k}_N = \mathbf{g}_N - BA^{-1}\mathbf{f}_N$. A system of this form needs to be solved at each time step. Note that the entries of the matrix S will, in general, be time-dependent.

The matrix S is made positive definite by imposing a zero volume condition on the pressure field to eliminate its arbitrariness [7]. The system (14) is solved for the vector of pressure unknowns using a nested preconditioned conjugate gradient (PCG) method.

The solution of systems of algebraic equations of this form, derived from a spectral element discretization of the constant viscosity Stokes problem, is well-documented in the literature [13] and suitable preconditioners have been advocated which yield efficient numerical algorithms. However, for the solution of realistic journal bearing problems in which $R_B - R_J$ is small the system of equations is extremely ill-conditioned and the spectrum of S is not so well-behaved [5]. A comprehensive study of preconditioners for this problem is given by Gwynllyw and Phillips [5] in the case of the statically loaded bearing.

The motion of the journal is determined by solving the equation of motion for the journal. To increase the stability of the calculation this equation is solved implicitly to determine the position and translational velocity of the journal [8].

5. NUMERICAL RESULTS

5.1 Phan-Thien Tanner Model

Here we investigate the effect of changes in the relaxation time on the path of the journal. In Figure 2 we compare the clearance plots when the angular velocity of the journal is $\omega = 20$ rad/s for $\lambda = 2 \times 10^{-4}$ s and $\lambda = 2 \times 10^{-6}$ s. This clearance plot demonstrates the well-known phenomenon of half-speed whirl associated full-film, constant viscosity lubricants and for the lubricant with the larger relaxation time there is a greater propensity for the journal to make contact with the bearing. When the relaxation time is increased there is a reduced dependence of the hydrodynamic forces on the relaxation time. This can be seen in Figure 3 where the same parameters have been used as for Figure 2 except that $\omega = 400$ rad/s. We observe that there is no discernible difference between the paths of the journals.

When the pressure-dependent cut-off model for the viscosity is employed the journal dynamics are quite different. This is to be expected since it is well-known that cavitation has a beneficial influence on the stability of journal bearings. In Figure 4 we display the paths for $\lambda = 2\times10^{-4}$ s and $\lambda = 2\times10^{-6}$ s when $\omega = 20$ s. Although both situations result in a stable trajectory with the journal reaching an equilibrium point the lubricant with the larger relaxation time results in a physically more stable journal. We have thus shown that it is

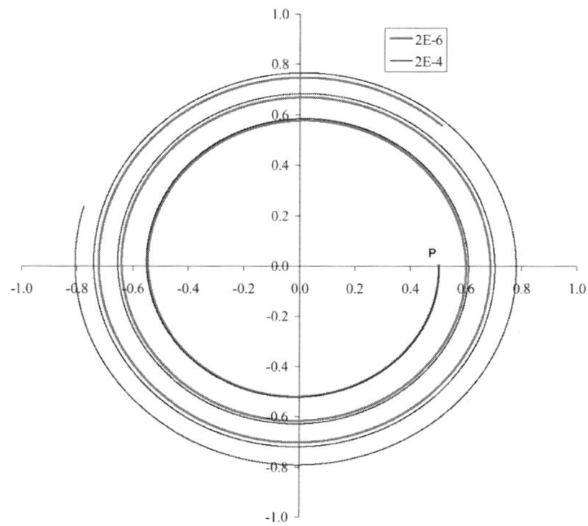

Figure 2. Comparison of clearance path plots of a slow DLJ system ($\omega = 20$) for two different relaxations times: (a) $\lambda = 2 \times 10^{-4}$ s and (b) $\lambda = 2 \times 10^{-6}$ s. Point P denotes the journal's initial position.

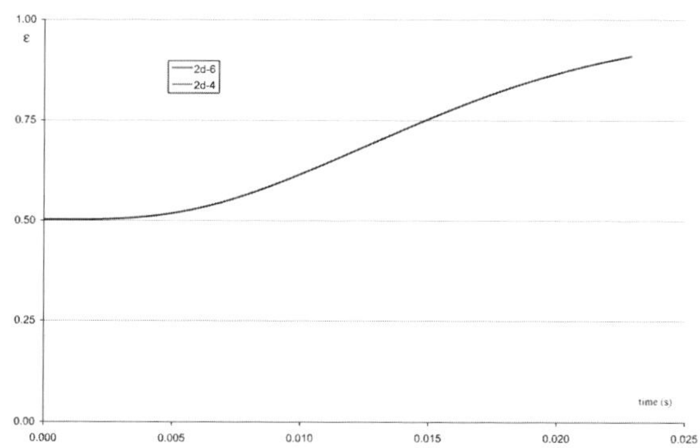

Figure 3. Comparison of clearance path plots of a fast DLJ system ($\omega = 400$) for two different relaxations times: (a) $\lambda = 2 \times 10^{-4}$ s and (b) $\lambda = 2 \times 10^{-6}$ s.

possible to predict viscoelastic effects using a simple constitutive relationship with a constant viscosity and that these effects are significant.

5.2 White–Metzner Model

We now turn our attention to the White–Metzner model and consider the effects of viscoelasticity on the performance of the lubricant and the dynamics of the journal when a constant applied load is imposed. This work has been

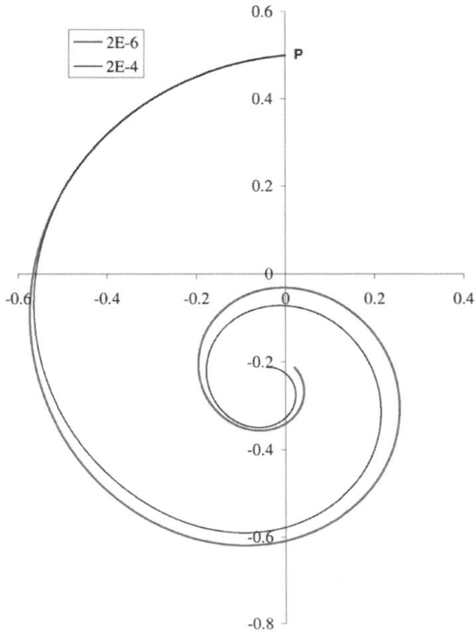

Figure 4. Comparison of DLJ paths using cavitation models with (a) $\lambda = 2 \times 10^{-6}$ s and (b) $\lambda = 2 \times 10^{-4}$ s. Point P denotes the journal's initial position.

reported previously [14] but we include a summary here in order to make comparisons with the PTT predictions. Two lubricants are considered for which $\lambda/\eta = 2 \times 10^{-4}$ Pa^{-1} and $\lambda/\eta = 4 \times 10^{-4}$ Pa^{-1}. A constant applied load of -10^5 N is applied and the journal rotates at an angular speed of 100 rad/s. In each case the journal very quickly reaches its equilibrium position, for which its eccentricity ratio is approximately 0.684, after 0.2 s. This can be seen in Figure 5 in which the evolution of the eccentricity ratio is shown for both lubricants. The dotted line represents the path of the journal for the lubricant with the higher elastic content, for which the final eccentricity ratio is lower. This figure clearly demonstrates that increased elasticity increases the load-bearing capacity and hence the minimum oil film thickness.

When the angular velocity of the journal is increased to 500 rad/s the journal takes longer to reach the equilibrium position (about 0.5 s), but again increased elasticity results in a larger minimum oil film thickness (see Figure 6). The final eccentricity ratios are 0.19807 and 0.19806, respectively, for the lubricants with the lower and higher elastic contents. Compared with the effect of increasing elasticity in the PTT model these changes are quite insignificant.

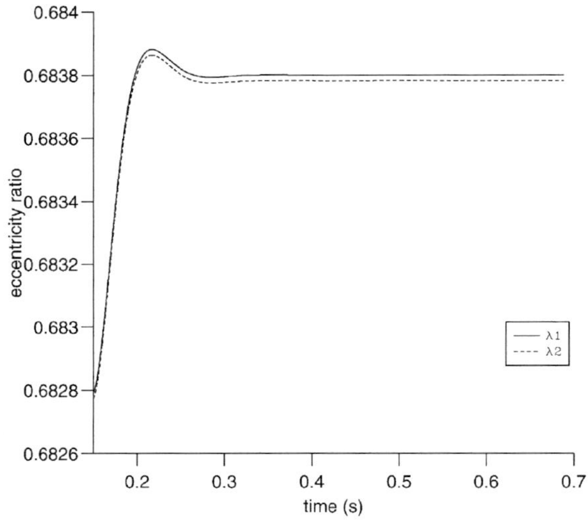

Figure 5. The evolution of eccentricity ratio in response to a constant applied load of -10^5 N and a journal angular velocity of 100 rad/s where $\lambda/\eta = 2 \times 10^{-4}$ Pa^{-1} and $\lambda/\eta = 4 \times 10^{-4}$ Pa^{-1}. Calculations by Phillips et al. [14].

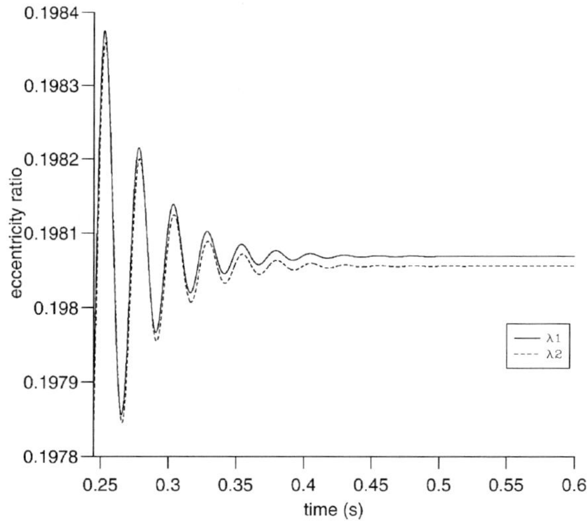

Figure 6. The evolution of eccentricity ratio in response to a constant applied load of -10^5 N and a journal angular velocity of 500 rad/s where $\lambda/\eta = 2 \times 10^{-4}$ Pa^{-1} and $\lambda/\eta = 4 \times 10^{-4}$ Pa^{-1}. Calculations by Phillips et al. [14].

6. CONCLUSIONS

In this paper we have described a model for predicting the behaviour of dynamically loaded journal bearings. In particular, we have investigated the effects of viscoelasticity on the minimum oil film thickness within a journal bearing. The spectral element method has been used for the spatial discretization of the problem and the movement of the mesh has been facilitated by means of the ALE formulation. We have shown that for both of the constitutive models discussed in this paper that increased elasticity results in a larger minimum oil film thickness. For the first time we have shown that a differential constitutive equation with a constant viscosity (the PTT model) is able to predict the positive influence of viscoelasticity on the performance of journal bearings.

REFERENCES

[1] T. W. Bates, B. Williamson, J. A. Spearot and C. K. Murphy. A correlation between engine oil rheology and oil film thickness in engine journal bearing. Technical Report 860376, Society of Automotive Engineers, 1986.

[2] A. R. Davies and X. K. Li. Numerical modelling of pressure and temperature effects in viscoelastic flow between eccentrically rotating cylinders. *J. Non-Newtonian Fluid Mech.*, 54:331–350, 1994.

[3] M. I. Gerritsma and T. N. Phillips. Discontinuous spectral element approximations for the velocity-pressure-stress formulation of the Stokes problem. *Int. J. Numer. Meth. Eng.*, 43:1401–1419, 1998.

[4] W. J. Gordon and C. A. Hall. Construction of curvilinear coordinate systems and applications to mesh generation. *Int. J. Numer. Meth. Eng.*, 7:461–477, 1973.

[5] D. Rh. Gwynllyw, A. R. Davies and T. N. Phillips. A moving spectral element approach to the dynamically loaded journal bearing problem. *J. Comput. Phys.*, 123:476–494, 1996.

[6] D. Rh. Gwynllyw, A. R. Davies and T. N. Phillips. On the effects of a piezoviscous lubricant on the dynamics of a journal bearing. *J. Rheol.*, 40:1239–1266, 1996.

[7] D. Rh. Gwynllyw and T. N. Phillips. On the conditioning of the mixed spectral element method for the Stokes problem. *Comput. Meth. Appl. Mech. Engng.*, 2004, submitted.

[8] D. Rh. Gwynllyw and T. N. Phillips. Some issues regarding the generation and movement of meshes for the spectral element method. *Int. J. Numer. Meth. Fluids*, 2004, submitted.

[9] H. H. Hu. Direct simulation of flows of solid-liquid mixtures. *Int. J. Multiphase Flow*, 22:335–352, 1996.

[10] H. H. Hu, D. D. Joseph and M. J. Crochet. Direct simulation of fluid particle motions. *Theor. Comput. Fluid Dyn.*, 3:285–306, 1992.

[11] H. H. Hu, N. A. Patankar and M. Y. Zhu. Direct numerical simulations of fluid-solid systems using the arbitrary Lagrangian–Eulerian technique. *J. Comput. Phys.*, 169:427–462, 2001.

[12] J. F. Hutton, B. Jones and T. W. Bates. Effects of isotropic pressure on the high temperature high shear viscosity of motor oils. Society of Automotove Engineers, Paper No. 830030, 1987.

[13] Y. Maday, A. Meirion, A. T. Patera and E. M. Rønquist. Analysis of iterative methods for the steady and unsteady Stokes problem: Application to spectral element discretizations. *SIAM J. Sci. Comput.*, 14:310–337, 1993.

[14] T. N. Phillips, R. E. Need, A. R Davies, B. P. Williamson and L. E. Scales. The effect of viscoelasticity on the performance of dynamically loaded journal bearings. Technical Report 982639, Society of Automotive Engineers, 1998.

[15] B. P. Williamson, K. Walters, T. W. Bates, R. C. Coy and A. L. Milton. The viscoelastic properties of multigrade oils and their effect on journal-bearng characteristics. *J. Non-Newtonian Flud Mech.*, 73: 115–126, 1997.

SESSION 5

BEHAVIOUR OF TRANSVERSE RIDGES PASSING THROUGH A CIRCULAR EHL CONJUNCTION

Motohiro Kaneta, Hiroshi Nishikawa and Kenji Matsuda
Deapartment of Mechanical and Control Engineering, Kyushu Institute of Technology, 1-1, sensuicho, Tobata, Kitakyushu 804-8550, Japan

Abstract The transient effects of transversely oriented moving ridges having different heights on point contact EHL films are examined by direct observations using the optical interferometry technique. It has been found that if cracks, which may produce rolling contact fatigue, initiate at the surface having ridges, they occur easily near the midplane along the entraining direction when the velocity of the rough surface is smaller than that of the smooth surface. Wear of contact surfaces and scuffing occur initially near the region where the horseshoe shaped constriction is generated in smooth-surface contacts.

Keywords: elastohydrodynamics, optical interferometry, roughness, slide-roll ratio, wear, scuffing, rolling contact fatigue.

1. INTRODUCTION

A clear understanding of the interaction between asperity-film thickness in the elastohydodynamic lubrication (EHL) regime is very important in order to improve the working performance, function and reliability of many machine elements. Therefore, a great deal of investigation has been carried out on rough surface EHL problems by numerous researchers from different points of view. One of the important results is that the film factor, Λ, defined as the ratio of the minimum EHL film thickness existing under smooth surfaces to the composite root-mean-square roughness of the contact surfaces is a principal parameter characterizing the asperity interaction, wear and fatigue phenomena of rolling contacts[e.g.1,2,3]. Kaneta et al.[4] have found in

R.W. Snidle and H.P. Evans (eds), IUTAM Symposium on Elastohydrodynamics and Micro-elastohydrodynamics, 189–200.

an EHL point contact that the vertical deformation of asperities always occurs when the film factor defined by the ratio of central film thickness calculated for smooth surfaces to root-mean-square roughness is less than 3, although the degree is very small under pure rolling conditions. Li et al.[5] pointed out that the surface fatigue life in the region of $\Lambda<3$ is expressed by a function of the ratio of the root-mean-square roughness to the average radius of asperity summit and Λ.

It has been found that local changes in the film thickness caused by a moving surface irregularity travel through the EHL conjunction approximately at the average speed of the surfaces regardless of the speed at which the surface irregularity itself moves [e.g.6-13]. Tao et al.[14] have pointed out through their numerical simulation taking real rough surface profiles into account that pressurized valleys make a significant contribution to load support, and the fluid film breaks down due to cavitation between asperities and local contact of the surfaces. Furthermore, it has been pointed out that a consideration of non-Newtonian lubricant behaviour is required in order to predict asperity deformation accurately [e.g.12,15].

In this study the transient behaviour of transverse ridges, which have different height, passing through a point EHL conjunction and the behaviour of EHL films affected by the ridges are examined by using the optical interferometry technique in order to investigate the basic mechanisms of rolling contact fatigue, wear and scuffing in rolling/sliding EHL contacts.

2. EXPERIMENTAL PROCEDURE

2.1 Apparatus

The experimental apparatus used in this study is basically the same as that used previously[4,9,10,16]. The contacting bodies are a precision steel ball of 25.4 mm diameter and an optical glass disk of 180 mm diameter and 12 mm thickness, whose lower surface is sputtered with a semi-reflecting chromium layer. The reduced elastic modulus, E', is equal to 117 GPa. The steel ball was mounted between four taper rollers. The ball and the disk were driven separately so that the slide-roll ratio, Σ ($=2(u_D-u_B)/(u_D+u_B)$), can be adjusted to any desired value, where u_D and u_B are the surface velocities of the glass disk and steel ball, respectively.

The duochromatic optical interferometry technique[17] was used to measure the shape and thickness of the lubricating film. A Xenon flash lamp with a flash duration of 20 μs and a metal halide lamp were used as light sources and the fringe pattern was recorded with a 35 mm camera and a high speed digital camera (2000 fps) attached to a microscope. In photographing with the 35 mm camera, the flash of the Xenon lamp was triggered by a signal from the rotary encoder attached to the end of the shaft to which the steel ball was fastened. The exposure time was fixed by a single discharge from the Xenon lamp, which was short enough to freeze movement. Consequently, the transient behaviour of EHL films influenced by the transverse asperities passing through the contact region was recorded at a spacing of 55.4 μm. The comparison between a series of interferograms recorded with the high speed digital camera and that with the 35 mm camera proved that the transient effect induced by the moving ridges can be evaluated using interferograms recorded with the 35 mm camera.

2.2 Specimens

Five types of patterns of transversely oriented ridges were produced onto one part of the ball surface as model asperities. Their stylus traces are shown in Fig. 1. Type A has two ridges having the same height. Type B has two ridges having different height; one is about twice as large in height as the other. Types C, D, E have 9 ridges, but the height of one or two ridges is approximately two times larger than the others. The ridges having different heights were produced by two stages of sputter deposition of chromium through a foil mask pasted on the surface of the steel ball. As the first step, the ridges with the same height and wavelength were made using a mask having nine slits, which have a length of 1 mm, a width of 30 μm, and a wavelength of 90 μm. As the second step, the height of one or two ridges was increased using the same mask, where non-necessary slits of the mask were covered with adhesive tape. The maximum height of ridges was about 0.6 μm and the heights of other asperities were about 0.3μm. The generatrix of each ridge was arranged so as to be perpendicular to the oil entraining vector.

2.3 Experimental conditions

The oil used was a mineral bright stock (BS) having a kinematic viscosity of 1388 mm^2/s at 21 °C and 376 mm^2/s at 40 °C, a specific gravity of 875 kg/m^3 at 21 °C, 863 kg/m^3 at 40 °C and a pressure-viscosity coefficient, *a*, of 23.7 GPa^{-1} at 21 °C and 19 GPa^{-1} at 40 °C. A load of w=39.2 N, which gives a maximum Hertzian pressure of 0.54 GPa and a Hertzian diameter of 0.37 mm, was used. The oil temperature in the inlet to the contact was 20.5 ± 1.0 °C. These conditions correspond to the dimensionless load parameter $W=w/E'R^2=2.08\text{x}10^{-6}$ and the materials parameter $G=\alpha E'=2630$.

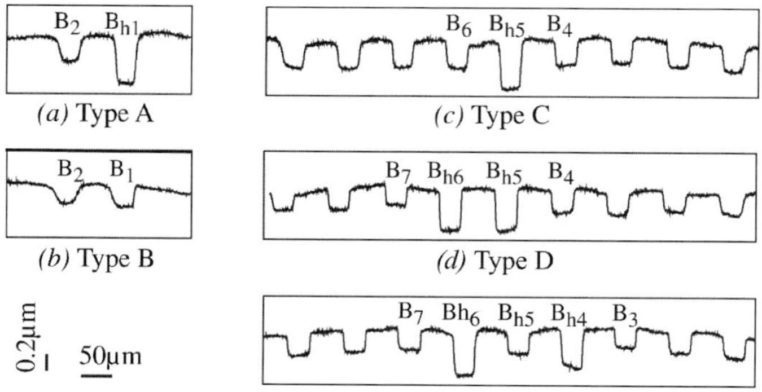

Figure 1. Stylus traces of ridges.

3. RESULTS AND DISCUSSION

3.1 Variation of Film Profile

Figure 2 shows an interferogram observed with type A specimen at the moment when the valley between both ridges is located near the centre of the EHL conjunction, in pure rolling, i.e., Σ=0 (u_D=23.1 mm/s and u_B=23.1 mm/s) giving $U=\eta_0(u_D+u_B)/2E'R=2.0\text{x}10^{-11}$. The corresponding mid-plane film profile along the entraining direction is shown, together with the film profiles for smooth surfaces. The direction of oil entraining velocity is from the left to the right. Arrows shown in the film profile show the position of the ridges. As has been pointed out[9,18] each ridge entering the contact area is deformed by the squeeze or micro-squeeze action and consequently, the leading side of the ridge forms a divergent gap during the passage through the EHL conjunction. The effect of the ridge is localized to a region around the ridge itself.

Figures 3 and 4 show interferograms and the corresponding midplane film profiles in the direction of motion observed at various positions when the type A ridges pass through the EHL conjunction under conditions of Σ=1 (u_D=34.7 mm/s and u_B=11.6 mm/s) and Σ=-1 (u_D=11.6 mm/s and u_B=34.7 mm/s), respectively, with $U=2.0\text{x}10^{-11}$. For Σ=1, where the speed of the glass disk is three times larger than that of the steel ball with ridges, the average speed of the fluid in the EHL conjunction is two times larger than the speed of the ridges. Therefore, each ridge forms a convergent gap by the hydrodynamic wedge or micro-wedge action (see B_1 ridge in Figure 3(a) and

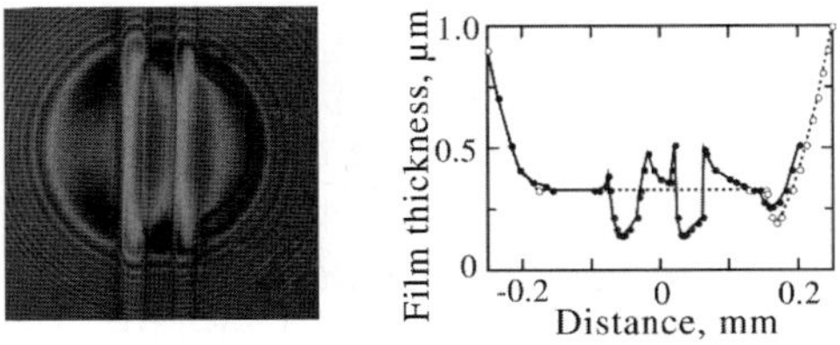

Figure 2. Interferogram and midplane film profile (Type A ridges, Σ=0, U=2.0×10^{-11}).

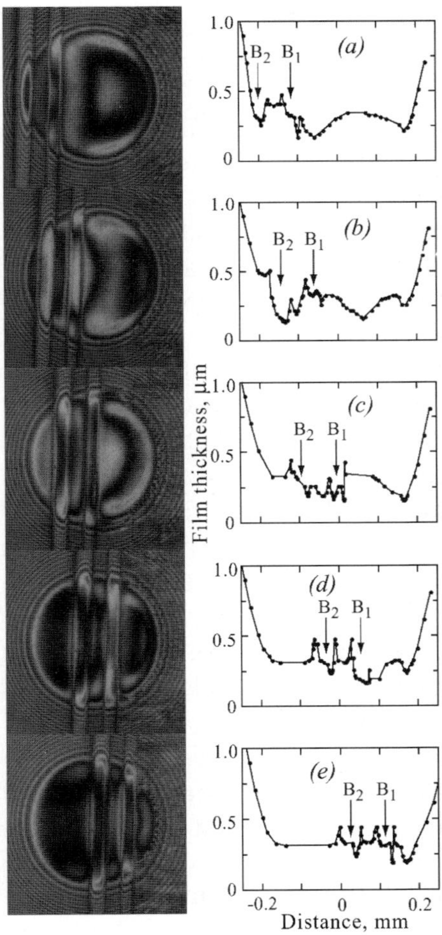

Figure 3. Interferograms and midplane film profiles during the passage of type A ridges (Σ =1, U=2.0×10^{-11}).

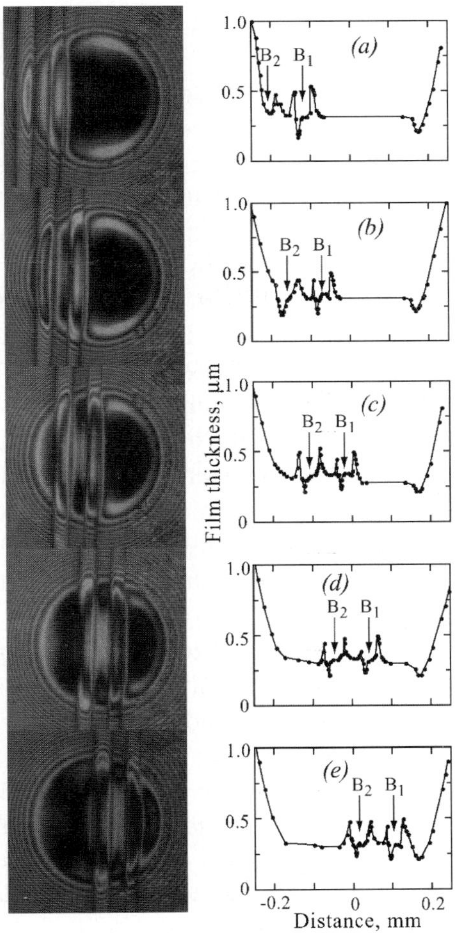

Figure 4. Interferograms and midplane film profiles during the passage of type A ridges (Σ =-1, U=2.0×10^{-11}).

B_2 ridge in Figures 3(b) and 3(c)). Furthermore, a band shaped thin film, which is produced by the B_1 ridge entering the contact area and located just downstream of the B_1 ridge, travels through the contact area at the average speed of the surfaces, while holding almost the original width but bending downward. A band shaped thin film produced by the B_2 ridge (see Figure 3(b)), in process of time, moves at the middle of both ridges (see Figure 3(c)), and just downstream of the B_1 ridge (see Figure 3(d)). That is, thick films are reduced by this thin film. It should be noted from Figure 3(d) that film thicknesses at the centre of each ridge and at the middle of both ridges are almost the same as the smooth-surface film thickness. Such phenomena mean that the conservation of mass in the flow is held, i.e., their film thicknesses are influenced by the smooth-surface film thickness upstream of the B_2 ridge.

In the case of Σ=-1 shown in Figure 4, the leading side of each ridge forms a divergent gap (reverse horseshoe shaped constriction), because the motion of the oil in the contact area relative to the ridge is opposite to the motion of the entraining fluid. Even in this case, film thicknesses at the centres of both ridges existing inside the contact area and at the middle of both ridges are almost the same as the smooth-surface film thickness formed at the downstream side of the leading ridge. However, each ridge is hardly influenced by the other ridge. This is because the area of influence induced by each ridges is much smaller for Σ=-1 than for Σ=1. In the case of Σ=1, the average speed of the oil is two times as fast as the speed of each ridge. In the case of Σ=-1, however, the average speed of the oil is 2/3 of the ridge speed. However, when $|\Sigma|$ increases under conditions of $\Sigma<0$, the film upstream of a local film shape will be markedly influenced by its local film shape. This means that the film thickness fluctuation depends on the wavelength of the roughness.

Figure 5 shows mid-plane film profiles observed with the types C, D and E ridges under conditions Σ=0 and U=$2.0\text{x}10^{-11}$. Each ridge is depressed vertically a little, and forms a divergent shape due to the micro-squeeze action; each trough or valley forms a convergent shape. Their configurations are frozen during the passage through the EHL conjunction. The midplane minimum film thickness at each ridge including the highest ridge is almost the same.

As seen from Figures 2, 3 and 4, and also pointed out in the previous paper[4,16], there is a big difference in film profile between Σ=0 and $\Sigma\neq0$. Figure 6 shows mid-plane film profiles observed under conditions Σ=1. The mid-plane minimum film thickness at each ridge is almost the same regardless of the height of ridge. This means that the highest ridge takes a larger vertical deformation and is pressed much more into the surfaces. The

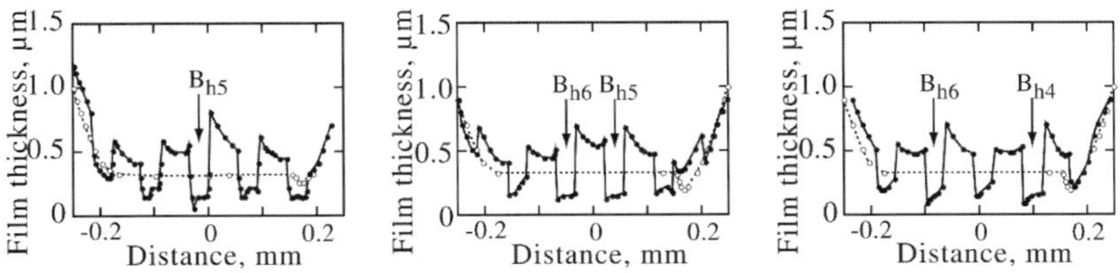

(a) Type C ridges *(b)* Type D ridges *(c)* Type E ridges

Figure 5. Midplane film profiles (Σ =0, U=2.0×10^{-11}).

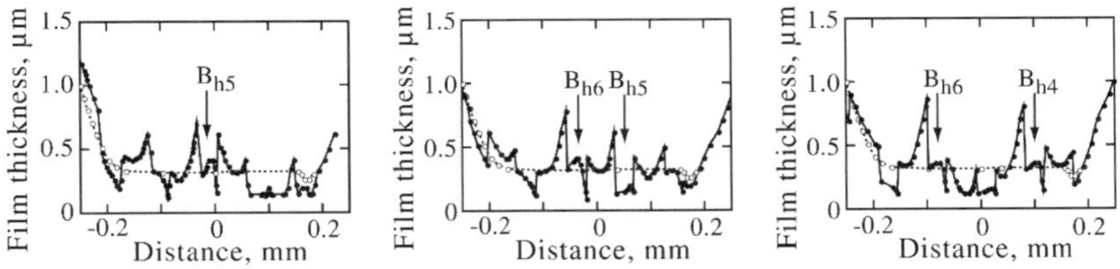

(a) Type C ridges *(b)* Type D ridges *(c)* Type E ridges

Figure 6. Midplane film profiles (Σ =1, U=2.0×10^{-11}).

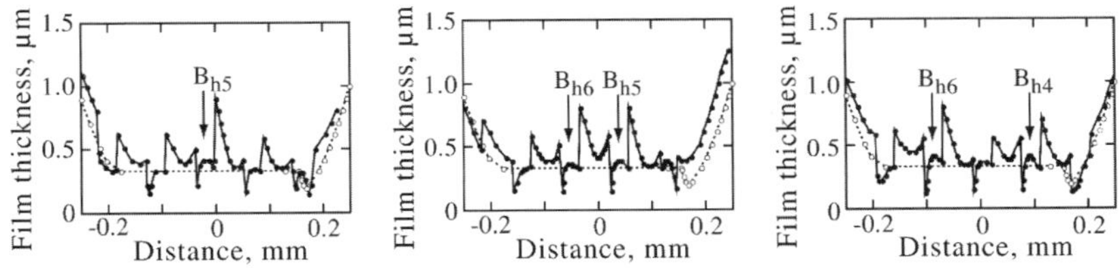

(a) Type C ridges *(b)* Type D ridges *(c)* Type E ridges

Figure 7. Midplane film profiles (Σ =-1, U=2.0×10^{-11}).

film thickness profile, however, fluctuates complicatedly but with a regularity[16] as each ridge passes through the EHL zone.

Figure 7 shows the results for Σ=-1. The dashed lines shown in Figure 7 correspond to the smooth-surface results. Each ridge forms a reverse horseshoe shaped constriction due to micro-squeeze and micro-wedge actions. The mid-plane minimum film thickness at each ridge is almost the same regardless of the height of ridge. As has been described, each ridge including the highest ridge hardly affects the film profile upstream of the ridge itself.

3.2 Vertical Deformation of Ridges

The vertical deformation in the midplane of each ridge passing through the EHL conjunction was evaluated by the film thickness differences, Δh_u and

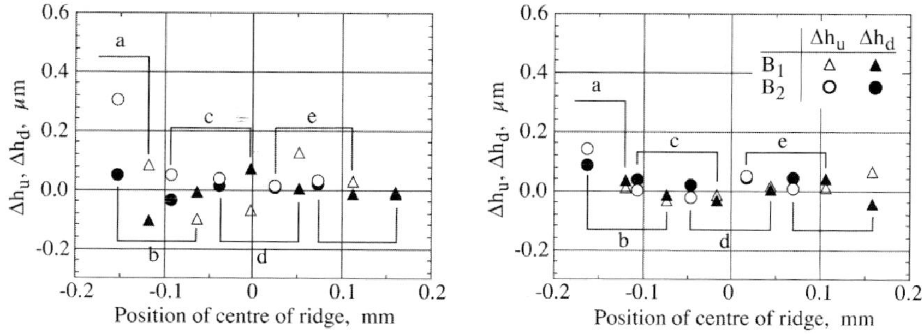

(a) Σ=1, corresponding to Fig. 3 *(b)* Σ=-1, corresponding to Fig. 4

Figure 8. Vertical deformation of B_1 and B_2 ridges passing through EHL conjunction.

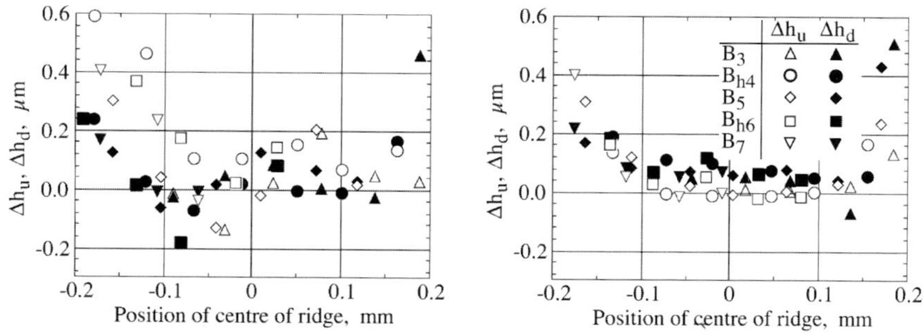

(a) Σ=1 *(b)* Σ=-1

Figure 9. Vertical deformation of ridges passing through EHL conjunction (Type E ridges, $U=2.0\times10^{-11}$).

Δh_d. It is difficult to measure the film thickness near the root of each ridge accurately, since the fringe changes very sharply. Therefore, the film thickness differences, Δh_u and Δh_d, were defined by subtracting the film thickness, h_c, at the centre of the ridge from the film thickness, h_u and h_d, at 30 μm up and down from a ridge centre, respectively. That is, $\Delta h_u = h_u - h_c$, and $\Delta h_d = h_d - h_c$. $\Delta h=0$ means that the film thickness at the centre of the ridge is the same as that of the adjacent valley or trough.

The results for the type A ridges are shown in Figure 8. The abscissa plots the position of the centre of the ridge passing through the EHL conjunction. The solid marks correspond to Δh_u and the open marks Δh_d. The alphabetical symbols in Figures 8(a) and 8(b) correspond to those in Figures 3 and 4, respectively. In the case of Σ=0, the actual vertical deformation of the ridge is very small. However, it should be noted that each ridge is forced into the surfaces as shown in Figure 2, though the amount is not so large compared with the cases of Σ=1 and Σ=-1. In Σ=1, when the B_1 ridge exists at the inlet side, $\Delta h_{uB1}>0$ and $\Delta h_{dB1}<0$, because of thin film produced when the ridge enters the contact area. When the B_1 ridge moves further, $\Delta h_{uB1}<0$ and $\Delta h_{dB1}>0$. This is because the entering of the B_2 ridge produces a thin film upstream of the B_1 ridge and the thin film caused by the B_1 ridge moves away from the vicinity of the B_1 ridge. When the thin film brought in by the B_2 ridge moves just to the downstream side of the B_1 ridge, Δh_{uB1} becomes positive though Δh_{dB1} keeps around 0. On the other hand, the B_2 ridge keeps almost a constant deformation due to the effect of the smooth-surface film thickness existing on the upstream side of the ridge itself, though the B_2 ridge is temporarily influenced by the thin film produced by itself when the B_2 ridge enters the contact area. In the case of Σ=-1, Δh_u and Δh_d are hardly dependent on the position of the ridges in the EHL conjunction, though both ridges are compressed and are forced into both surfaces.

That is, the root of the B_1 ridge for Σ=1 and the corresponding mating smooth surface receive a larger shear stress cycle compared with that of the B_2 ridge and that of each ridge in the case where Σ=-1.

Figure 9 shows the results obtained with the type E ridges. It should be noted from the comparison between Σ=1 and Σ=-1 that the fluctuation of Δh as a function of the position of each ridge is much larger for Σ=1 than for Σ=-1. Furthermore, as seen from Figure 11, although the ridges existing a certain distance from the mid-plane along the entrainment direction deform a little vertically, they are not largely pressed into the surfaces compared with those at the mid-plane. These facts suggest that rolling contact fatigue occurs easily near the mid-plane along the entraining direction when the velocity of the rough surface is smaller than that of the smooth surface, if cracks initiate at the surface having ridges. It is of course necessary to conduct detailed stress analyses to confirm this.

3.3 Minimum Film Thickness

In the present experiments, direct contact of both surfaces could not be observed. The minimum film thickness took place in the places where the highest asperity overlapped with the side-lobes of the macroscopic horseshoe-shaped constriction. Figure 10 shows the film profiles along the middle lines of the ridge perpendicular to the entrainment direction at a position where the ridge B_{h4} of the type E ridges locates near the center of the contact. Σ=1 gives the smallest minimum film thickness (say 0.017 μm under the condition of $U=2.0\text{x}10^{-11}$) and Σ=0 the largest minimum film thickness (0.060 μm). The minimum film thickness for Σ=0 was 0.039 μm.

Figure 11 compares the film profiles at a distance 0.157 mm from the midplane along the entraining direction, on whose line the smooth-surface film thickness takes the minimum value, with those at the mid-plane, at a position where the ridge B_{h4} of the type E ridges locates near the centre of

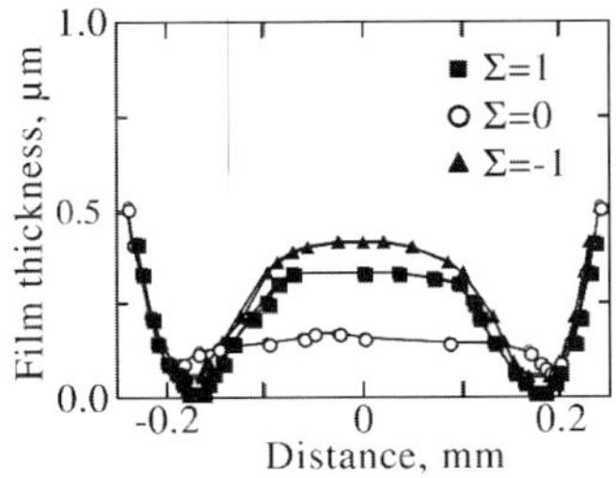

Figure 10. Film profiles perpendicular to oil entraining direction along middle line of ridge B_{h4} corresponding to Figure 11A.

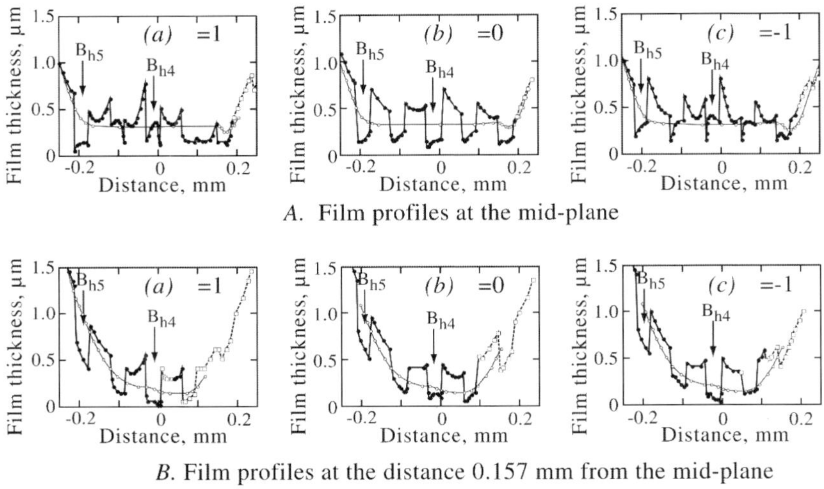

Figure 11. Film profiles along the entraining direction at the moment when ridge B_{h4} locates near the centre of the contact (Type E ridges, $U=2.0\times10^{-11}$).

the contact. The dotted line, which connects measured points, indicates that air bubbles occur in that region. The thin solid line connecting open marks shows the corresponding smooth-surface film profile. The film profile does not depend significantly on the slide-roll ratio Σ. The vertical deformation of each ridge is small compared with that of the mid-plane.

These facts suggest that wear of contact surfaces and scuffing will be initiated at the region where the horseshoe shaped constriction is generated in smooth-surface contacts.

4. CONCLUSIONS

The effects of transversely oriented ridges passing through the EHL conjunction on micro-EHL films and vertical deformation of each ridge have been examined by direct observations using the optical interferometry technique. The main results obtained under the present experimental conditions are as follows:

1. When the velocity of the rough surface is smaller than that of the smooth surface, each ridge affects local film profiles downstream of the ridge; in particular the effect of the highest ridge is remarkable.
2. When the speed of the rough surface is larger than that of the smooth surface, local films are hardly influenced by ridges including the highest ridge; the extent of the effects depends on the wavelength of asperities.
3. Rolling contact fatigue occurs easily near the mid-plane along the entraining direction when the velocity of the rough surface is smaller than that of the smooth surface, if cracks initiate at the surface having ridges.
4. Wear of contact surfaces and scuffing occur initially at the region where the horseshoe shaped constriction is generated in smooth-surface contacts.
5. Wear and scuffing occur easily in the case where the speed of the smooth surface is larger than that of the rough surface.

ACKNOWLEDGMENTS

The authors wish to express their sincere thanks to N. Tani, T. Takayama, K. Kunihiro and M. Sakamoto of Kyushu Institute of Technology for their kind cooperation in this study. This work was supported by Grant-in-Aid for Scientific Research (No. 15002009) from the Ministry of Education, Culture, Sports, Science and Technology, Japan.

REFERENCES

1. Tallian, T.E., McCool, J.L. and Sibley, L.B., Partial Elastohydrodynamic Lubrication in Rolling Contact, *Proc. Inst. Mech. Eng.* **180**(3B), 169–184 (1965/1966).
2. Poon, S.Y. and Hains, D.J., Friction Behavior of Lubricated Rolling-Contact Elements, *Proc. Inst. Mech. Eng.* **181**(1), 363–376 (1966/1967).
3. Johnson, K.L., Greenwood, J.A. and Poon, S.Y., A Symple Theory of Asperity Contact in Elastohydrodynamic Lubrication, *Wear* **19**, 91–108 (1972).
4. Kaneta, M., Sakai, T. and Nishikawa, H., Effects of Surface Roughness on Point Contact EHL, *Tribology Trans.* **36**(4), 605–612 (1993).
5. Li. D.F., Kauzlarich, J.J. and Jamison, W.E., Surface Roughness Effects on Fatigue in Partial EHD Lubrication, *ASME Journal of Lubrication Technology* **97**, 530–537 (1976).
6. Chang, L., Cusano, C. and Conry, T.F., Effects of Lubricant Rheology and Kinematic Conditions on Micro-Elastohydrodynamic Lubrication, *ASME Journal of Tribology* **111**, 344–351 (1989).

7. Venner, C.H. and Lubrecht, A.A., Numerical Simulation of a Transverse Asperity in a Circular EHL Contact under Rolling/Sliding, *ASME Journal of Tribology* **116**, 751–761.
8. Ai, X. and Cheng, H.S., The influence of Moving Dent on Point EHL Contacts, *Tribology Trans.* **37**(2), 323–335 (1994).
9. Kaneta, M., Sakai, T. and Nishikawa, H., Optical Interferometric Observations of the Effects of a Ridge on Point Contact EHL, *ASME Journal of Tribology* **114**, 779–784 (1992).
10. Kaneta, M. and Nishikawa, H., Local Reduction in Thickness of Point Contact EHL Films Caused by a Transversely Oriented Moving Groove and Its Recovery, *ASME Journal of Tribology* **116**, 635–639 (1994).
11. Guangteng, G., Cann, P.M., Olver, A.V. and Spikes, H.A., Lubricant Film Thickness in Rough surface, Mixed Elastohydrodynamic Contact, *ASME Journal of Tribology* **122**, 65–76 (2000).
12. Felix-Quinonez, A., Ehret, P. and Summers, J.L., New Experimental Results of a Single Ridge Passing Through an EHL Conjunction, *ASME Journal of Tribology* **125**, 252–259 (2003).
13. Glovnear, R.P., Choo, J.W., Olver, A.V. and Spikes, H.A., Compression of a Single Transverse Ridge in a Circular Elastohydrodynamic Contact, *ASME Journal of Tribology* **125**, 275–282 (2003).
14. Tao, J., Hughes, T.G., Evans, H.P. and Snidle, R.W., Elastohydrodynamic Response of Transverse Ground Gear Teeth, in *Proc. 28th Leeds-Lyon Symposium on Tribology*, pp. 447–458 (2002).
15. Venner, C.H., Kaneta, M., Nishikawa, H. and Jacod, B., Effects of Waviness on the Film Thickness in a Circular EHL Contact under Rollin/Sliding, in *Proc. International Tribology Conf.*, Nagasaki, 2000, pp. 631–636 (2000).
16. Foord, C.A., Wedeven, L.D., Westlake, F.J. and Cameron, A., Optical Elasto-hydrodynamics, *Proc. Inst. Mech. Eng.* **184**(28), Part 1, 487–505 (1969/1970).
17. Kaneta, M. and Cameron, A., 1980, "Effects of Asperities in Elastohydrodynamic Lubrication," *ASME Journal of Lubrication Technology*, **102**, 374–379 (1980).
18. Kaneta, M., Tani, N., and Nishikawa, H., 2002, "Optical Interferometric Observations of the Effects of Moving Transverse Asperities on Point Contact EHL Films," Proc. 29th Leeds-Lyon Symposium on Tribology, pp.101-109.
19. Holmes, M.J.A., Evans, H.P., Hughes, T.G. and Snidle, R.W., Transient Elastohydrodynamic Point Contact Analysis Using a New Coupled Differential Deflection Method, Part 2: Results, *Proc. Inst. Mech. Eng. Part J, J. Engineering Tribology* **217**, 305–321 (2003).

EXPERIMENTAL INVESTIGATION ON THE PRESSURE DISTRIBUTION FOR PURE SLIDING EHL CONTACTS WITH DENTED SURFACES

P. Vergne and F. Ville
LaMCoS, UMR CNRS 5514, INSA de Lyon, Bâtiment Jean d'Alembert, 20, avenue Albert Einstein, 69621 Villeurbanne cedex, France

Abstract Dented contacts have been studied in an EHL ball on disk device under pure sliding conditions. Contact temperature therefore lubricant rheological parameters and sliding speed influences have been investigated. Dents whose diameter was around one fifth of the Hertzian contact diameter have been maintained in the centre of the contact. According to the operating conditions, the measured pressure distributions show very different shapes, from a nearly hydrodynamic profile to almost two discontinuous peaks. The influence of the dent shoulders is also visible, underling the strong elastohydrodynamic effects generated by the contact micro geometry. The experimental findings are then compared to analytical results obtained under dry conditions. Some possible consequences for fatigue life prediction are proposed thanks to a measured 3-dimensional pressure map.

Keywords: elastohydrodynamic lubrication, dented contacts, Raman Microspectroscopy, pressure distribution, fatigue life prediction.

1. INTRODUCTION – BACKGROUND

The tribological behavior of contacting surfaces and hence the lifetime of mechanisms depends on the topography of the surfaces. The minimum film thickness and the maximum pressure can be considered as valuable criteria to estimate the durability of machine elements.

Much work has been focused on the consequences of dented surfaces in lubricated contacts. Film thickness measurements have been carried out by Cusano and Wedeven[1] in pure sliding conditions and extended by Kaneta et al.[2] to rolling - sliding conditions. The knowledge of the actual pressure field

R.W. Snidle and H.P. Evans (eds), IUTAM Symposium on Elastohydrodynamics and Micro-elastohydrodynamics, 201–213.

acting on a dented surface is critical for failure prediction. It is used to calculate the subsurface stress tensor in order to estimate the fatigue life reduction. Numerical models where the critical parameter is the pressure have been developed[3] to take into account the dent's influence on the lifetime reduction. Simulations of starved EHL contacts under pure rolling conditions[4] have explained the advantage of making small pits on one of the contacting surfaces. In this situation, film thickness and pressure distributions in the contact are strongly modified and showed large gradients. However, the assumptions made to conduct these calculations can be considered to limit their applicability.

Pressure measurements by Raman Microspectroscopy (RM) have been firstly carried out in static contacts by Gardiner et al.[5], then by Mansot[6]. In 1989, Gardiner et al.[7] applied for the first time RM to a rolling EHL contact. They obtained pressure distributions that qualitatively agreed with theoretical predictions. Recently, RM has been applied to map the pressure field in highly loaded EHL contacts[8-9]. This technique seemed convenient to study lubricated dented contacts for two reasons: it presents a high spatial resolution and it is well adapted to large pressure variations, as those expected to occur. A preliminary work[10] has confirmed that this tool was appropriate: a huge pressure peak was observed at the place where the lubricant flows out from the dent, leading to very high pressure gradients.

This paper presents recent original results obtained after improving the experimental procedures. Contact temperature therefore lubricant rheological parameters and sliding speed influences have been investigated. Varying these parameters allowed covering a wider range of operating conditions than those studied in Coulon et al.[10].

2. OPERATING CONDITIONS

Raman Effect

The Raman effect consists of an inelastic scattering related to the vibrational energies of the molecules, each Raman peak corresponding to a specific vibrational frequency. When inter-atomic distances decrease due to an applied pressure, Raman band shifts induced by vibrational frequency changes occur. Frequency shifts are recorded and transformed into pressure thanks to a calibration factor.

Jubault et al.[8-9] have improved the application of RM to obtain pressure distributions in highly loaded smooth lubricated contacts. Their results were in good agreement with the predictions of EHL theory. The presence of the Petrusevich pressure spike and its variations with load and speed were clearly shown. The effects of severe conditions (high loads, high speeds, large sliding ratio) on the contact thermal regime were also discussed.

Lubricant

Due to the very weak intensity of the scattered signal in RM, the lubricant must satisfy at least two conditions: to be a very efficient Raman scatterer and to be able to generate relatively thick films, as the scattered signal is directly proportional to the exposed volume and so to its thickness.

The 5P4E polyphenyl used here presents an absorption band in the near UV. When using incident light in the visible spectrum pre-resonance Raman effect occurs and enhances the signal to noise ratio. The 5P4E vibrational frequency shift with respect to pressure has been already calibrated: Laplant et al.[11] reported a linear relationship between wave number and pressure in the pressure range of interest in lubrication. Furthermore, we showed that the calibration factor is not temperature dependent[8].

In addition, 5P4E rheological properties have been extensively studied[12-13]: its viscosity and piezoviscosity coefficient are high. Consequently it can form sufficiently thick films: a minimum film thickness of 0.1 µm is required[8] to run reliable measurements with the current device.

Dent Geometry

In order to compare experimental and numerical results, the dent geometry should be defined according to the same parameters. An analytical expression approximates the dent geometry:

$$r(x,y) = d_{th} \exp\left(-K\frac{x^2+y^2}{4\phi^2}\right)\cos\left(\frac{\pi\sqrt{x^2+y^2}}{\phi}\right) \quad (1)$$

As shown in Eq.1, the dent is analytically characterized by its depth noted d_{th}, its diameter φ and the height of its shoulder determined by K.

The geometry used in the experiments is generated with a Rockwell indenter. Dented surfaces have been run in ($>10^{+4}$ cycles, pure rolling, 0.5 m/s, 40 °C, 90 N) to limit plastic deformation during pressure measurements, the majority of plastic deformation occurring after a reduced number of contact cycles[14]. The effective dent characteristics after running in were: d_{th} = 3 µm, φ = 75 µm and K = 15 leading to the following shoulder dimensions: height of 0.3 µm and overall diameter of 200 µm.

Experimental Set-up

Details on the experimental set-up are reported in [10]. A ball on disk tribometer is used: the steel ball that has been previously dented remains stationary and the sapphire disk is driven at the given sliding speed. The

Raman microscope supported by two independent X and Z motorized stages is placed above the disk. The Y direction is used to focus the incident beam into the lubricating film. The uncertainty on the spatial co-ordinates is less than 5 µm.

The monochromator gives a spatial dispersion less than 0.4 cm^{-1}/pixel. The central frequency of a Raman band is obtained by fitting the experimental data by a Gaussian curve, providing a resolution of typically ± 0.1 cm^{-1}, yielding to a pressure determination within ± 0.03 GPa.

Operating Conditions

All tests were realized under a normal load of 90 N that gives a Hertzian pressure of 1.33 GPa and a contact width (360 µm) almost five times the dent diameter (75 µm). Experimental conditions are summarized in Table 1, where M and L represent the dimensionless parameters defined by Moes[15] and film thicknesses for smooth contacts are deduced from the Hamrock and Dowson[16] formula for point contacts. Isothermal conditions are assumed by choosing adequate operating parameters.

Table 1. Operating conditions.

M	50	64	103	149	185
L	56	46	39	35	29
Sliding speed (mm/s)	40	75	40	24	40
Temperature (°C)	15	20	20	20	25
h_c (µm)	1.54	1.15	0.76	0.54	0.42
h_{min} (µm)	0.85	0.64	0.42	0.29	0.23

3. EXPERIMENTAL RESULTS

For comparison purposes, pressure measurements were conducted on both dented and smooth contacts and compared to Hertzian profiles in Figures 1–4. Three lines are also drawn: the upper dotted line represents the dent diameter φ, the medium dotted line the distance between two shoulder summits diametrically located and the lower full line the shoulder overall diameter. During the experiments it is assumed that the dent centre coincides with the contact centre.

The most hydrodynamic case is reported in Figure 1: L is high (56) and M is moderate (50) leading to a thick film prediction (Table 1). Measurements show that pressure initiates typically one dry contact radius (≈ 180 µm) upstream of the theoretical Hertzian zone. The measured maximum pressure of the smooth contact remains very close to the Hertzian one and in the exit the pressure drops faster than for the Hertzian case. These points agree with former results[9] which showed that at low or medium load, pressure profiles were significantly modified when sliding was increased.

The dented contact shows similar pressure profile in the inlet zone but a significant deviation is observed when the shoulder area is reached, when x > -100 μm in Figure 1. It deviates from the smooth case then shows a plateau (1.4 GPa) and increases very sharply when the shoulder maximum height is overtaken. A second hydrodynamic effect occurs over a zone lying from the left summit of the shoulder to 10–15μm after the right summit. The maximum pressure (2.6 GPa) is almost twice the Hertzian prediction for a smooth contact. A steep pressure decrease down to the ambient value is finally observed.

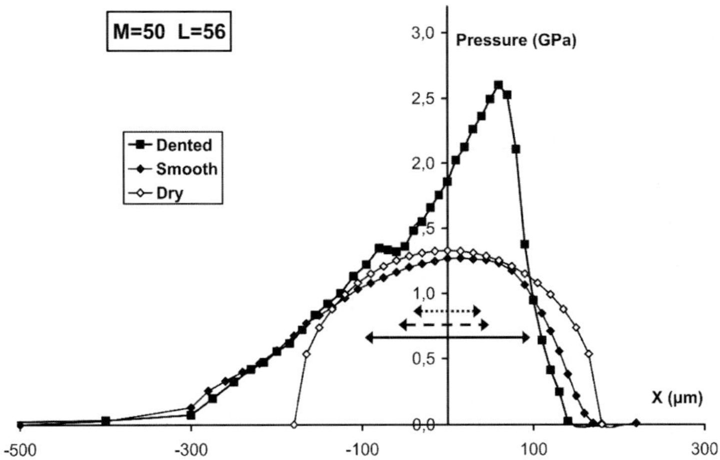

Figure 1. Pressure profiles when M=50 and L=56.

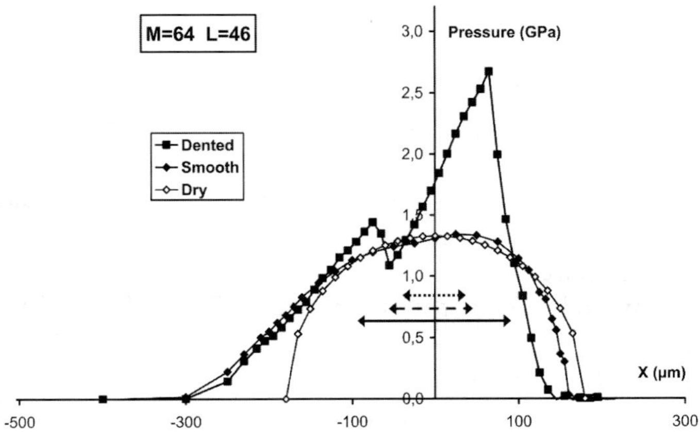

Figure 2. Pressure profiles when M=64 and L=46.

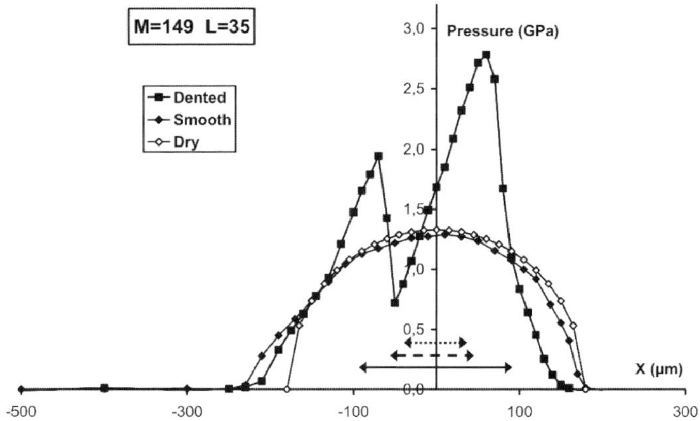

Figure 3. Pressure profiles when M=149 and L=35.

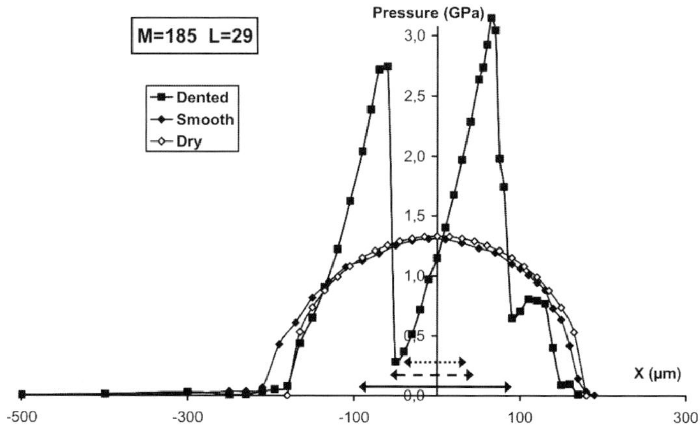

Figure 4. Pressure profiles when M=185 and L=29.

When M increases and L decreases several changes occur, from the contact inlet to the exit (respectively from left side to right side in Figures 1–4):

- For both smooth and dented contacts, the point where the pressure rise becomes significant moves towards the border of the Hertzian zone that is reached for the most severe operating condition, M=185 and L=29 (Figure 4).
- A first pressure peak appears and grows in dented contacts, the maximum pressure location being closer and closer to the left summit of the shoulder. It reaches 1.5 GPa for M=64, L=46 (Figure 2) and finally 2.75 GPa for M=189 and L=29 (Figure 4). Subsequently, a very steep pressure drop over a very narrow length is found. Its amplitude increases due to the simultaneous increase of the first pressure peak and decrease of the following pressure collapse (0.3 GPa in Figure 4).

- The amplitude of the second pressure rise also increases, the maximum pressure reaches 3.15 GPa in Figure 4 (2.6 GPa in Figure 1).
- The only point that seems unchanged among these observations concerns the location of the second pressure peak. Whatever the operating conditions are, it starts at x=-50 μm (corresponding to the top of the left side shoulder) and finishes at x=+60−65 μm. The existence of this relative dissymmetry can be explained by the dent shoulder deformation caused by severe operating conditions. Pure sliding conditions, huge pressure variations located onto a small area and two superposed constrictions occur in the exit zone: a classical constriction due to the contact itself and a second resulting from the presence of the shoulder.
- Finally, the exit smooth contacts give pressure values very close to Hertzian profiles. Dented contacts show huge pressure falls leading once again to very large pressure gradients. Furthermore in the last case reported here (Figure 4), one observes the appearance of a pressure shoulder located just after the second pressure peak.

To summarise these experimental findings, is should be noted that the presence of a dent induces major modifications of the pressure profiles whatever the operating conditions are. The influence of local features such as the dent shoulder has very important consequences. Two intense pressure peaks appear, the maximum pressure value being equal or greater than twice the Hertzian value calculated for smooth conditions. Huge pressure gradients are observed just after the pressure maxima positions and a noticeable pressure remains in the dent. This demonstrates that no cavitation occurs after the first shoulder and that the greater part of the initial lubricant volume is still entrapped inside the dent.

Typically we observe a pressure drop of 2.5 GPa over a distance of 10−20 μm. The presence of such gradients strongly modifies the stress field in the solid subsurface and will have important consequences on damage risk and hence on fatigue life. Furthermore, the important changes in the pressure profiles suggest that a measured 3-dimensional pressure field would be helpful to show how normal load is distributed over the whole contact area.

4. COMPARISON WITH NUMERICAL RESULTS

Pressure distributions of dented dry contacts have been analytically determined by Coulon et al.[17]. Depending on the applied load, three contact configurations could be obtained for a given dent geometry:

- Continuous contact area. The dent is completely absorbed by the elastic deformation of the surfaces and the pressure peak is proportional to the dimensionless dent slope (defined in [17]).
- Intermediate configuration (Figure 5). In this situation, the pressure peak still depends upon the dimensionless dent slope.
- Discontinuous contact area with load only supported by the dent shoulders and the contact area is a ring.

Using a new set of dimensionless parameters, Coulon et al.[17] proposed an abacus to determine the pressure peak as a function of load and dent geometry. For the experimental conditions presented before, the maximum pressure peak can be evaluated using this dry contact analytical approach. It has been estimated to be 2.4 times the maximum Hertzian pressure which corresponds in our case to 3.19 GPa.

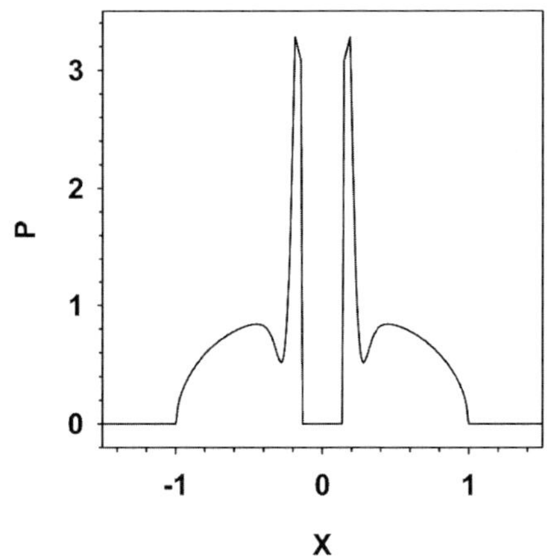

Figure 5. Analytically determined pressure distribution from Coulon et al.[17].

Using classical dry contact software to calculate the pressure field a maximum pressure of 3.12 GPa is found (Figure 6) under similar conditions[10].

It has been noticed before that the maximum pressure peaks obtained experimentally are in the range of 2.6−3.15 GPa. There is a satisfying agreement between the analytical, numerical and experimental results for the maximum pressure value. The evolution of the pressure profile with decreasing L and increasing M has shown that it becomes more symmetrical. It tends towards the dry contact pressure distribution, even if the first peak is significantly overestimated by the dry contact analysis and if pressure in the dent central region is necessarily nil in this later approach. The presence of a pressure shoulder on the right side of the contact in Fig.4 confirms this tendency. It strongly suggests that the minimum film thickness would be much thinner than predicted (0.2 μm) over a quite large area in the exit zone.

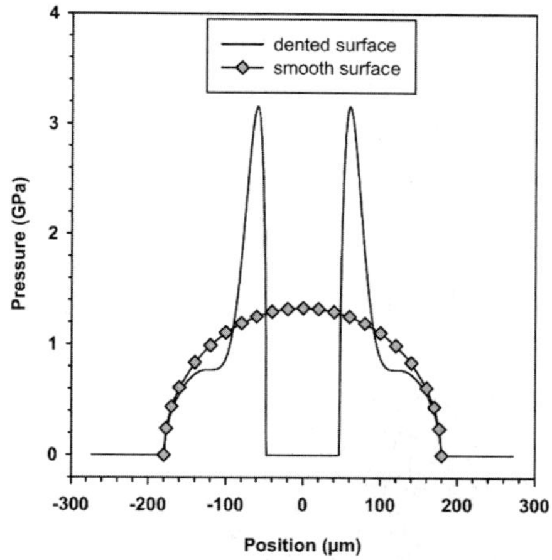

Figure 6. Pressure profiles in dry contacts from Coulon et al.[10].

5. CONSEQUENCES ON FATIGUE LIFE

The presence of geometrical defects like dents on contacting surfaces modifies the pressure distribution. It generates elastic pressure peaks when over-rolled, increasing the stress level. The maximum stresses occur closer to the surface, leading to surface initiated spalls (Figure 7). The maximum shear stress, taken as a fatigue criterion, could be considered as the driving force creating damage by particle detachment.

The effect of a dent on fatigue life can be studied according to its geometry to obtain a damage risk abacus. A lower bound of the infinite life is defined by the Hertzian pressure considered as the maximum value that induces no local plastic deformation within the material[18]. The maximum shear stress is then compared to the shear elastic limit, calculated applying elasticity theory. This endurance criterion has been plotted versus dent dimensions by Coulon et al.[3]. For a given Hertzian pressure, the dent geometry for which the local maximum shear stress is slightly less than the shear elastic limit can be determined.

In this paper, it has been previously noticed that the measured maximum pressure peak corresponds to the calculated one. So the maximum shear stress encountered in the contacting bodies is rather well evaluated by the theoretical studies and the endurance criterion proposed can be validated. This endurance criterion is based on a local limit and is representative of the contact severity. Unfortunately, with such a criterion, the prediction of the fatigue life reduction is not possible: a quantitative study of the influence of dents requires stressed volume considerations. It has been demonstrated[3] that the stressed volume depends directly on the pressure distribution and thus, in our case, on the dent geometry and operating parameters. Within the studied operating conditions, it appears that the measured pressure field is not symmetrical. So the induced actual stressed volume will be different from the calculated one. Moreover, longitudinal pressure profiles do not provide all the input data required to compute a representative 3-dimensional stress field under the contact surface. As the risk is based on the stressed volume, it can be stated that the risk evaluation could be optimized if 3-dimensional pressure measurements are conducted.

Figure 7. Spall due to a dent.

This new experiment has been carried out applying the medium operating conditions presented in Table 1, M=103 and L=39. Almost 350 pressure measurements have been carried out at x and z regular intervals of 15 μm: mainly longitudinal profiles have been evaluated as well as a limited number of transverse profiles to check the symmetry of the 3-dimensional distribution. The results are reported in Figures 8a and 8.b, in the form of top and perspective views. This experiment gives a valuable confirmation of the accuracy of pressure measurements made by RM: the pressure integration over the whole contact area gives the experimentally applied load within 2%. The 3-dimensional plots show that overpressures (i.e. pressures higher than 1.33 GPa, the Hertzian value for smooth conditions) form a deformed ring of external and internal diameters close to 220 μm and 60 μm. The ring shape is stretched toward the contact exit where the highest pressure values are found: a transversally oriented band (90 μm long, 15 μm wide) supports pressures equal to or greater than 2.6 GPa. In the contact centre, the results show a deep fall surrounded by pressures equal to or greater than 1.9 GPa. These observations underline that high pressures and very important pressure gradients occur over a significant area. The consequences on the subsurface stress field can be now evaluated to improve risk estimation and finally fatigue life prediction.

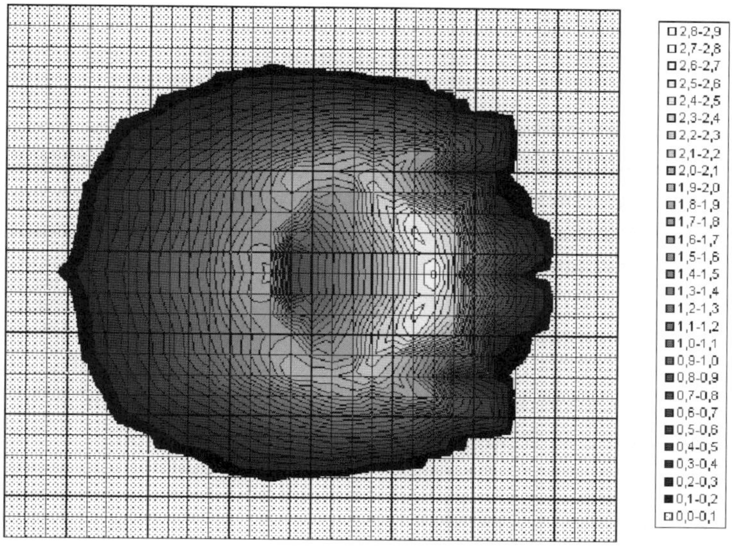

Figure 8a. Top view of a 3-D pressure field measured for M=103 and L=39.

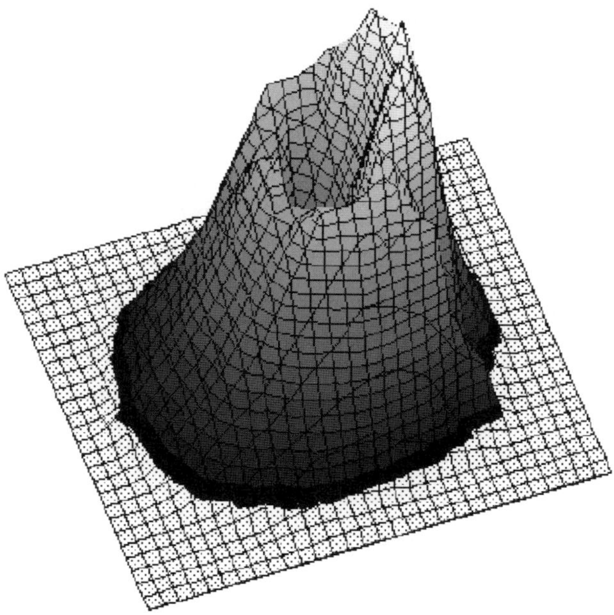

Figure 8b. Perspective view of a 3-D pressure field measured for M=103 and L=39.

6. CONCLUSION

In this paper, dented contacts have been experimentally studied under pure sliding conditions. Dents whose diameter was around one fifth of the

Hertzian contact diameter have been maintained in the centre of the contact: temperature therefore lubricant rheological parameters and sliding speed influences have been investigated. According to the operating conditions, the measured pressure distributions show very different shapes, from a nearly hydrodynamic profile to almost two discontinuous peaks. Compared to the smooth cases, these major modifications result from strong elastohydrodynamic effects generated by the contact micro geometry. It is shown that the presence of the dent shoulder has a critical influence on the pressure field. The experimental results are compared to analytical and numerical ones obtained using dry contact simulations. A satisfying agreement between the different approaches is obtained. Finally a measured 3-dimensional pressure map gives a comprehensive view of an actual case and suggests future work to optimize fatigue life prediction.

ACKNOWLEDGEMENT

The authors would like to thank A.A. Lubrecht (LaMCoS, INSA de Lyon) for his contribution in suggesting this experimental work and his support on damage risk evaluation and fatigue prediction.

REFERENCES

1. Cusano C., Wedeven L.D., J. of Lubr. Tech., July 1982, **104**, 365–375.
2. Kaneta M., Kanada T., Nishikawa H., Proc. of the 23rd Leeds-Lyon Symp. on Tribology, D. Dowson et al. (eds), Elsevier Science, 1997, pp. 69–79.
3. Coulon S., Ville F., Lubrecht A.A., Proc. of the 28th Leeds-Lyon Symp. on Tribology, D. Dowson et al. (eds), Elsevier Science, 2002, pp. 283–294.
4. Dumont M.L., Lugt P.M., Tripp J.H., J. of Trib. (2002), **124**, 2, 358–366.
5. Gardiner D.J., Baird E., Gorvin A.C., Marshall W.E., Dare-Edwards M.P., Wear (1983), **91**, 111–114.
6. Mansot J.L., PhD Thesis (in French), University of Lyon I, 01/12/1986, N.86-50.
7. Gardiner D.J., Baird E., Craggs C., Dare-Edwards M.P., Bell J.C., Lubrication Science (1989), **1**, 301–313.
8. Jubault I., Mansot J.L., Vergne P., Mazuyer D., J. of Tribology (2002), 124, 1, 114–120.
9. Jubault I., Molimard J., Lubrecht A.A., Mansot J.-L., Vergne P., Tribology Letters (2003), **15**, 4, 421–429.
10. Coulon S., Jubault I., Lubrecht A.A., Ville F., Vergne P., Tribology Int. (2004), **37**, 111–117.
11. Laplant F., Hutchinson E. J., Ben-Amotz D., J. of Trib. (1997), **119**, 4, 817–822.
12. Bezot P., Hesse-Bezot C., Berthe D., Dalmaz G., Vergne P., J. of Trib. (1986), **108**, 4, 579–583.
13. Yasutomi S., Bair S., Winer W. O., J. of Trib. (1984), **106**, 2, 291–302.
14. Howell M.B., Rubin C.A., Hahn G.T., J. of Trib., 2004, **126**, 413–421.
15. Moes H., Wear (1992), **159**, 57–66.

16. Hamrock B.J., Dowson D., J. of Lub. Tech. (1976), **98**, 223–229.
17. Coulon S., Ville F., Lubrecht A.A., J. of Trib. (2002), **124**, 1, 220–223.
18. Lamagnere P., Fougeres R., Lormand G., Vincent A., Girodin D., Dudragne G., Vergne F., J. of Trib. (1998), **120**, 3, 421–426.

SESSION 6

EHL FILM THICKNESS BEHAVIOUR UNDER HIGH PRESSURE – COMPARISON BETWEEN NUMERICAL AND EXPERIMENTAL RESULTS

Martin Hartl[1], Ivan Křupka[1] and Dong Zhu[2]
[1]*Faculty of Mechanical Engineering, Brno University of Technology, Technicka 2896/2, 616 69 Brno, The Czech Republic*
[2]*Eaton Corporation, 26201 Northwestern Highway, Southfield, MI 48037, U.S.A.*

Abstract This paper is devoted to the study of the behaviour of thin lubricant films under high contact pressures. Central film thickness values obtained from optical measurements are compared with numerical solutions from Zhu and Hu's model. The comparison indicates that both the numerical solutions and the experimental measurements are in good agreement with the Hamrock-Dowson formula at moderate loads and speeds. However, at low speeds and heavy loads accelerated non-linear film thickness reductions shown on the log-log scale are observed in the simulation results and confirmed by the experimental data.

Keywords: thin lubricant films, central film thickness, high contact pressure.

1. INTRODUCTION

Compact design of modern heavy-duty high efficiency mechanical systems demands improved understanding of the contact and lubrication of critical components under severe and extreme conditions. The mixed elastohydrodynamic lubrication (EHL) model developed by Zhu and Hu provides a means to simulate and predict the EHL performance in extended ranges of operating conditions. Recently presented numerical results have shown that the relationship between the film thickness and the speed may not always follow the simple power law, especially under severe conditions [1]. As the EHL film thickness measurement techniques are greatly improved, it is now possible to measure extremely thin films with nano-scale accuracy at significantly increased loads and reduced speeds.

Several experimental studies have been carried out at high contact pressures in the second half of the 20th century. They were oriented on relatively thick lubricant films, mainly due to the limitations of the conventional interferometry

R.W. Snidle and H.P. Evans (eds), IUTAM Symposium on Elastohydrodynamics and Microelastohydrodynamics, 217–228.

used. Gohar [2] reached contact pressures up to 3.5 GPa using a sapphire disc and a tungsten carbide ball and Gentle et al. [3] studied the thick film behavior for contact pressures above 2 GPa with the same combination of materials. Dalmaz [4] studied lubricant films within elliptical contacts between a small steel barrel and a sapphire disc at contact pressures from 1 to 2 GPa.

In the last decade techniques based on the evaluation of chromatic or monochromatic interferograms have enabled the study of the behavior of very thin lubricant films down to a few nanometers [5–10]. However, the sapphire disc that is required for higher contact pressures has been used rarely because of its high cost and optical properties (birefringence). Smeeth et al. [11] used a tungsten carbide ball and a small sapphire window inserted in a steel disc to obtain central film thickness values at very high contact pressures up to 3.4 GPa. This approach was further improved by Smeeth and Spikes [12] who obtained film thickness profiles using a sapphire disc and modified ultra thin film interferometry. They compared measured central and minimum film thickness values with theoretical predictions at contact pressures as high as 3.5 GPa. Recently, Křupka et al. [13] used thin film colorimetric interferometry (TFCI) to evaluate the central and minimum film thickness values within lubricated contact formed between a sapphire disc and a steel ball (maximum Hertz pressures from 0.511 to 1.315 GPa).

In the present study, the TFCI technique is used to evaluate the film thickness distribution within lubricated contacts. Its recent modification [14] enables us to study the behaviour of the central film thickness down to a few nanometers and compare obtained results with data from numerical solutions.

2. EXPERIMENTAL APPARATUS AND CONDITIONS

The experimental apparatus (Figure 1) consists of a conventional optical test rig, a microscope imagining system and a control unit. The principal parts of the test rig are a steel or tungsten carbide (WC) ball and a glass or sapphire disc. The ball is 25.4 mm in diameter with a root mean square surface roughness of about 4 nm. The lower surface of the disc is covered with a thin chromium layer. The disc is in a pure rolling contact with the ball that is driven by an AC servomotor. The lower part of the ball is submerged in a lubricant reservoir to ensure a fully flooded condition. The microscope imaging system consists of an industrial microscope, a halogen light source, and a color video camera.

Film thickness values within lubricated contacts were obtained from chromatic interferograms by thin film colorimetric interferometry. This measurement technique can provide detailed information about lubricant film shape down to a few nanometers within the whole lubricated contact. As the refractive index of lubricant changes with the density, evaluated film thickness values must be corrected [15].

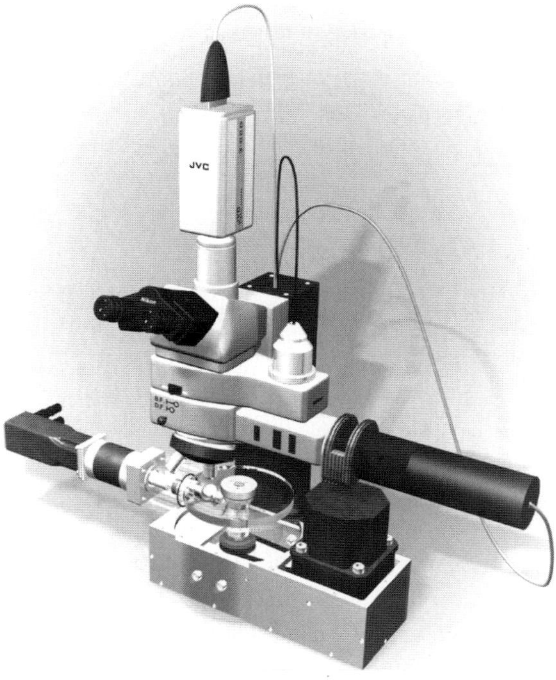

Figure 1. Experimental apparatus.

Measurements were carried out with paraffinic and naphthenic base oils under pure rolling conditions. Hexadecane was used to validate experimental results. The properties of the materials of the contact bodies and lubricants used in this study are given in Tables 1 and 2, respectively. Test conditions are summarised in Tables 3 and 4.

3. EXPERIMENTAL TECHNIQUE VALIDATION

Two series of experiments were carried out to check the validity of the present technique. At first, central film thickness values were measured with hexadecane at max. contact pressure of 512 MPa. It is obvious from Figure 2 that the results obtained exhibit the typical behaviour for this fluid observed also by other researchers, i.e. the linear behaviour down to a few nanometers. As these results were obtained using a glass disc, in the second series of experiments the influence of a sapphire disc on measured film thickness values was checked. Figure 3 shows the comparison of central film thickness values for naphthenic base oil N2400 obtained using both glass and sapphire discs at max. contact pressure of 512 MPa. It can be seen that the sapphire disc has no significant influence on obtained data. Hence, it can be used for obtaining accurate values of central film thickness at high contact pressures.

Table 1. Properties of materials of contact bodies.

	Glass disc	Sapphire disc	Steel ball	WC ball
Modulus of elasticity, GPa	87	420	207	641
Poisson's ratio	0.208	0.29	0.3	0.26

Table 2. Properties of lubricants at 40°C.

	Viscosity, Pas	Pressure-viscosity coefficient, GPa^{-1}
Hexadecane (23°C)	0.004	13
Paraffinic Base Oil SR600	0.1	23
Naphthenic Base Oil N2400	0.421	35

Table 3. Test conditions for the steel ball, the sapphire disc and SR600 oil.

Contact load, N	5, 35, 130
Max. Hertz pressure, GPa	0.52, 0.979, 1.517
Rolling speeds, ms^{-1}	0.00014–0.374
Temperature, °C	40±0.4
Hamrock and Dowson dimensionless parameters	
Speed parameter U	3.72×10^{-15}–1.03×10^{-11}
Load parameter W	0.98×10^{-7}–2.54×10^{-6}
Materials parameter G	6654

Table 4. Test conditions for the WC ball, the sapphire disc and SR600 oil.

Contact load, N	85, 280
Max. Hertz pressure, GPa	1.954, 2.908
Rolling speeds, ms^{-1}	0.00018–0.066
Temperature, °C	40±0.4
Hamrock and Dowson dimensionless parameters	
Speed parameter U	0.99×10^{-15}–3.66×10^{-12}
Load parameter W	0.83×10^{-6}–2.89×10^{-6}
Materials parameter G	20968

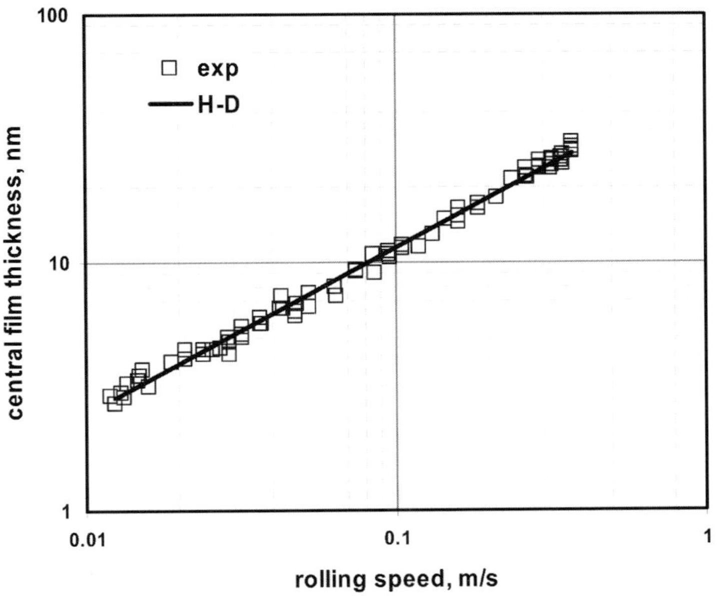

Figure 2. Central film thickness vs. rolling speed for hexadecane.

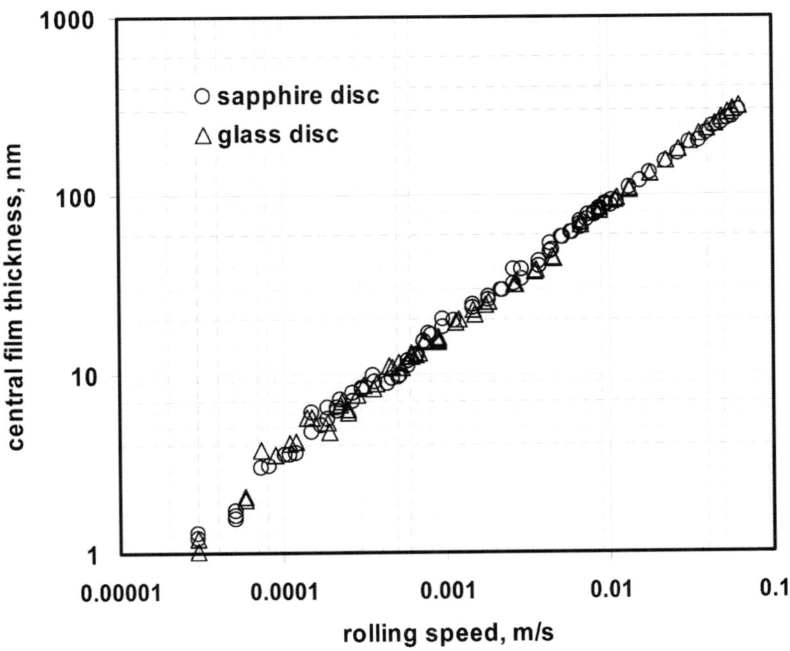

Figure 3. Central film thickness vs. rolling speed for naphthenic base oil.

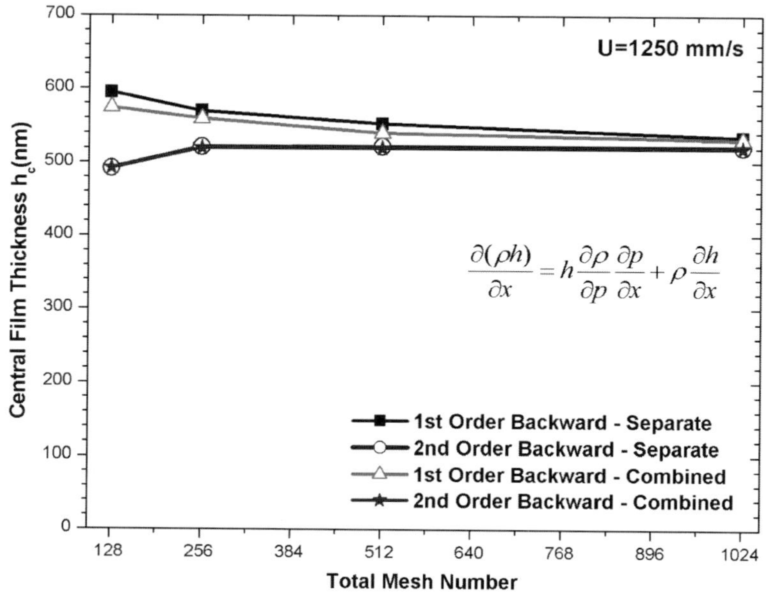

Figure 4. Mesh size effect on central film thickness for rolling speed of 1250 mm/s.

4. NUMERICAL SOLUTION

The numerical solution method used in the present study has been described in detail by Ai [16], Zhu and Hu [17, 18], so it will not be repeated here. However, as the lubricant film becomes very thin, numerical accuracy may become more critical. In order to investigate the effect of possible discretization errors, different numerical schemes and mesh sizes have been used and results compared. Figure 4 shows a typical thick film case (central film thickness about 530 nm), in which the mesh size and differential scheme used for the wedge term in the Reynolds equation do not seem to have any significant influence on the solution. A coarse mesh, such as 256×256 (mesh size $\Delta X = \Delta Y = 0.01172$), yields about the same film thickness as that with the dense mesh of 1024×1024 ($\Delta X = \Delta Y = 0.002929$). The different differential schemes do not significantly affect the solution either.

However, for ultra-thin film cases at low speeds and/or heavy loads, the mesh size and the differential scheme become more important. Figure 5 shows a typical thin film case (central film thickness around 15–17 nm). It can be seen that the first order backward scheme (separate) yields a thicker film, which gives an upper limit for all the solutions at the same mesh size (but the second order schemes are supposed to be more accurate). As the mesh size is continuously reduced, the solutions from the first order schemes slightly go down and those from the second order schemes go up, all approaching a constant.

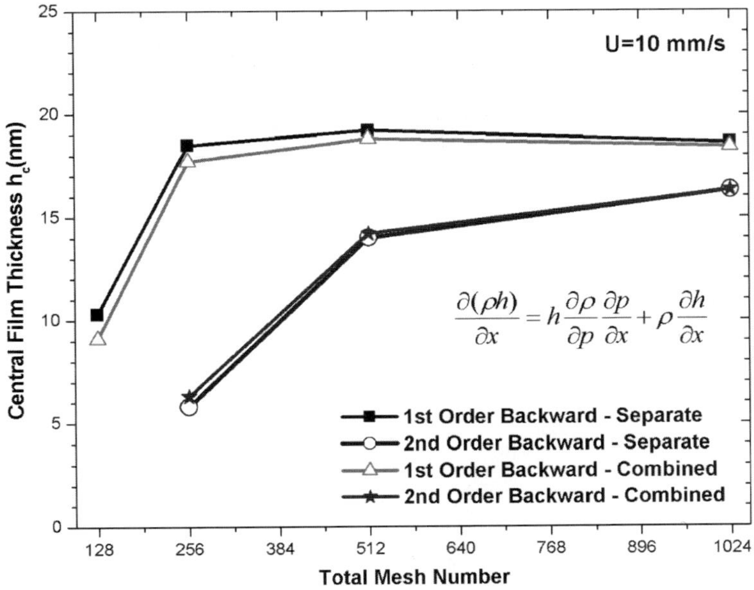

Figure 5. Mesh size effect on central film thickness for rolling speed of 10 mm/s.

In principle, for ultra-thin film cases, a very dense mesh with a second order scheme is always desirable. However, the dense mesh of 1024 × 1024 ($\Delta X = \Delta Y = 0.002929$) requires a huge amount of computing time (one case with a convergence criterion of $P_{\text{error}} < 10^{-9} \sim 10^{-8}$ may take several days on a Pentium IV 3.2 GHz computer). Actually, the difference between the solutions from the 512 × 512 mesh and the 1024 × 1024 mesh appears to be quite limited. It is believed, based on the observation of the solution trend, that it may not be necessary to infinitely reduce the mesh size. In the present work the majority of cases are analyzed with the second order backward scheme and the 512 × 512 mesh. A limited number of ultra-thin film cases (film thickness as low as 3.2 nm) are also checked with the 1024 × 1024 mesh. The difference is found to be within 10–17%, which does not appear to have significant influence on the basic trend of the results.

5. RESULTS

Experimental results presented in this paper were obtained while the load was kept constant and the rolling speed was changed with a pre-chosen interval to ensure very thin lubricant film conditions. When a sapphire disc and steel ball were used, the applied loads of 5, 35 and 130 N correspond to maximum Hertz pressures of 0.52, 0.979 and 1.517 GPa, respectively. The use of the WC ball

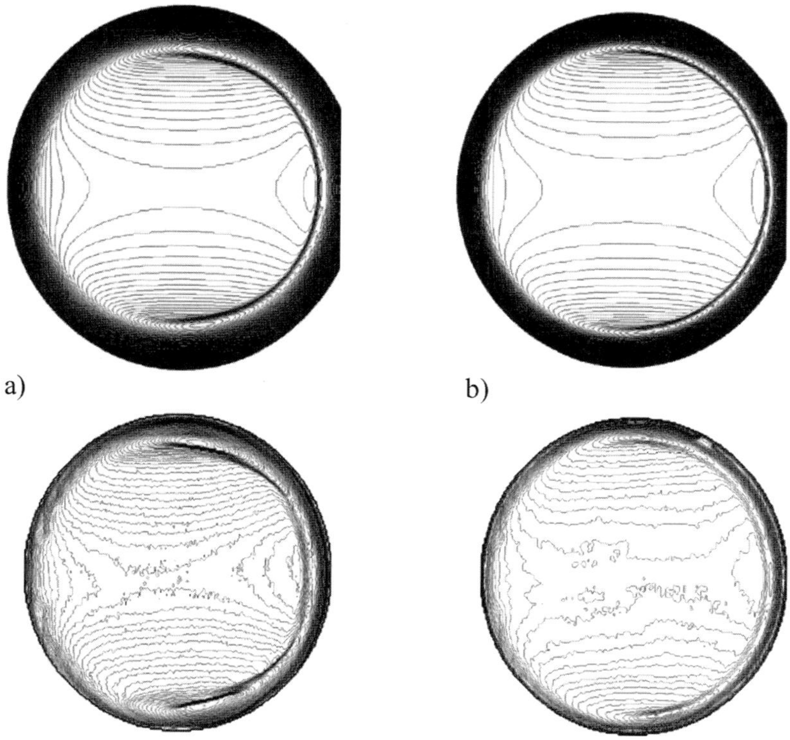

Figure 6. Calculated and measured film thickness contour plots for paraffinic base oil, max. Hertzian pressure of 1.517 GPa and rolling speed of 0.372 m/s (a) and 0.133 m/s (b).

enabled us to reach a max. Hertz pressure of 1.954 and 2.908 GPa (load of 85 and 280 N, respectively).

The film thickness distributions within the whole lubricated contact were obtained for each combination of operating conditions from which it was possible to determine precise values of the central film thickness. Obtained results were compared with data provided by the numerical solution and they are summarized in Figures 6 to 11. Figure 6 shows calculated and measured film thickness contour plots for paraffinic base oil, max. contact pressure of 1.517 GPa and rolling speed of 0.372 m/s and 0.133 m/s. It can be seen that there is a very good qualitative agreement between the measured and calculated film thickness distributions within the lubricated contact. In Figures 7 to 11 logarithmic plots of central and minimum film thickness against rolling speed are shown. In these figures, values calculated from the Hamrock and Dowson central and minimum film thickness equations are displayed as a reference.

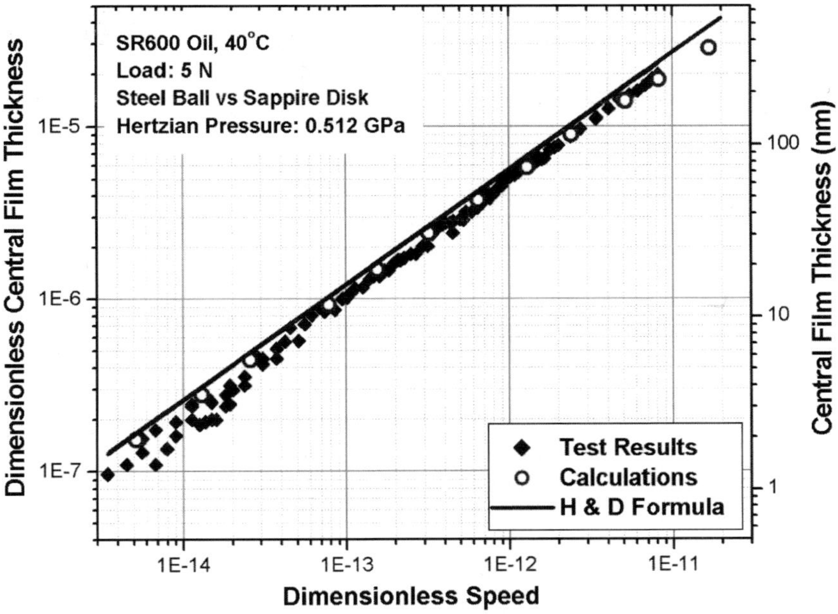

Figure 7. Central film thickness vs. speed for max. Hertzian pressure of 0.512 GPa.

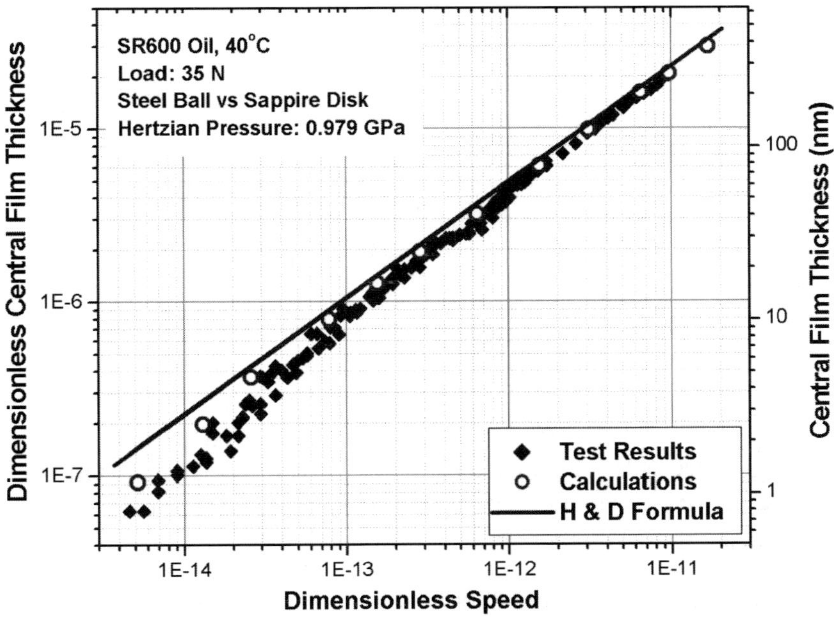

Figure 8. Central film thickness vs. speed for max. Hertzian pressure of 0.979 GPa.

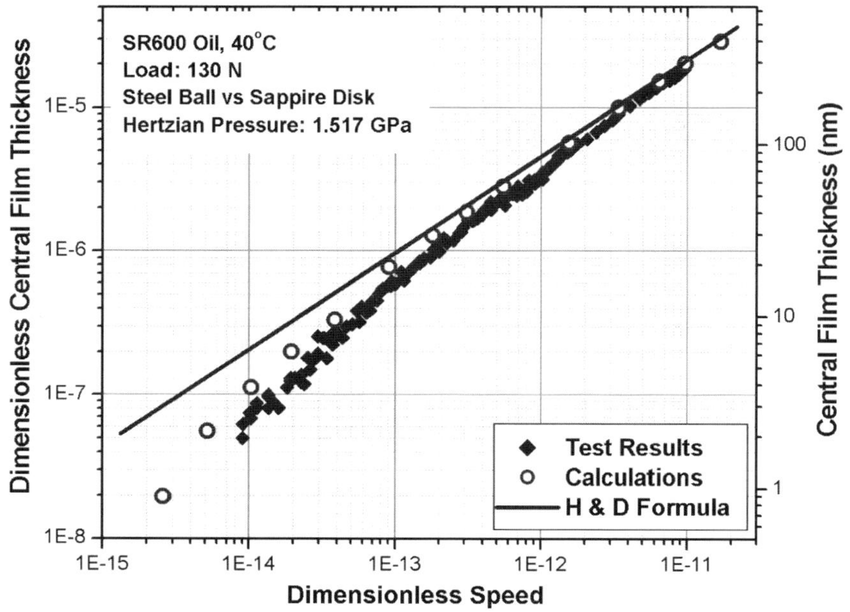

Figure 9. Central film thickness vs. speed for max. Hertzian pressure of 1.517 GPa.

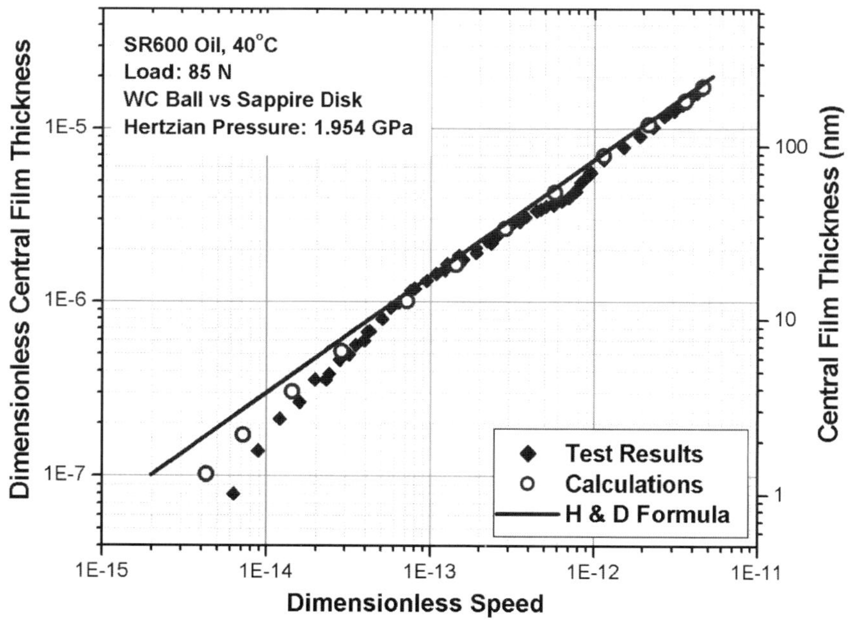

Figure 10. Central film thickness vs. speed for max. Hertzian pressure of 1.954 GPa.

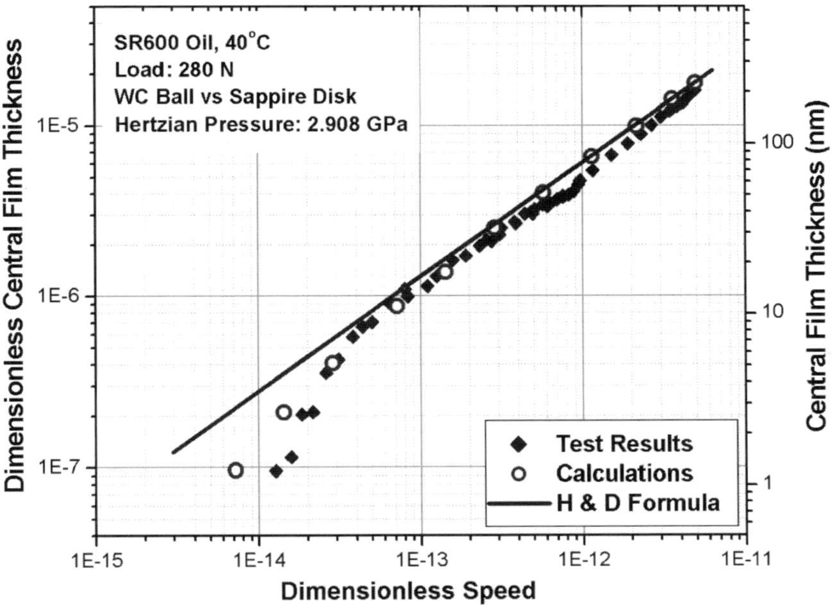

Figure 11. Central film thickness vs. speed for max. Hertzian pressure of 2.908 GPa.

ACKNOWLEDGMENTS

This research was supported by the Czech Science Foundation under grant No. 101/03/0525. The test oils were supplied by the Sun Oil Company Belgium. The numerical results presented in Figures 4 and 5 were obtained in collaboration with Dr. Y.C. Liu and Prof. Q. Jane Wang of Northwestern University, USA, and Dr. W.Z. Wang and Prof. Y.Z. Hu of Tsinghua University, China.

REFERENCES

[1] D. Zhu, Elastohydrodynamic Lubrication in Extended Parameter Ranges – Part 1: Speed Effect, *Tribology Transactions* **45**, 540–548 (2002).

[2] R. Gohar, Oil Film Thickness and Rolling Friction in Elastohydrodynamic Point Contact, *Trans. of the ASME, Journal of Lubrication Technology* **93**, 371–382 (1971).

[3] C.R. Gentle, R.R. Duckworth and A. Cameron, Elastohydrodynamic Film Thickness at Extreme Pressures, *Trans. of the ASME, Journal of Lubrication Technology* **97**, 383–389 (1975).

[4] G. Dalmaz, Simulating Angular Ball Bearing Lubricated Elliptical Contacts – Film Thickness and Traction Measurement, *Tribology International* **11**, 273–279 (1978).

[5] G.J., Johnston, R. Wayte and H.A. Spikes, The Measurement and Study of Very Thin Lubricant Films in Concentrated Contacts, *Tribology Transactions* **34**, 187–194 (1991).

[6] L. Gustafsson, E. Höglund and O. Marklund, Measuring Lubricant Film Thickness with Image Analysis, *Proceedings Institution of Mechanical Engineers, Part J: Journal of Engineering Tribology* **208**, 199–205 (1994).

[7] P.M. Cann, H.A. Spikes and J. Hutchinson, The Development of a Spacer Layer Imaging Method (SLIM) for Mapping Elastohydrodynamic Contacts, *Tribology Transactions* **39**, 915–921 (1996).

[8] J.B. Luo, S.Z. Wen and P. Huang, Thin film Lubrication 1. Study on the Transition between EHL and Thin Film Lubrication Using a Relative Optical Interference Intensity Technique, *Wear* **194**, 107–115 (1996).

[9] M. Hartl, I. Křupka, R. Poliščuk, M. Liška, J. Molimard, P. Vergne and M. Querry, Thin Film Colorimetric Interferometry, *Tribology Transactions* **44**, 270–276 (2001).

[10] F. Guo and P.L. Wong, A Multiple-Beam Intensity-Based Approach for Thin Lubricant Film Measurement in Non-Conformal Contacts, *Proceedings Institution of Mechanical Engineers, Journal of Engineering Tribology* **216**, 281–291 (2002).

[11] M. Smeeth, P.M. Cann and H.A. Spikes, Measurement of Elastohydrodynamic Film Formation in Rolling Contacts at Very High Pressure, in *Lubricants and Lubrication, Proceedings of the 21st Leeds–Lyon Symposium on Tribology*, pp. 497–502 (1995).

[12] M. Smeeth and H.A. Spikes, Central and Minimum Elastohydrodynamic Film Thickness at High Contact Pressure, *Trans. of the ASME, Journal of Tribology* **119**, 291–296 (1997).

[13] I. Křupka, M. Hartl, R. Poliščuk and M. Liška, Experimental Study of Central and Minimum Elastohydrodynamic Film Thickness by Colorimetric Interferometry Technique, *Tribology Transactions* **43**, 611–618 (2000).

[14] M. Hartl, I. Křupka and M. Liška, Direct Observation of the Behaviour of Real Surface Roughness Passing through EHD Point Contact, in *Transient Processes in Tribology, Proceedings of the 30th Leeds–Lyon Symposium on Tribology*, pp. 419–428 (2004).

[15] B.O. Jacobson and P. Vinet, A Model for the Influence of Pressure on the Bulk Modulus and the Influence of Temperature on the Solidification Pressure of Liquid Lubricants, *Trans. of the ASME, Journal of Tribology* **109**, 709–713 (1987).

[16] X. Ai, Numerical Analyses of Elastohydrodynamically Lubricated Line and Point Contacts with Rough Surfaces by Using Semi-System and Multigrid Methods, Ph.D. Dissertation, Northwestern University, Evanston, IL (1993).

[17] Y.Z. Hu and D. Zhu, A Full Numerical Solution to the Mixed Lubrication in Point Contacts, *Trans. of the ASME, Journal of Tribology* **122**, 1–9 (2000).

[18] D. Zhu and Y.Z. Hu, A Computer Program Package for the Prediction of EHL and Mixed Lubrication Characteristics, Friction, Subsurface Stresses and Flash Temperatures Based on Measured 3-D Surface Roughness, *Tribology Transactions* **44**, 383–390 (2001).

CORRELATION BETWEEN MODEL TEST DEVICES AND FULL BEARING TESTS UNDER GREASE LUBRICATED CONDITIONS

H. Baly[1], G. Poll[1], P.M. Cann[2] and A.A. Lubrecht[3]
[1]*Institute for Machine Elements, Engineering Design and Tribology, Hannover University, Germany*
[2]*Tribology Section, Department of Mechanical Engineering, Imperial College London, U.K.*
[3]*Laboratoire de Mécanique des Contacts et des Solides, UMR-CNRS 5514, INSA de Lyon, France*

Abstract This paper compares friction and film thickness measurements from a full-scale bearing test device and optical ball-on-disc results. The lubricants used were commercial greases with a range of base oil viscosities and thickeners formulated for high-speed applications. The bearing tests operated in the semi-starved regime over most of the speed range. The film thickness evolution shows the similarity in starvation curves, as well as the influence of the running-in speed, on subsequent runs. Indirectly the paper shows the influence of spin on relubrication and that lubricant reflow occurs primarily in the vicinity of the contact.

Keywords: rolling element bearings, grease lubrication, friction, starvation.

1. INTRODUCTION

In most areas of Tribology model laboratory tests are routinely used to provide an understanding of fundamental mechanisms and to develop new lubricant formulations. Examples include the many and varied tests for fuel lubricity [1], gear oil additives [2] and engine oils [3]. These simple laboratory devices do not replace application testing, however they do provide an assessment of the lubricant performance under controlled conditions. It is of course vitally important to establish a relationship between the laboratory test results and performance in field service. Such a correlation has been reported for many model test devices [1, 4] and these tests are incorporated into research and development strategies for new lubricants. This is not the case however for grease lubrication of rolling element bearings. One exception is a recent paper presented by Nakajima and co-authors where a combined laboratory and field

R.W. Snidle and H.P. Evans (eds), IUTAM Symposium on Elastohydrodynamics and Micro-elastohydrodynamics, 229–240.

service approach is used to develop new grease formulations for steel rolling mill bearings [5]. There are very few papers in the open international literature reporting grease lubrication performance in bearing tests [6, 7]. Therefore, the majority of the published grease lubrication studies have used laboratory machines [8–12]. Hence a legitimate question is whether the model test device can simulate full bearing behaviour. Furthermore the open literature provides few answers relating film thickness or friction measured in bearing tests to the results from the model device studies. However the operational film thickness is an important parameter as it affects both surface damage and fatigue [13].

One conclusion from both types of test is that the most important parameter determining grease film thickness is the inlet supply condition. Grease has a yield stress, and once it has been pushed aside by the passage of the rolling element, it will not readily flow back to replenish the track. Reflow, both on the free surfaces and around the ball, is surface tension driven and is retarded by the lubricant viscosity [14]. If the time interval between overollings is too short, replenishment will be incomplete. This can result in starvation of the contact and a severe reduction in film thickness [12]. Earlier workers have shown that grease film thickness in bearings is often less than predicted using classical (fully flooded) equations and the base oil viscosity; this suggests that the bearing operates in the starved regime [6]. Starvation occurs most readily at low temperatures, high rotational speeds or high-aspect ratio contacts. It is an important cause of lubricant-related failure in many bearing systems [15] but also the key to operation with continuously low friction. In some bearing types additional relubrication mechanisms occur, for example spin and cage effects [15, 16]. Particular operating conditions such as heavy vibration can also resupply the contact [15].

The question remains – can a model test device operating under laboratory test conditions successfully predict the performance of a grease in a bearing under industrial conditions? Some of the authors have claimed that the answer is affirmative providing the major failure mechanism (starvation in this case) is "scaled" correctly from the application to the test machine [17]. This paper addresses the relationship between full-scale bearing tests and laboratory simulation.

2. EXPERIMENTAL PROGRAMME

A series of tests has been carried out with a bearing device to establish film thickness and torque with different greases for a range of speed conditions. The results have been compared to measurements from tests run in a bearing simulation device.

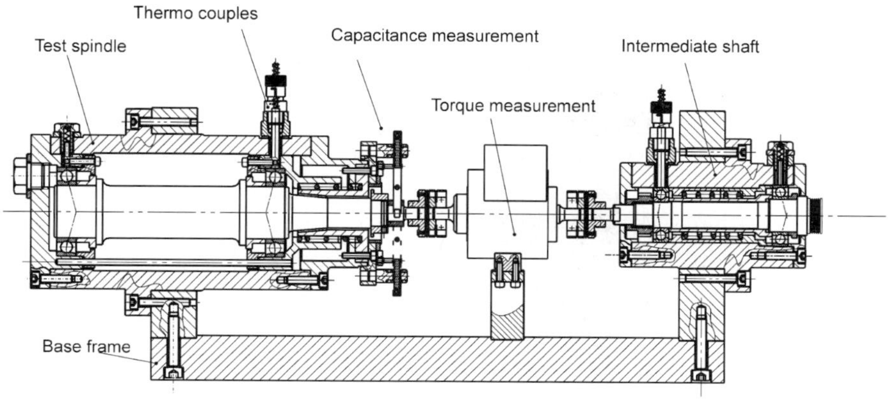

Figure 1. Bearing test device.

2.1 Bearing Test Device

Figure 1 shows the test rig with purely axially loaded angular contact ball bearings in the test spindle at the Institute for Machine Elements, Engineering Design and Tribology of Hannover University. The frictional torque in the two test bearings is measured through the torque meter. Film thickness in the ball/raceway contact is measured by a capacitance method.

The test bearings are angular contact ball bearings with phenolic cages under an axial load of 160 N. The speed range was 0–18000 rpm. Four different greases were tested, the first three were designated A, B and C with a mineral base oil of relative viscosity: $\eta_A = 2$, $\eta_B = 3$ and $\eta_C = 6$. Grease A has a polyurea thickener, whereas greases B and C have a lithium-complex-thickener. A fourth grease designated D contained a barium-complex-thickener with ester base oil.

2.2 Model Bearing Test Device

The effect of operating parameters on starvation was studied at the Tribology Section, Imperial College in a ball-on-disc device supplied by PCS Instruments. Thin film optical interferometry [18] was used to measure central film thickness as a function of speed for a range of test conditions. A diagram of the test rig is shown in Figure 2. The device uses a steel ball loaded against a glass disc to simulate a single bearing contact. In most cases, the contact pressures and rotational speeds are much lower than found in practice.

All tests were run at a bulk temperature of 60°C and a maximum Hertz pressure of 0.5 GPa. In this rig configuration a ball is driven by the disc in nominal pure rolling. The ball is supported on three angled bearings so that the rotational axis of the ball coincides with the disc. Two series of tests were carried

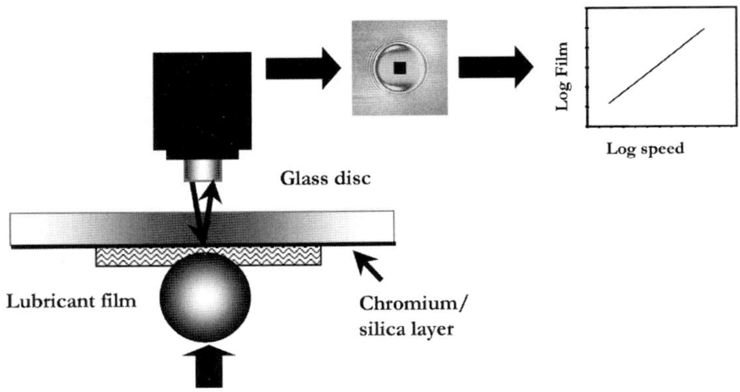

Figure 2. Model bearing test device.

out; in both cases a small amount of grease (2–3 g) was initially distributed around the disc and the rig temperature allowed to stabilise before measurements taken.

Series 1: Tests were run without an external lubricant resupply mechanism. Speed was slowly increased until starvation occurred. The speed was then decreased and held constant at 5 mm/s to allow time for the lubricant to redistribute. The speed was then increased as before. The speed cycle was repeated twice.

Series 2: Tests were run with the bearing support offset at an angle. This in effect introduces ball spin and is known to be a replenishment mechanism in some bearings. The speed cycle was the same as Series 1.

3. BEARING FRICTION AND FILM THICKNESS RESULTS

3.1 Bearing Tests – Friction and Torque

Results from the bearing test device are plotted in Figures 3–6. Figure 3 shows a typical friction torque versus speed curve for the three greases with different base oil viscosity at very low axial load. Under these conditions, the measured friction is mainly due to losses in the lubricant.

The duration of the measurements at each rotational speed depended on the measured friction torque and temperature. The measurements were continued until steady state conditions were established. During this process, there were several speed cycles between zero and maximum speed, until the measured friction between two subsequent cycles no longer differed. Then, at each speed level a measurement of the friction torque was taken after the torque had stabilized. The measurement cycle was repeated at least twice. Preceding these

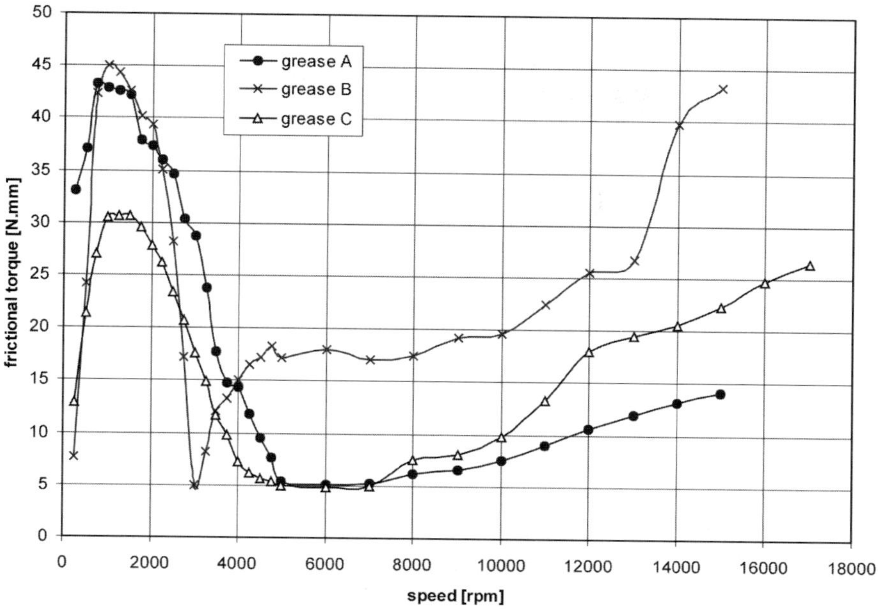

Figure 3. Measured frictional torques for three different greases at low axial load.

measurements, a running-in procedure was observed. During this process, the speed was increased in steps of 1000 rpm up to the maximum speed. At each speed level friction torque and temperature were allowed to stabilize.

Three friction regimes can be identified depending on the operating speed:

- At low speeds: as predicted by classical theory, the friction increases with speed.
- At medium speeds: beyond 500 to 1000 rpm, however, this tendency is inverted and the friction drops with speed.
- At high speeds: from 3000 to 7000 rpm upwards, the friction increases again, but at a much lower rate. In the high speed regime it is likely that other mechanisms dominate the friction results.

Figure 4 shows the measured film thicknesses corresponding to Figure 3.

The film thickness increases as predicted up to speeds of 500 to 1000 rpm, but then starts to drop. Below 500 rpm film thickness follows the classical fully flooded predictions and scales with base oil viscosity [19]. From 500 to 1000 rpm starvation occurs and the film thickness decreases with increasing speed. Between 1000 and 5000 rpm the film thickness is approximately constant. At speeds beyond 5000 rpm other phenomena dominate associated with the dynamics of the bearings used.

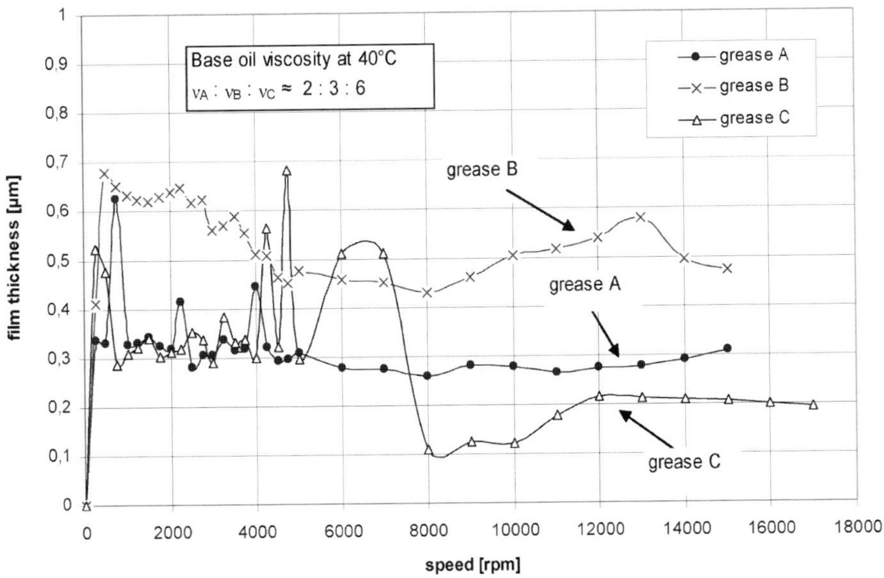

Figure 4. Measured film thickness as a function of speed for three different greases.

3.2 Bearing Friction and Maximum Speed

The results plotted in Figures 5 and 6 show the influence of the maximum speed attained during the running-in process of the ball bearings. In both cases the friction and film thickness decrease when the higher speed was used during the running-in period. This implies that starvation is more prone to occur after running at high speeds.

3.3 Bearing Friction and Number of Rolling Elements

Figures 7 and 8 show the influence of the number of balls and thus the time elapsed between subsequent overrollings. The axial load was adapted in such a manner that the load per ball remained constant. Not surprisingly, the friction torque was lower with fewer balls up to 7000 rpm; albeit not by the same factor as the number of balls. Above this speed the friction curve is similar, most likely due to dynamical effects invoked previously.

Figure 8 shows only slightly lower film thickness levels with fewer balls over parts of the speed range. If the film thickness depended on the replenishment between overrollings, as predicted by [20] the 8-ball experiment should show a twice-thicker film. This is obviously not observed.

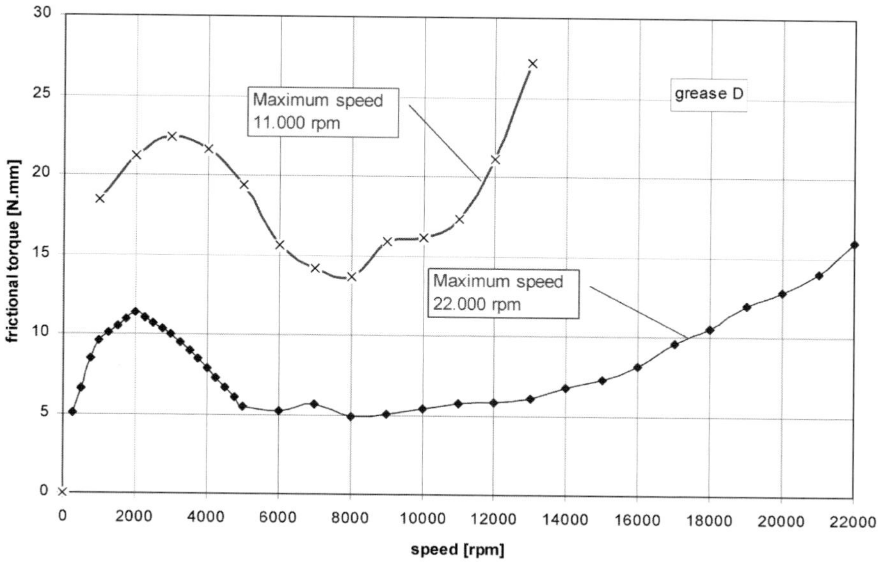

Figure 5. Influence of the maximum speed reached on the friction torque.

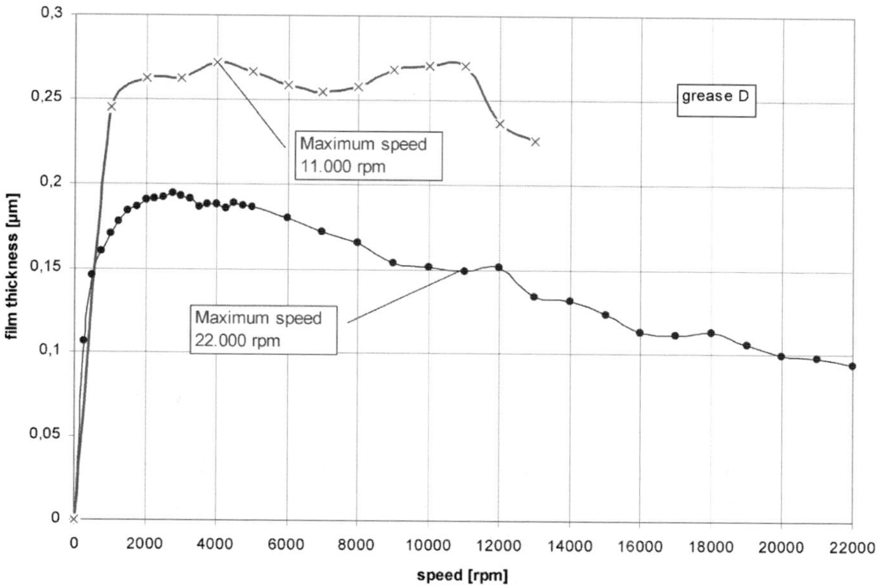

Figure 6. Influence of the maximum speed reached on the film thickness.

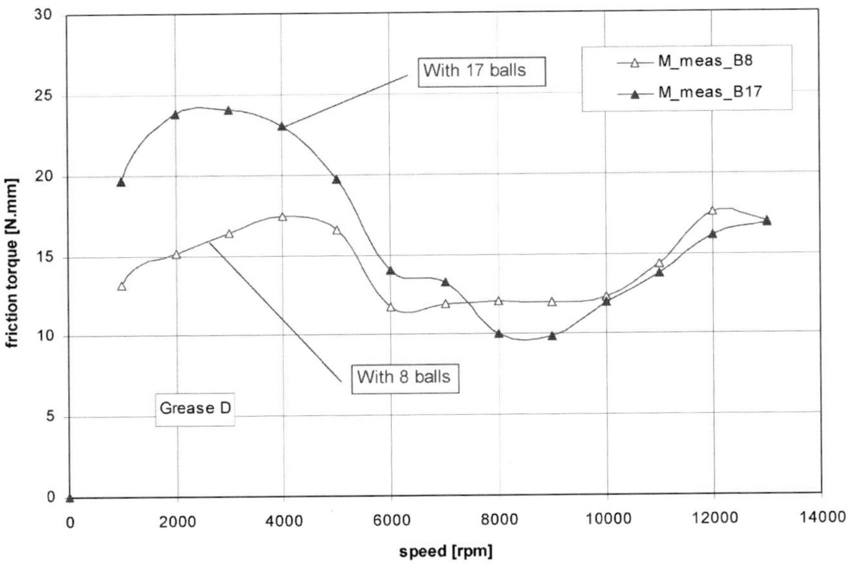

Figure 7. The influence of the number of balls on the measured friction torque.

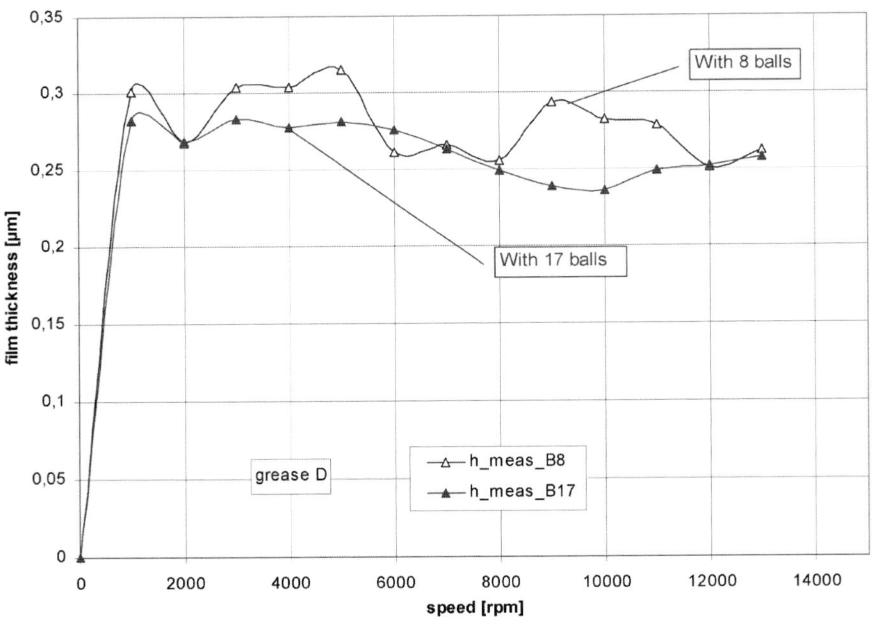

Figure 8. The influence of the number of balls on the measured film thickness.

3.4 Model Test Results

Figures 9–11 show representative results from the model bearing device. Central film thickness as a function of speed is plotted for grease A in Figure 9. Four

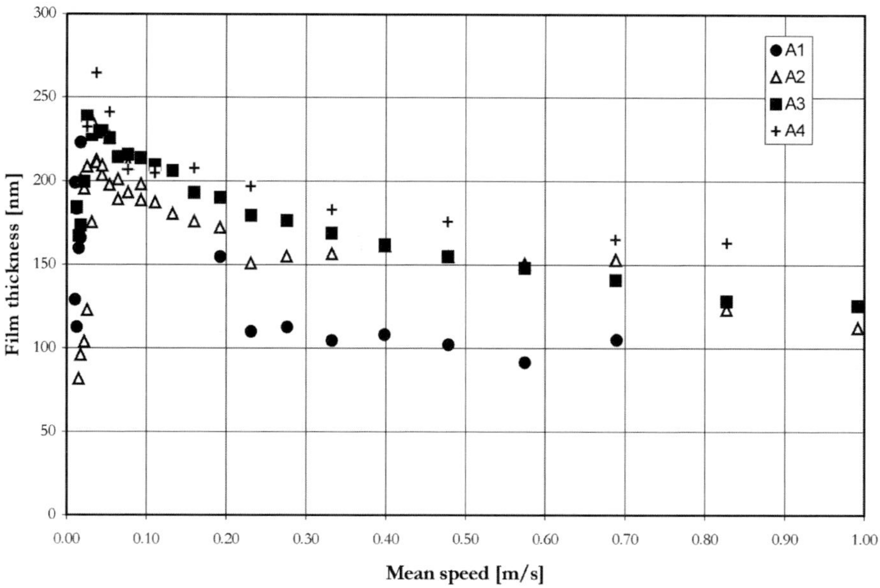

Figure 9. Film thickness as a function of speed – starvation curve, grease A.

successive speed increase curves are shown. Initially film thickness increases with speed, however at speeds above 0.02 m/s starvation occurs and the film thickness drops. It is interesting that the starvation speed (defined as the speed at which the film thickness starts to decrease) increases, with each subsequent speed cycle, from 0.017 to 0.04 m/s. The film thickness in the starved regime also increases from 100 to 160 nm. This would imply that lubricant replenishment is improving as the test continues due to shear degradation of the grease close to the contact [21].

The effect of running-in speed on starvation and film thickness levels can be seen in Figure 10. Two test curves are shown for grease B, B1 was taken after running-in at 0.01 m/s, curve B3 after running at 1 m/s. A classical starvation curve was obtained for B1, however B3 was severely starved throughout the speed sweep.

The effect of ball spin on film thickness is shown in Figure 11. Speed sweeps were run in a similar fashion to the results shown in Figures 9 and 10, however very different lubrication behaviour was obtained.

When spin was present grease A gave fully flooded results up to the maximum speed of 2.7 m/s, this compares to a starvation speed of 0.04 m/s obtained without spin (Figure 9). The first speed curve with grease B was fully flooded up to 1.5 m/s, subsequent sweeps starved at 0.5 m/s. Obviously spin provides

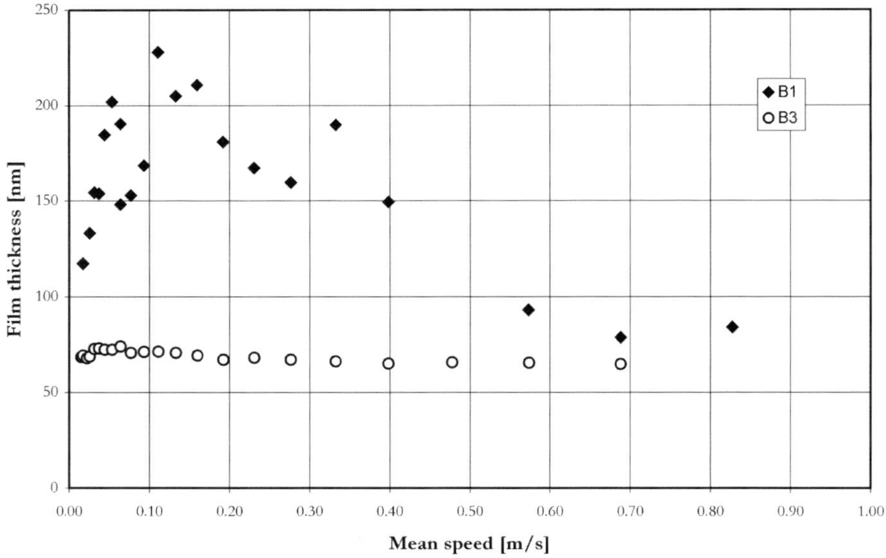

Figure 10. Film thickness as a function of speed – effect of running-in speed, grease B. Before B1 the test was run for 5 minutes at 0.01 m/s and B3 was measured after 5 minutes at 1.0 m/s.

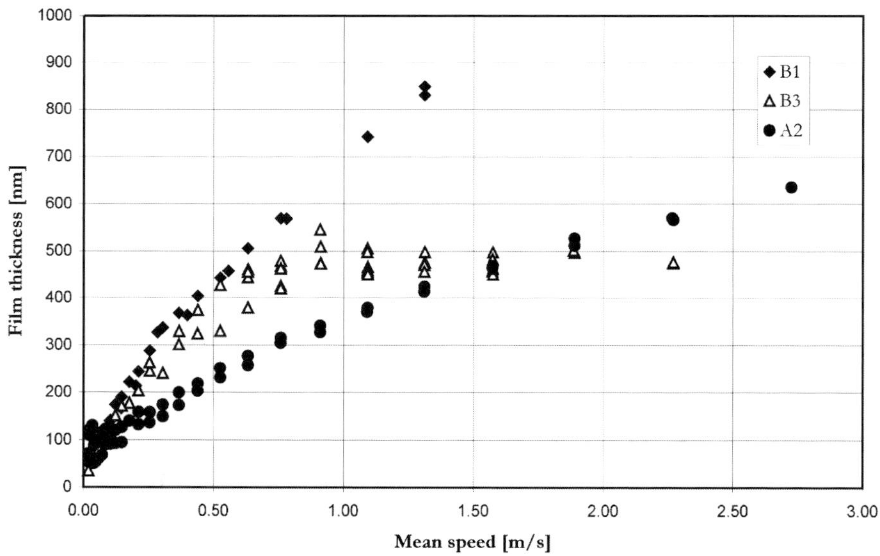

Figure 11. Film thickness as a function of speed – effect of ball spin, grease A and B. B1 first speed curve, B3 third speed curve. A2 second speed curve.

an important lubricant supply mechanism, similarly the presence of a cage can also significantly increase starvation speed [16].

4. DISCUSSION

The friction and film thickness measurements in Figures 3 and 4 clearly show that the bearings operate in the semi-starved regime over most of the speed range. A similar pattern is observed in the model test results in Figure 9 although there is a shift in the starvation speed. These results were taken in the absence of any bearing-related lubrication supply. However, Figure 11 shows that spin alone can increase the starvation speed by a factor of ten.

Furthermore, a similar effect of the running-in speed on the lubricant supply, is observed in the full bearing (Figures 5 and 6) and model tests (Figure 10). As a consequence of the high speed, the grease is expelled from the vicinity of the track, and hence subsequent tests will be heavily starved.

Figures 7 and 8 show the influence of the number of balls on friction and film thickness.

In the starved regime the film thickness is not influenced by the number of balls. This confirms the hypothesis used in [17] that the main relubrication mechanism is local to the contact.

5. CONCLUSION

Bearing test rigs can provide valuable information on bearing behaviour under realistic operating conditions. In turn, the advantage of laboratory test devices is the possibility to study in detail all the mechanisms and processes involved through direct observation, whereas measurements in complete bearings only deliver overall information. Therefore, the two methods are complementary, not alternatives. Without laboratory tests it is not possible to understand the mechanism, but measurements on complete bearings are still needed to generate industrial data. For example, the velocity at which the onset of starvation is observed, depends very much on the presence of a cage and at higher speeds of cage vibration It is necessary to extend the theory developed in [17] to bearing geometries, operating conditions and include the presence of the cage.

ACKNOWLEDGEMENT

The authors wish to thank FVA e.V. (Research Association for Drive Systems) for the kind permission to use data from a joint research project.

REFERENCES

[1] C. Bovington, R. Caprotti, K. Meyer and H.A. Spikes, "Development of a laboratory test to predict lubricity properties of diesel fuels and its application to the development of

highly refined diesel fuels", *Tribo Test*, **2**, 93–112 (1995).

[2] R.P.S. Bisht and S. Singhal, "Laboratory technique for the evaluation of automotive gear oils of API GL-4 level", *Tribo Test*, **6**, 69–77 (1999).

[3] S. Gunsel, M. Smeeth, S. Korcek and H.A. Spikes, "Elastohydrodynamic friction and film forming properties of lubricant base oils", *STLE Trib. Trans.*, **49**, 569–579 (1999).

[4] W. Dan Ping, S. Korcek and H.A. Spikes, "Comparison of the lubricity of gasoline and diesel fuels', *SAE Special Publications Gasoline Performance and Deposit Control Additives*, **1214**, 51–59 (1996).

[5] S. Nakajima, K. Izaki, K. Takemura, Y. Kimura and P.M. Cann, "Development of new urea grease for steel continuous casting machine under boundary lubrication", *NLGI Spokesman*, **68**(8), 8–16 (2004).

[6] A.R. Wilson, "The relative thickness of grease and oil films in rolling bearings", *Proc. I. Mech. E.*, **193**, 185–191 (1979).

[7] B.R. Williamson and D. Miller, "Condition monitoring of grease lubricated rolling element bearing", *NLGI Spokesman*, **63**, 8–15 (1999).

[8] L. Wedeven, Ph.D. Thesis, University of London (1975).

[9] F. Rastegar and W.O. Winer, "On the traction and film thickness behaviour of greases in concentrated contacts", *NLGI Spokesman*, August, 162–170 (1986).

[10] H. Åström, J.O. Ostenson and E. Höglund, "Lubricating grease replenishment in an elastohydrodynamic point contact", *ASME J. Trib.*, **115**, 501–506 (1993).

[11] P.M. Cann, "Starvation and reflow in a grease lubricated elastohydrodynamic contact", *STLE Trib. Trans.*, **39**, 698–704 (1996).

[12] S. Hurley and P.M. Cann, "Starved lubrication of EHL contacts – Relationship to bulk grease properties", *NLGI Spokesman*, **64**, 15–23 (2000).

[13] E. Zaretsky (ed.), *STLE Life Factors for Rolling Bearings*, STLE Publ. SP-34 (1992).

[14] B. Jacod, F. Publier, P.M. Cann and A.A. Lubrecht, "Track replenishment in and out of contact", in *Proceedings 25th Leeds–Lyon Symposium on Tribology*, pp. 483–492 (1999).

[15] P.M. Cann and A.A. Lubrecht, "Analysis of grease lubrication in rolling element bearings", *Lubrication Science*, **11**, 227–245 (1999).

[16] B. Damiens, P.M. Cann and A.A. Lubrecht, "Influence of cage clearance on bearing lubrication", *STLE Trib. Trans.*, **47**, 2–6 (2004).

[17] P.M. Cann, B. Damiens and A.A. Lubrecht, "The transition between fully flooded and starved regimes", *Tribology International*, submitted.

[18] G.J. Johnston, R.W. Wayte and H.A. Spikes. "The measurement and study of very thin lubricant films in concentrated contacts", *Trib. Trans.*, **34**, 187–195 (1991).

[19] B.J. Hamrock and D. Dowson, "Isothermal elastohydrodynamic lubrication of point contacts", *Journal of Lubrication Technology, Trans. ASME*, **99**, 264–276 (1977).

[20] Y.P. Chiu, "An analysis and prediction of lubricant starvation in following contact systems", *ASLE Trans.*, **17**, 22–28 (1974).

[21] J.-S. Mérieux, S. Hurley, A.A. Lubrecht and P.M. Cann, "Shear-degradation of grease and base oil availability in starved EHL lubrication", in *Proceedings 26th Leeds–Lyon Symposium on Tribology*, pp. 581–588 (2000).

EXPERIMENTAL AND THEORETICAL APPROACHES TO THIN FILM LUBRICATION PROBLEMS

I. Lee-Prudhoe[1,†], C.H. Venner[2], P.M. Cann[3] and H. Spikes[3]
[1] *PCS-Instruments, London W3 7SZ, U.K.*
[2] *University of Twente, P.O. Box 217, 7500 AE Enschede, The Netherlands*
[3] *Imperial College, Tribology Section, London SW7 2BX, U.K.*

Abstract The last fifty years have seen tremendous advances in the field of thin film lubrication. This is particularly true of steady-state Elastohydrodynamic lubrication (EHL) where the accurate measurement of film thickness combined with advanced numerical modelling methods has been very successful. However there are still many challenges to be faced in areas of practical importance for example transient operation, rough surfaces, lubricant failure and two-phase fluids. Such systems present many experimental problems and it is often difficult to obtain detailed measurements of film formation under such conditions. This insight is essential if we are to develop physically realistic models to describe these processes and predict lubrication performance. The aim of this paper is to report on the continued development of experimental methods to study thin film lubrication and to compare the results from this work to current models. The paper will also discuss the limitations of the existing experimental and modeling capabilities and the new challenges to be tackled.

Keywords: elastohydrodynamic lubrication, theoretical predictions, starved lubrication, grease lubrication, additised films.

1. INTRODUCTION

One of the most important research topics in Tribology is that of EHL and thin film lubrication, which has seen tremendous advances through the parallel, and complementary, development of modelling and experimental capabilities. In the late 1950s Dowson and Higginson reported numerical solutions to the EHL problem [1, 2]. This work showed for the first time details of the film thickness variation in the contact and the relationship between film thickness

†Deceased 3 March 2003.

R.W. Snidle and H.P. Evans (eds), IUTAM Symposium on Elastohydrodynamics and Micro-elastohydrodynamics, 241–255.

and operating conditions for line contacts. In the early 1960s these advances were matched by the development of new experimental techniques by Gohar and Cameron [3, 4]. Optical interferometry was used to measure and map EHL films and thus verify many of the model predictions. In the next few years these methods were applied to a wide range of lubrication conditions of practical importance, this included starved lubrication [5], grease [6] and rough surfaces [7]. Initially the technique was limited to films of greater than 100 nm, however this restriction was also overcome by the spacer layer method of Westlake and Cameron [8]. However, because of technical problems it was a number of years before it could be routinely exploited.

Thus by the end of the 1970s experimental techniques had largely overtaken modelling capabilities through the improved simulation of real conditions. Numerical methods suffered from instability of algorithms, slow convergence and limited computer speeds and as a result were restricted to simple steady state line contact problems. However in 1977, Dowson and Hamrock [9, 10] published the first solutions for the circular contact problem, which was also extended to the starved condition.

The 1980s saw significant improvements to the numerical techniques. Firstly, there were developments in algorithms for line contact problems [11]. Also non-Newtonian rheology was also incorporated into the film thickness solvers [12]. The research at this time was still mainly concerned with line contacts. Because of computational difficulties only few studies of point contacts appeared. Evans and Snidle [13] successfully applied an inverse approach for highly loaded point contacts and Chittenden et al. [14] studied the elliptic contact problem. However, the major advances made in this period were through the application of multi-grid techniques to line [15] and point contact [16] problems. For the first time it was possible to obtain point contact solutions with high resolution on a small computer in an acceptable time. The considerable improvement to spatial resolution allowed local surface features to be studied [17]. As a result, the modelling again outstripped our experimental capabilities. Further refinements were achieved in the stability and speed of the computations and in the evaluation of elastic deformation [18]. These developments were aided by faster hardware, which allowed evaluation of more complex problems, for example the time dependent point and line contacts with moving surface features [19]. In recent years these techniques have become widely adopted in the field [20].

The surge in modelling expertise outstripped experimental capabilities. However the early 1990s saw further development of the thin film technique [21]. The consequent improvement in film thickness resolution and sensitivity ($\sim$ 1 nm) opened up new areas of study and the technique has been applied to wide range of problems [22]. Automated imaging techniques were also developed for both normal [23, 24] and thin film optical interferometry [25,

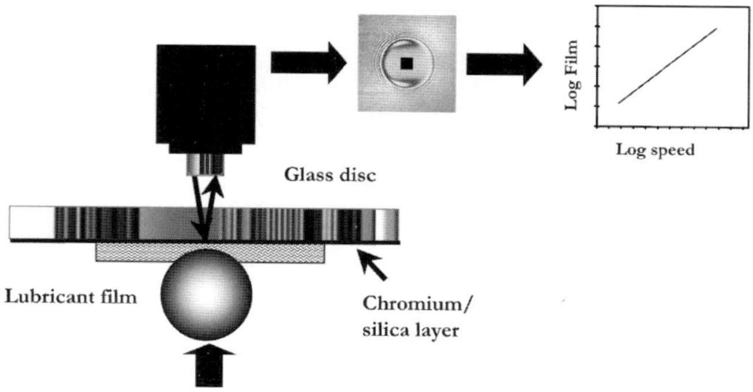

Figure 1. Diagram of thin film test device.

26], which gave detailed maps of the Hertzian contact. Recent work has used multiple-reflection techniques to measure sub-nanometre lubricant films [27].

The parallel experimental and numerical development has been directed towards solving problems of increasing realism and reducing film thickness. The ability to study thin film problems is important as this is linked to surface damage and failure, and many new applications (magnetic discs, MEMS) operate under thin film conditions. Recently, attention has turned to increasingly complex problems that simulate more closely lubrication in real life. For example the prediction of film properties or failure with rough surfaces, transient or starved lubrication and boundary films. Such knowledge is of great practical importance and the development of valid models requires physical insight into lubrication mechanisms particularly under very thin film conditions where the fluid film continuum assumptions break down. The aim of this paper is to report on the continued development of optical interferometric methods to study thin film lubrication and to compare the results from this work to current models. The discussion will also identify areas where the existing models break down and experimental methods can only provide limited information. It is in these areas that future research efforts need to be concentrated.

2. EXPERIMENTAL METHODS

EHL film thickness measurements were made using the ultra [21] and SLIM [25] spacer layer methods. The experimental set-up is illustrated in Figure 1.

Results from other studies carried out in the Tribology Section at Imperial College are also presented with the appropriate reference given in these instances. Results are presented for a number of test conditions, which range from the simple case of smooth steady state operation to transient rough surfaces and film reaction formation. In each case for which models are available

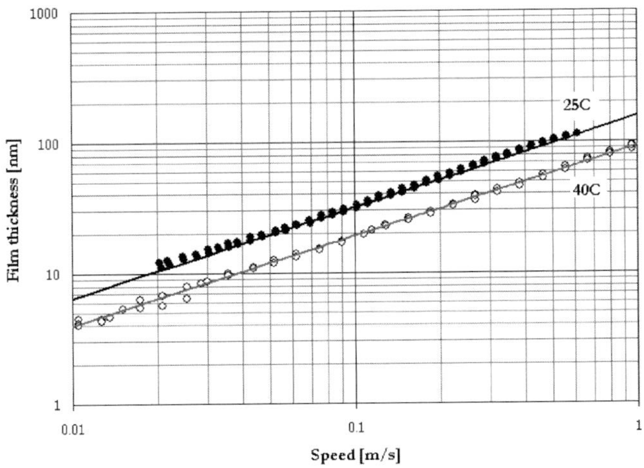

Figure 2. Central film thickness as a function of mean speed for a simple hydrocarbon, comparison of measured and computed values.

the experimental results are compared with the numerical predictions, these were obtained with multigrid solvers of the modelling equations [20].

3. RESULTS & DISCUSSION

3.1 Classical, Steady State

The correlation between measured and theoretical EHL film thickness is demonstrated in the following figures. The comparison starts with the simplest test conditions; fully flooded base oil (non polar) for steady state operation and smooth surfaces. Figure 2 shows central film thickness as a function of mean speed for two temperatures for a synthetic hydrocarbon. The experimental results are plotted as symbols and the computed results as a curve-fitted line.

Computed and measured EHL film shapes are compared in Figure 3, where line profiles (cross section, perpendicular to rolling direction) are plotted for two speeds. In both figures very good agreement is observed even in the thin film regime. This illustrates that for simple non-polar fluids the classical continuum approach is valid down to a few nanometers [26–28]. However, the film thickness and shape can be modelled successfully only when the fluid rheological and chemical properties are known and there is an understanding of the physical mechanisms. For example, the classical models fail when the lubricant contains polar or polymer components which is the case in many practical applications. Thus film measurements can show a significant deviation from the classical predictions (using the bulk oil properties) particularly at low speeds or in the boundary regime [29–32].

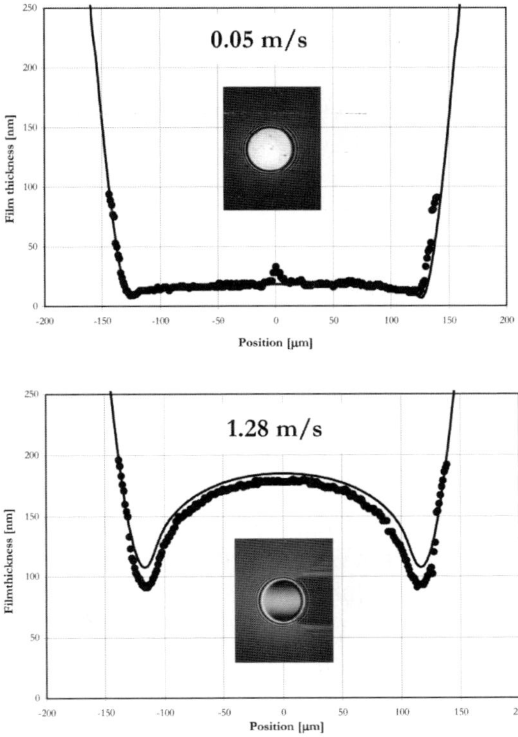

Figure 3. Film profiles (Y-direction) measured and computed for two speeds.

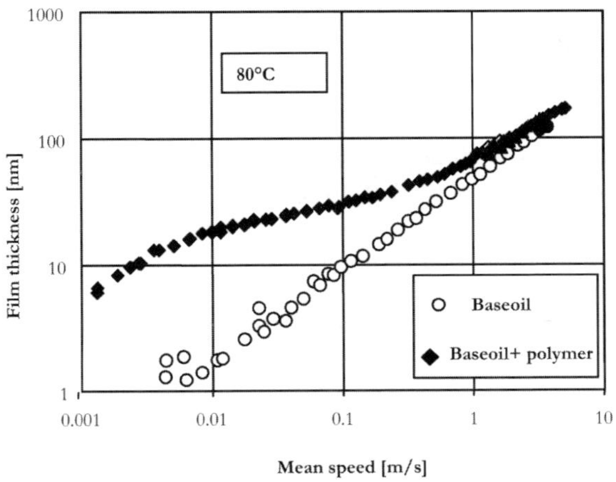

Figure 4. Effect of polymer additives on EHL film thickness from Topolevec Miklozic [31].

An example is shown in Figure 4 where film thickness is plotted for a base oil containing a polymer additive [31]. Film thickness in the near surface regime will be determined by the local lubricant properties; here it is a layer properties of polymer-rich molecules. In many cases the chemical and physical properties will be modified by the presence of the rubbing surfaces or the effect of shear and will thus differ from the bulk lubricant, such considerations are not included in classical EHL theory.

Problems are also encountered in modelling complex rheological fluids [33]. Many lubricants exhibit non-Newtonian behaviour at the high shear and strain rates associated with the EHL inlet region. This can result in a much lower EHL film thickness than predicted if a low-shear rate viscosity is assumed. The most common approach to non-Newtonian behaviour is to use a classical EHL model and replace the viscosity with some effective viscosity that is derived from the constitutive equation describing the shear stress shear strain relation of the lubricant. With the resulting models it is quite easy to show in a theoretical study the effects of non-Newtonian behaviour on the film thickness or the friction. However, it is often difficult to select the most appropriate rheological model for a given lubricant and to obtain values for the parameters in the constitutive equation. Experimental measurements of film thickness profiles and friction do not contain enough information to really validate the rheological predictions. This requires comparison of cross-film quantities such as the velocity profile and acquiring such information for EHL conditions is a major experimental challenge.

3.2 Non-Classical, Steady State

One assumption inherent to the classical EHL models is that the inlet to the conjunction is filled with lubricant. In many applications this is not the case and this results in lubricant starvation of the contact. This is a problem for many systems particularly where repeated contact occurs within very short time scales so that the lubricant is unable to replenish the rolling track. The condition commonly occurs in rolling element bearings particularly at high speeds, low temperatures, or with grease lubrication. Lubricant starvation has been the topic of many theoretical and experimental studies [6, 34, 35] and one of the aims is to be able to predict when starvation will occur for a given lubricant and operating conditions. Starvation is related to the amount of lubricant available, the viscosity and the speed. Passage of the rolling element expels lubricant from the track and this is not completely replenished before the next overrolling. Track replenishment was ascribed to surface tension driven reflow occurring out of contact [34, 35], thus the time between overrollings was identified as a critical parameter in determining the onset of starvation.

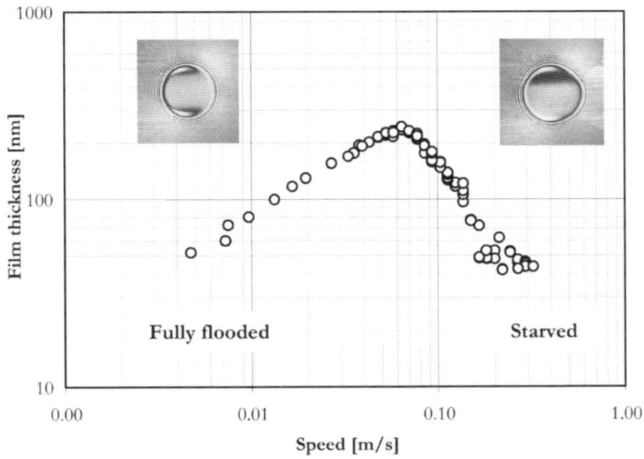

Figure 5. Lubricant starvation – log-log film thickness as a function of speed.

A typical starvation curve is shown in Figure 5 for a high viscosity fluid. At low speeds the results follow the fully flooded predictions. However, at high speeds the film thickness is much reduced. Interferometric images are also shown for the fully flooded and starved regimes. For severely starved contacts the EHL film shape is heavily distorted with the minimum film thickness in the centre of the contact.

Starved lubrication was modelled initially with the classical EHL approach but taking the inlet boundary in the computations as less than the fully flooded distance [6]. However, in severely starved cases the boundary reaches or enters the Hertzian contact region and this cannot be modelled. Experimental observations of the air-oil meniscus have shown it to be curved rather than straight as assumed in the models. In recent work a more physically realistic model has been developed by Chevalier [36, 37]. The parameter controlling starvation is now the amount of lubricant present on the surfaces in the inlet region. The starvation boundary is treated as a free boundary that is allowed to move into the Hertzian contact. Figure 6 shows an experimental interferometric image from a starved contact, the film cross section is also shown together with the predicted profile. Clearly the model accurately predicts the phenomena observed in the experiments. However, this still is only a qualitative agreement. The problem remains of being able to quantify how much lubricant is present in the inlet and to model the mechanisms controlling track replenishment. Early papers [34] cited surface tension driven reflow back into the track whilst out of contact. This mechanism has been observed for tests where a very high viscosity fluid was used and there is a lot of lubricant present in the ridges at the side of the track [38]. However there is little evidence of this for systems where starvation is induced by a reduction in the amount of lubricant present

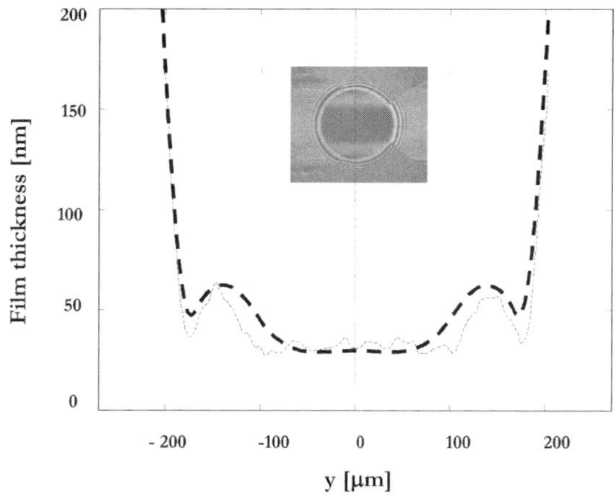

Figure 6. Interferometric image from starved EHL film; measured (· · ·) and predicted (- - -) cross-section from Chevalier et al. [37].

so that the ridges are a few microns in height. Under these conditions it is possible that lubricant flow and replenishment only occurs close to the conjunction [39]. Further experimental work needs to be carried out to measure the lubricant distribution in the track and inlet zone and to quantify replenishment. The flow of lubricant around the contact could be modelled by CFD analysis, first aiming at the reproduction of the characteristic butterfly shape formed around the contact for simple base oils.

3.3 Transient

It is usual to presume steady state operation for EHL calculations, however this is overly simplistic for many real life applications. Changes in speed, load, direction of relative motion and thermal cycling occur during operation of many machine elements, and naturally the effects of surface roughness moving through the contact are transient. These topics are of much interest in relation to vibrations and noise of bearings. It is only recently that the experimental and modelling capabilities to study such changes have been developed.

Several studies have been published concerning the effect of speed and load on variation on film thickness [40, 41]. The results show that variations of the operating conditions can induce significant local changes in the film thickness in the Hertzian contact region. The details of these changes are related to the dynamic behaviour of the loading system. However, when these properties are known the changes can be predicted very accurately as is illustrated in Figure 7.

The study of time dependent effects in EHL due to surface roughness has received much more attention both experimentally [7, 43, 44] and theoretically [17–19, 45]. As a result of advances in computer algorithms [20] and hard-

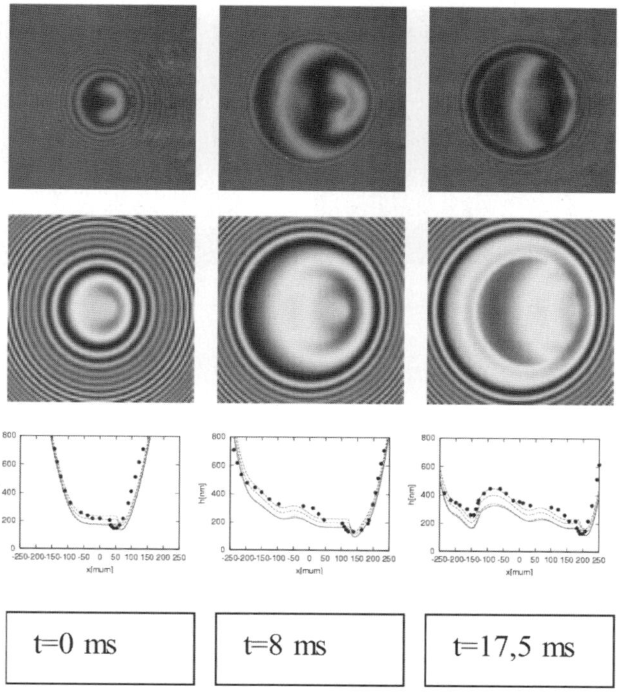

Figure 7. Film thickness at different times after sudden change in the load from 4 to 60 N. Top: measured interferometric image. Centre: interferometric image of computed result. Bottom: Computed centreline profiles for different temperatures (lines) and values from measurement (dots). Experimental results by Sakamoto et al. [41].

ware full time dependent simulations with surface features moving through the contact has become possible even for point contacts. In the thick film regime good agreement between experiment and modelling predictions is seen [42], as is also illustrated in Figure 8.

However for thin lubricant films where the feature dimension is greater than the film height the agreement is not so good. An example is shown in Figure 9 where the computed and measured film profiles are shown for a ridge passing through the centre of the contact. For the case shown the central film thickness is about 40 nm. The amplitude of the bump is 150 nm and width about 40 micrometers. In the numerical simulation an amplitude of 100 nm was used. For large amplitudes the model indicates that contact will occur and this is not incorporated in the model. This difference explains why the valley following the ridge in the experimental result is much deeper than in the numerical result.

The extension of the EHL models to account for (local) contact to occur is still a topic on which the opinions diverge. The main problem is to ensure that the transition from separation to contact is done in a physically accept-

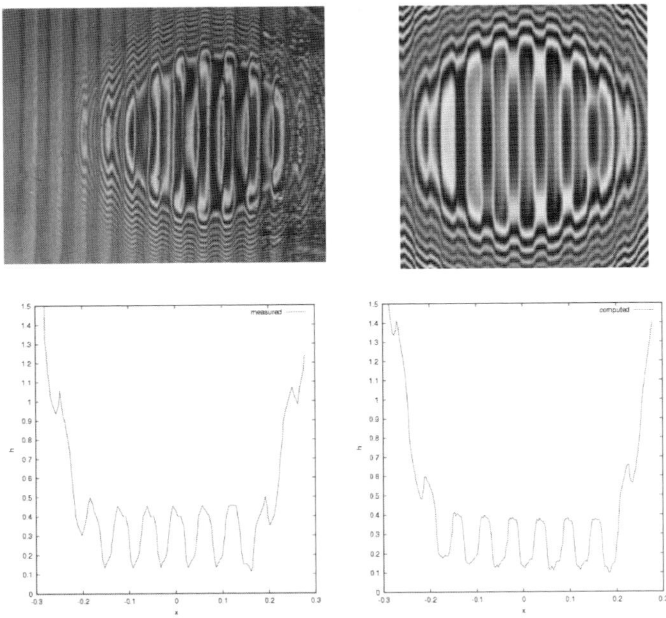

Figure 8. Computed and measured film thickness for the case of an artificial waviness moving through the contact. Experimental results from Kaneta [42].

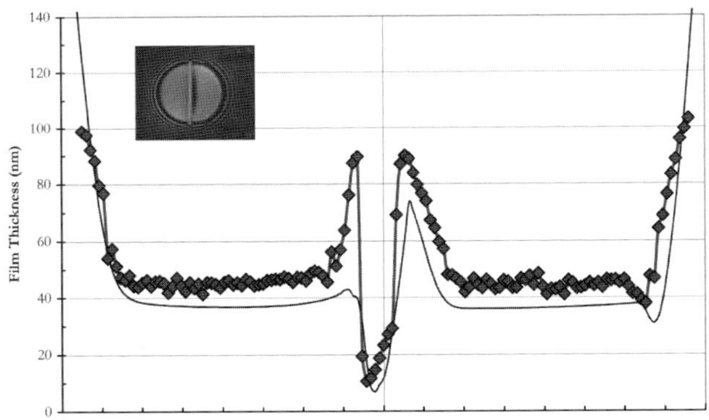

Figure 9. Ridge passing through an EHL contact (fully flooded, pure rolling) – measured and computed plus cross section from Choo [44].

able manner. Whatever model is used, for it to be valid it should be clearly demonstrated that it gives mesh size or grid independent results. Development and validation of such a model is a much-needed precursor for forays into the boundary or mixed lubrication regime. At present this regime can only be studied experimentally.

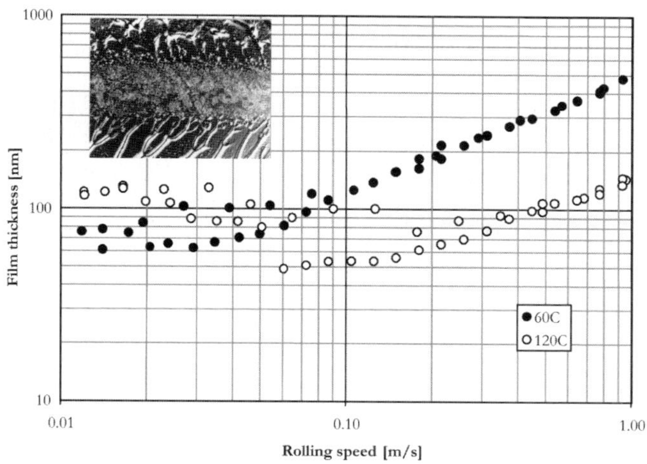

Figure 10. Fully flooded results for lithium grease – a phase contrast microphotograph of the rolled track and deposited layer is also shown.

3.4 Multiphase Fluids

In most applications it is assumed that the lubricant behaves as a single-phase Newtonian fluid, however there are many industrially important lubricants that are an exception. For example grease is a two-phase lubricant containing a solid thickener and base oil that is highly non-Newtonian and is prone to starvation. Models to predict EHL film thickness with grease normally assume fully flooded conditions and continuum flow behaviour [45]. Neither of these assumptions is physically realistic and thus predicting grease lubrication performance in bearings remains a challenge.

For fully flooded conditions where resupply of the contact is maintained grease typically shows two distinct film thickness/speed regimes (Figure 10). In the low speed regime the film thickness is often almost independent of speed and is due to the entrainment of thickener concentrations into the contact. These locally distort the EHL film and this has been linked to grease noise and local pressure fluctuations [47]. The grease structure is shear-degraded in the contact zone and thickener is deposited in the track forming a solid particulate layer shown in the phase contrast image from the glass disc. In many cases this film has low friction properties and maintains surface separation even at zero speed. Thus the lubricating performance of grease cannot be described solely by the base oil properties and this has important implications for many practical applications [48].

At higher speeds grease conforms more closely to fluid behaviour, however the rheology is complex and highly shear rate dependent. Moreover at higher speeds grease is also prone to starvation. The onset of starvation will be de-

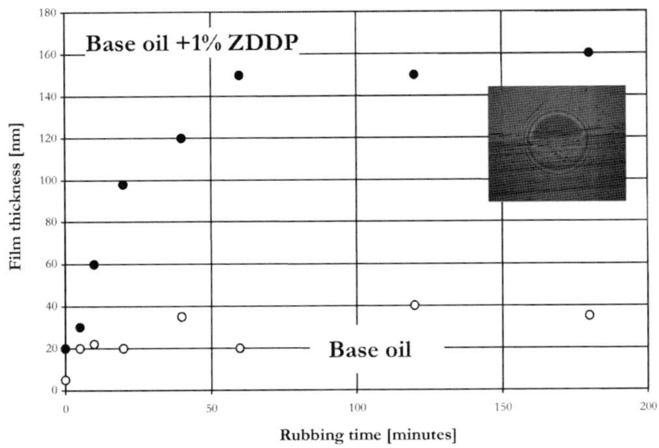

Figure 11. ZDDP reaction film formation at 100°C from Taylor [50].

termined by the properties of the grease (particularly base oil viscosity and shear-stability) and the design and operating conditions of the bearing. Similar effects of phase separation, layer deposition and starvation are also observed with emulsions [49]. Two-phase lubricants combine many of the problems described earlier in this paper, with complex chemical and rheological properties and starvation.

Clearly modelling two-phase lubricant mechanisms in engineering systems is challenging and requires an appreciation of the interaction between the lubricant properties and the machine design and operating conditions. The most important factor determining film thickness is lubricant replenishment of the contact and the role of the grease properties and bearing parameters in controlling this must be quantified if a predictive model is to be achieved.

3.5 Additised Fluids

Modelling of lubricant mechanisms in the thin film regime with additised lubricants represents one of the most difficult challenges in Tribology. There have been some attempts to incorporate additive effects into thin film lubrication models however this has been hampered by the lack of fundamental information concerning the formation and local properties of reacted layers.

One of the problems with additives studies is that rubbed metal surfaces are a necessary precursor to reaction film formation. This is not possible with conventional optical interferometry, however it is possible to observe and quantify additive film formation using out-of-contact imaging [50]. The additive film is formed in a ball-on-disc tribometer with metal test specimens. At different stages of the test the motion is halted and a glass spacer layer disc loaded against the track on the ball. The glass window displaces excess lubricant and it is now possible to use thin film optical interferometry to measure the additive film thickness and distribution [50]. An example of ZDDP reaction film growth is shown in Figure 11. An out-of-contact image is also shown; in this case the optical disc was placed only partly over the rubbed track. The ZDDP film is seen as interference colours in the bottom-half of the contact. It is evident that the reaction film is only formed in the rubbed track.

As the lubricant film thickness decreases the experimental and modelling problems increase. Most models break down due to non-continuity and the nature of the surface chemistry plays an increasingly important role. Thus it is not valid to apply a continuum approach. To be able to predict film formation or failure more information is needed on the chemical and physical properties of reacted layers particularly under high-pressure and shear conditions. Experimental research is needed to provide detailed information on the local adhesion, shear strength and rheology of surface films and the chemical kinetics of additive reactions.

4. CONCLUSIONS

The development of our understanding of EHL and thin film lubrication has benefited from a combined empirical and theoretical approach. Such collaboration becomes even more necessary as the complexity, and thus approximation to reality, of the conditions increases. At present we can only predict EHL film thickness with any degree of confidence for a very limited number of conditions. Once the system approaches the very thin film or failure regime the prediction methods are invalid. There are many experimental and modelling problems that need to be addressed for continued development. For example:

- Experimental – development of new techniques to give:
  - Depth profiling of film properties.
  - Measurement of near-surface rheology.
  - Local measurements of friction and pressure.
  - Chemical/physical properties of thin/reacted films in contact.
  - Chemical reaction kinetics in-contact.
  - Measurements with more realistic surfaces.
- Modelling requires the development of realistic physical models for
  - Local contact varying in time and location.
  - Sub contact scales.
  - Thin lubricant layers and reacted films, formation and behaviour.

- Near-surface rheology (shear stress/shear rate, shear strength).
- Surface modification during running-in.
- Thin film multiphase flow models.

ACKNOWLEDGEMENTS

The authors would like to thank the UK Government Agency EPSRC for travel funding for Dr Venner as part of their Platform Grant programme. The authors would also like to thank their colleagues at Imperial College and of other institutes for generously supplying results.

REFERENCES

[1] D. Dowson and G.R. Higginson, *J. Mech. Eng Sci.*, **1**, 6–12 (1959).

[2] D. Dowson and G.R. Higginson, *Elastohydrodynamic Lubrication, The Fundamentals of Roller and Gear Lubrication*, Pergamon Press, Oxford (1966).

[3] R. Gohar and A. Cameron, *Nature*, **200**, 458–459 (1963).

[4] A. Cameron and R. Gohar, *Proc. Roy. Soc. Lond. A*, **291**, 520–536 (1966).

[5] L.D. Wedeven, D. Evans and A. Cameron, *ASME, J. Lub. Tech.*, **93**, 349–363 (1971).

[6] L.D. Wedeven, Ph.D. Thesis, London University (1971).

[7] A. Jackson and A. Cameron, *ASLE Trans.*, **19**, 50–60 (1976).

[8] F. Westlake and A. Cameron, *Proc. Instn. Mech. Engrs.*, **182**, 75–78 (1967–1968).

[9] B.J. Hamrock and D. Dowson, *ASME, J. Lub. Tech.*, **98**, 223–229 (1976).

[10] B.J. Hamrock and D. Dowson, *ASME, J. Lub. Tech.*, **99**, 264–276 (1977).

[11] M.M. Kostreva, *J. Tribology, ASME Trans.*, **108**, 551–556 (1984).

[12] T.F. Conry, S. Wang and C. Cusano, *J. Tribology, ASME Trans.*, **109**, 648–658 (1987).

[13] H.P. Evans and R.W. Snidle, *J. Tribology, ASME Trans.*, **103**, 539–546 (1991).

[14] R.J. Chittenden, D. Dowson, J.F. Dunn and C.M. Taylor, *Proc. Roy. Soc. Lond. A*, **397**, 245–269.

[15] A.A. Lubrecht, G.A.C. Breukink, H. Moes, W.E. ten Napel and R. Bosma, in *Proc. 13th Leeds–Lyon Symposium on Tribology*, pp. 175–182 (1987).

[16] A.A. Lubrecht, W.E. ten Napel and R. Bosma, *J. Tribology, ASME Trans.*, **109**, 437–443 (1987).

[17] A.A. Lubrecht, Ph.D. Thesis, University of Twente, ISBN 90-9001583-3 (1987).

[18] C.H. Venner, Ph.D. Thesis, University of Twente, ISBN 90-9003974-0 (1991).

[19] C.H. Venner and A.A. Lubrecht, *J. Tribology, ASME Trans.*, **116**, 751–761 (1994).

[20] C.H. Venner and A.A. Lubrecht, *Multilevel Methods in Lubrication*, Elsevier Tribology Series, 37, D. Dowson (ed.), ISBN 0-444-5050-32 (2000).

[21] G.J. Johnston, R.W. Wayte and H.A. Spikes, *Trib. Trans.*, **34**, 187–195. (1991).

[22] H.A. Spikes and P.M. Cann, *Proc. Instn. Mech. Engrs. (Part J)*, **215**, 261–277 (2001).

[23] J. Lord, A. Jolkin, R. Larsson and O. Marklund, *J. Tribology, ASME Trans.*, **122**, 16–22 (2000).

[24] M. Hartl, I. Křupka and M. Liška, *STLE Trib. Trans.*, **42**, 361–368 (1999).

[25] P.M. Cann, J.H. Hutchinson and H.A. Spikes, *STLE Trib. Trans.*, **39**, 915–921 (1996).

[26] M. Hartl, I. Křupka, R. Poliščuk, M. Liška, J. Molimard, M. Querry and P. Vergne, *STLE Trib. Trans.*, **44**, 270–276 (2001).

[27] R.P. Glovnea, A.K. Forrest, A.V. Olver and H.A. Spikes, *Trib. Lett.*, **15**, 217–230 (2003).

[28] G. Guanteng and H.A. Spikes, *STLE Trib. Trans.*, **39**, 448–454 (1996).

[29] V. Anghel, C. Bovington and H.A. Spikes, *Lub. Sci.*, **11**, 313–335 (1999).

[30] G. Guangteng, M. Smeeth, P.M. Cann and H.A. Spikes, *Proc. Instn. Mech. Engrs.*, **210**, 1–15 (1996).

[31] K. Topolovec Miklozic, Ph.D. Thesis University of London (2003).

[32] G. Gao and H. Spikes, *Wear*, **200**, 336–345 (1996).

[33] S. Bair, *STLE Trans.*, **43**, 91–99 (2000).

[34] Y.P. Chiu, *ASLE Trans.*, **17**, 22–35 (1974).

[35] G. Gao, P.M. Cann and H.A. Spikes, *Wear* **153**, 91–105 (1992).

[36] F. Chevalier, A.A. Lubrecht, P.M.E. Cann, F. Colin and G. Dalmaz, in *Proc. 22nd Leeds–Lyon Symposium on Tribology*, pp. 213–223 (1996).

[37] F. Chevalier, A.A. Lubrecht, P.M.E. Cann, G. Dalmaz and F. Colin, *J. Tribology, ASME Trans.*, **120**, 126–133 (1998).

[38] J. Yin, K.I. Eissa, R. Gohar and P.M.E. Cann, *Lub. Sci.*, **12**, 3–30 (1999).

[39] P.M.E. Cann, B. Damiens and A.A. Lubrecht, *Trib. Int.*, **37**, 859–864 (2004).

[40] R.P. Glovnea and H.A. Spikes, *Lub. Sci.*, **15**, 311–320 (2003).

[41] M. Sakamoto, H. Nishikawa and M. Kaneta, in *Proceedings 30th Leeds–Lyon Symposium on Tribology*, pp. 391–399 (2004).

[42] Venner, C.H., Kaneta, M. and Lubrecht, A.A., in *Proceedings 26th Leeds–Lyon Symposium on Tribology*, pp. 25–36 (2000).

[43] G. Gao, P.M. Cann, H.A. Spikes and A. Olver, in *Proceedings 25th Leeds–Lyon Symposium on Tribology*, pp. 175–183 (1999).

[44] J.W. Choo, Ph.D. Thesis University of London (2004).

[45] C.J. Hooke, *Proc. Instn. Mech. Engrs. (Part C)*, **214**, 253–265 (2000).

[46] J.J. Kauzlarich and J.A. Greenwood, *ASLE Trans.*, **15**, 269–277 (1973).

[47] H. Astrom and C.H. Venner, *Proc. Instn. Mech. Engrs. Part (J), J. Eng. Tribol.*, **210**, 259–268 (1996).

[48] S. Nakajima, K. Izaki, K. Takemura, Y. Kimura and P. M. Cann, *NLGI Spokesman* (2004).

[49] A.A. Lubrecht, D. Mazuyer and P.M. Cann, *C.R. Acad. Sci. Paris, Série IV*, **2**, 717–728 (2001).

[50] L.J. Taylor, Ph.D. Thesis, University of London (2001).

EXTENSION OF CONVENTIONAL OPTICAL EHL TECHNIQUE

P.L. Wong[1] and F. Guo[2]
[1]*Department of Manufacturing Engineering and Engineering Management, City University of Hong Kong*
[2]*School of Mechantronic Engineering, Qingdao Polytechnic University, Qingdao, P.R. China*

Abstract The term "conventional" optical EHL technique refers to the simple mechanical set up of a steel ball rolling/sliding on a partially Cr-coated glass disc and illuminated using monochromatic light source, as first introduced by Cameron's group in Imperial College in the 1960s. The technique has been widely used in experimental studies of EHL since its introduction. It gains its popularity not only because of the provision of detailed mapping of the highly pressurized and stressed EHL contact but also its simplicity which, as we recognize, is the beauty of the technique. The film thickness was inferred from the known order of interference fringes, which limits the minimum thickness measurement down to only a quarter of the wavelength of the light source. Recently, the magnitude of both the film thickness measurement and its resolution has been extended to nano-scale, which is achieved by a full optical analysis of the multilayer optical system comprising the lubricated contact, such that the film thickness can be inferred from the fringe intensity instead. The mathematical derivation, at first sight, may be very complicated since it consists of multiplications of complex numbers and lengthy expressions. The aim of this paper is to summarize the derived algebraic equations such that one can easily make use of them. Even though they look lengthy, calculation is quite straightforward and can be readily completed using an ordinary spreadsheet, such as Excel. It is noted that the mechanical set up of the technique and the interferograms obtained are exactly the same as the ones introduced in the 1960s. Thus this current extension of the technique does not obliterate the beauty of its simplicity which was, after all, one of the key considerations in experimental studies as repeatedly stressed by the late Prof. A. Cameron.

Keywords: optical elastohydrodynamics, multi-beam interferometry, thin-film optics.

R.W. Snidle and H.P. Evans (eds), IUTAM Symposium on Elastohydrodynamics and Micro-elastohydrodynamics, 257–268.

1. INTRODUCTION

This paper describes how a conventional optical EHL technique can be used for the measurement of thin film and small local variations in film thickness at the nano-scale. Conventional optical EHL interferometry relies on two-beam interference theory. Film thickness is inferred from the known order of interference fringes. Hence, for film thickness lower than that of the minimum fringe order it is difficult to measure. The two-beam interferometric method is thus limited to film thickness of about 0.15 micron. The extension of the technique for ultra-thin film measurement by using a silica 'spacer' layer was initially proposed in 1970[1]. The idea requires a known thickness of silica to be coated on top of the partially reflective Cr coating. In the last decade, the idea of using a spacer layer has been rapidly developed[2]. Literature reveals that another idea for thin film measurements based on the fringe intensity was mentioned by Roberts and Tabor[3] just a few years after the first publication of the optical EHL technique in the 1960s. The intensity of interferograms changes from the minimum (darkest) to the maximum (brightest) or vice versa between two adjacent interference fringes due to the variation in thickness. The relative intensity is related to the absolute film thickness. One can easily obtain the film thickness from the relative intensity provided that their relation is known. The idea was later applied for thin-film measurements by Luo et al.[4] in the 1990s. They used a simple cosine function for the relation of intensity and film thickness changes based on two-beam interferometry. The problem is that the variation of interferometric intensity with thickness for a typical EHL contact bounded by optically absorptive metal surfaces, steel and Cr, cannot be simply described by a simple cosine function. An incorrect function produces erroneous film thickness measurements[5]. The relation between intensity and film thickness for a conventional optical EHL contact has been derived with consideration of multi-beam interference and the optically absorptive characteristics of Cr-coated and steel surfaces[6, 7]. This paper presents the intensity and film thickness relation in the form of algebraic equations. The film thickness can be inversely calculated by the equation of reflection intensity. Hence, thin-film thickness and small variations in film thickness can thus be inferred from interferograms obtained by the conventional optical EHL technique. Film forming properties of hexadecane and a dilute solution of polybutene (PB2400) in hexadecane measured by the technique and comparison to similar results obtained by other investigators are also presented.

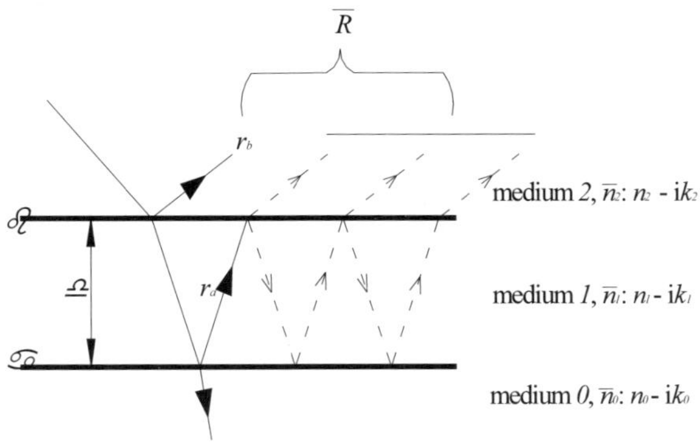

Figure 1. Multi-beam interferometry of a single layer.

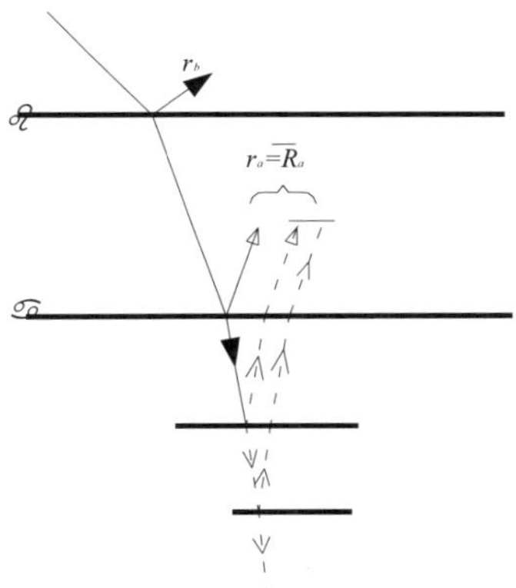

Figure 2. Multi-beam interferometry of a multi-layer system.

2. BASIC OPTICS OF MULTI-LAYER FILM

When a light beam strikes an interface of two media of different refractive indexes, it is divided into reflected and transmitted parts. Considering a layer of thickness, *d,* bounded on either side by semi-infinite layers of various refractive indexes as shown in Figure 1, the reflection *R* of the incident beam is the resultant of the multi-reflected and multi-transmitted elements at the two interfaces, *a* and *b*. For normal incidence, the (amplitude) reflection coefficient *R* can be expressed[8] as:

$$\bar{R} = \frac{r_b + r_a e^{-2i\delta}}{1 + r_b r_a e^{-2i\delta}} \tag{1}$$

where

$$\delta = \frac{2\pi}{\lambda}\overline{n}_1 d \tag{2}$$

r_b and r_a of Eq. (1) are the Fresnel reflection coefficient of interfaces, given respectively as

$$r_b = \frac{\overline{n}_2 - \overline{n}_1}{\overline{n}_2 + \overline{n}_1} \quad ; \quad r_a = \frac{\overline{n}_1 - \overline{n}_0}{\overline{n}_1 + \overline{n}_0} \tag{3}$$

Note that r_a is the Fresnel reflection coefficient as expressed in Eq. (3), only if there is no reflection from any layer below. If not, the reflected wave $\underline{r_a}$ from the interface a should be the resultant of all reflections (the reflection R_a) from the below layer (or layers), as schematically shown in Figure 2.

r_a and r_b are complex numbers and can be written respectively as

$$r_a = g_a + if_a \tag{4}$$

$$r_b = g_b + if_b \tag{5}$$

The exponential term in Eq. (1) can be rewritten as

$$\exp(-2i\overline{\delta}) = e^{-2\alpha}(\cos 2\gamma - i\sin 2\gamma) \tag{6}$$

where $\alpha = \frac{2\pi}{\lambda}k_1 d$ and $\gamma = \frac{2\pi}{\lambda}n_1 d$. Substituting Equations (4), (5) and (6) into Eq. (1), gives

$$\overline{R} = G + iF \tag{7}$$

where

$$G = \frac{g_b + e^{-2\alpha}(1+g_b^2+f_b^2)(g_a\cos 2\gamma + f_a\sin 2\gamma) + e^{-4\alpha}g_b(f_a^2+g_a^2)}{1+2e^{-2\alpha}(g_a g_b - f_a f_b)\cos 2\gamma + 2e^{-2\alpha}(g_a f_b + g_b f_a)\sin 2\gamma + e^{-2\alpha}(g_b^2+f_b^2)(g_a^2+f_a^2)} \tag{8}$$

and

$$F = \frac{f_b + e^{-2\alpha}(1 - g_b^2 - f_b^2)(f_a \cos 2\gamma - g_a \sin 2\gamma) - e^{-4\alpha} f_b (f_a^2 + g_a^2)}{1 + 2e^{-2\alpha}(g_a g_b - f_a f_b)\cos 2\gamma + 2e^{-2\alpha}(g_a f_b + g_b f_a)\sin 2\gamma + e^{-2\alpha}(g_b^2 + f_b^2)(g_a^2 + f_a^2)} \tag{9}$$

The reflectance $\Re$, defined as the percentage of energy of the incident beam being reflected magnitude of the reflectance is thus,

$$\Re = \left|\overline{R}\right|^2 = G^2 + F^2 \tag{10}$$

More detailed analysis can be found in Wong and Guo[5,6]. The following results were calculated using the aforementioned algebraic equations.

3. MULTI-BEAM INTERFEROMETRY OF TYPICAL OPTICAL EHL CONTACT

Typically, optical EHL tests are performed with a steel ball and a partially Cr-coated glass plane in conjunction with a specimen oil. The contact comprises two microscopically thin layers: the chromium coating and the lubricant film, bounded by the glass disc and the steel ball, as schematically shown in Figure 3. It is assumed that the optical absorption of the glass and oil over these microscopic dimensions is negligible, and hence their refractive indexes are described by only a real component. As chromium and steel are metals, their refractive indexes are taken as complex numbers. The optical parameters of the layers are abbreviated as in Table 1.

Table 1. Refractive indices of the EHL contact layers.

Glass	n_{glass}
Cr film	$n_{Cr} - \mathrm{i}\, k_{Cr}$
Oil	n_{oil}
Steel	$n_{steel} - \mathrm{i}\, k_{steel}$

The evaluation of the total reflectance from the multiple layers can be separated into two steps. The first and second steps refer respectively to layers of (1) Cr/oil/steel and (2) glass/Cr/oil. The calculation for the reflectance of each layer follows the aforementioned protocol.

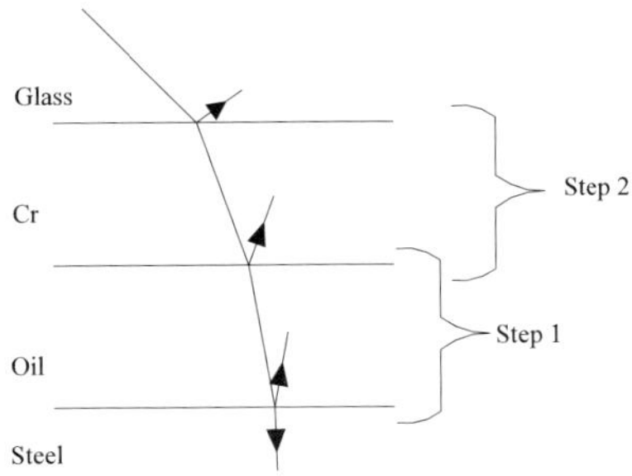

Figure 3. The optical layers in an EHL contact.

First step:
The reflectance from the oil thin layer, bounded by the chromium and steel media, is calculated with the numbering of media and interfaces as specified in Table 2.

Table 2. Numbering of media and interfaces.

Layer	Medium	Interface	
Cr	*2*	Cr/oil	*b*
Oil	*1*	Oil/steel	*a*
Steel	*0*		

As oil/steel is the lowest interface, r_a is thus equal to the Fresnel reflection coefficient of the oil/steel interface and is expressed as

$$r_a = g_a + if_a = \frac{\overline{n}_{oil} - \overline{n}_{steel}}{\overline{n}_{oil} + \overline{n}_{steel}} \tag{11}$$

It gives,

$$g_a = \frac{(n_{oil}^2 - n_{steel}^2) - k_{steel}^2}{(n_{oil} + n_{steel})^2 + k_{steel}^2} \tag{12}$$

$$f_a = \frac{2n_{oil}k_{steel}}{(n_{oil} + n_{steel})^2 + k_{steel}^2} \tag{13}$$

Similarly, r_b is the Fresnel reflection coefficient of the chromium/oil interface which gives

$$g_b = \frac{(n_{Cr}^2 - n_{oil}^2) + k_{Cr}^2}{(n_{Cr} + n_{oil})^2 + k_{Cr}^2} \tag{14}$$

$$f_b = \frac{-2n_{oil}k_{Cr}}{(n_{Cr} + n_{oil})^2 + k_{Cr}^2} \tag{15}$$

Combining Eqs. (12) to (15) together with the thickness of the oil film into Eqs. (8) and (9), yields the reflectance of the compound layer of chromium/oil/steel, as expressed

$$\overline{R}_{step1} = G_{step1} + iF_{step1} \tag{16}$$

Second step:
The reflectance of the chromium layer bounded by the glass disc and the oil film (in fact, a compound substrate below) is calculated with the numbering of media and interfaces as specified in Table 3.

Table 3. Numbering of media and interfaces.

Layer	Medium	Interface	
Glass	*2*	glass/Cr	*b*
Cr	*1*	Cr/oil	*a*
Oil	*0*		

The Fresnel reflection coefficient of the glass/chromium interface is now given by r_b and yields

$$g_b = \frac{(n_{glass}^2 - n_{Cr}^2) - k_{Cr}^2}{(n_{glass} + n_{Cr})^2 + k_{Cr}^2} \tag{17}$$

$$f_b = \frac{2n_{glass}k_{Cr}}{(n_{glass} + n_{Cr})^2 + k_{Cr}^2} \tag{18}$$

and from Eq. (16),

$$g_a = G_{step1} \tag{19}$$

$$f_a = F_{step1} \tag{20}$$

Substituting Eqs. (17) to (20) together with the known thickness of Cr film into Eqs. (8) and (9), gives the total reflection coefficient of the whole system. The ratio of reflected energy to the incident energy or alternatively the ratio of the intensity of the reflected to incident beam can be obtained by $\left|\overline{R}\right|^2$.

Table 4. Optical parameters of the optical EHL contact.

Glass	$n = 1.5$
Cr film	$n = 2.80$, $k = 1.0$ $d = 22$ nm
Air	$n = 1.0$
Steel	$n = 2.21$, $k = 2.31$
Wavelength	λ=600 nm

4. VALIDATION

The aforementioned protocol was used to simulate a stationary contact of a steel ball and chromium coated glass plane, in conjunction with air under a load of 0.93 N. The deformed ball shape is known from Hertzian theory so that the variation of the gap between the steel surface and the chromium coated glass along the radius can be obtained. The system's optical and material data used in the calculation are listed in Table 4. The refractive index of the steel ball was interpolated from the data provided in Palik[9]. However, the refractive index of a thin chromium film could not be found from the literature and was therefore determined by comparing the measured interference intensity distributions of two identical air wedges: one bounded with two glass planes, and the other with a chromium coated glass and an ordinary glass plane. The corresponding reflectance at different radii of the Hertzian contact was calculated. The theoretical results were normalized with the maximum and minimum intensities measured and plotted together with the experimental data in Figure 4. The curve of the calculated results show very good agreement with the experimental data points. The theoretical Hertzian profile, being the input data for the intensity calculation, is also plotted in Figure 4 for reference.

5. EHL FILM MEASUREMENTS

The first significant publication on the use of a spacer layer for the measurement of very thin lubricant films in EHL contacts was from Johnston

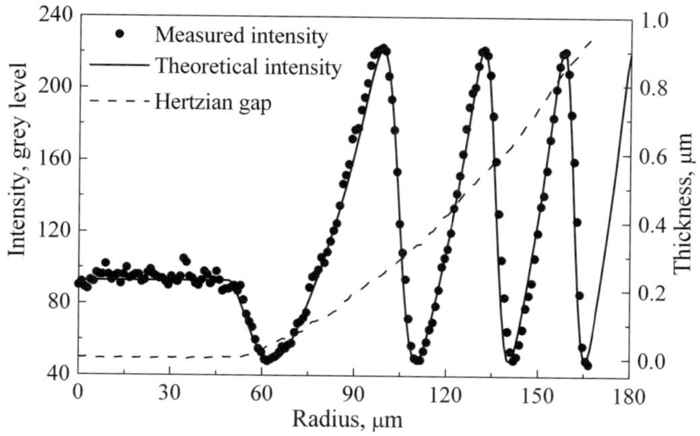

Figure 4. A steel ball of 25.4 mm in diameter loaded on a Cr coated glass plate in air, Cr film thickness = 22 nm, load = 0.93 N.

et al.[10] in 1991. They reported some anomalies in the relationship between film thickness and speed Both purified hexadecane and a saturated solution of stearic acid in hexadecane were tested and they demonstrated two different behaviours under slow speed ranges. Purified hexadecane gave no apparent residual film but the stearic acid solution appeared to give a rather uniform boundary film which persisted down to very low speeds. The phenomenon was attributed to an absorbed film of stearic acid on one or both surfaces. being found with film thickness of less than 15 nm.

In order to demonstrate the validity of the fringe intensity method for thin film lubrication study, similar tests on hexadecane and its mixture with a polymer, polybutene PB2500, were performed using a conventional optical EHL test rig. Polybutene is tacky and readily generates a boundary film. The viscosity of the tested fluid were measured at ambient temperature of 20°C using a Brookfield rotational concentric viscometer. The viscosity of the hexadecane is 4.10 cP and that of the solution of hexadecane/PB2400 (14.8% by weight) is 8.73 cP. The disc and ball were driven by separate motors with a worm-gear-reduction system such that sufficiently low speeds could be achieved by the current test rig to see whether similar phenomena can be observed. The two specimen fluids were tested four times each with the ball loaded on different radii of the disc and the results plotted in Figure 5 were an average of six readings. The load for all the tests was 3 N. Generally, the hexadecane/PB2400 solution was able to generate thicker films than the pure hexadecane. Most importantly, the two phenomena reported by Johnston et al. were also illustrated here. The pure hexadecane has an increased dependence of film thickness on rolling speed below 1 nm and there is a boundary film (or immobile film) of thickness of 2 nm formed

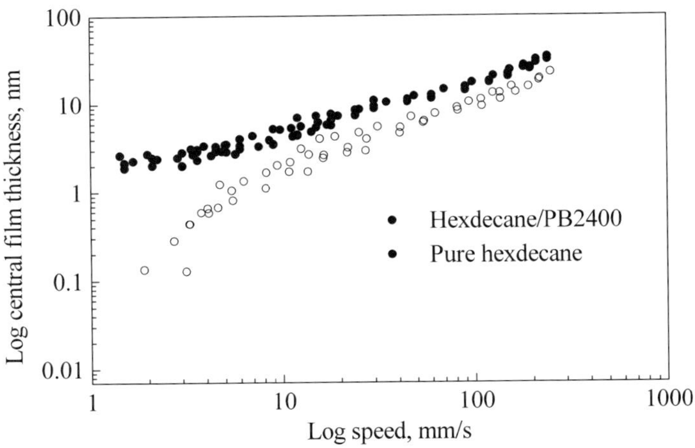

Figure 5. Film thickness variation with entrainment speed, load = 3 N.

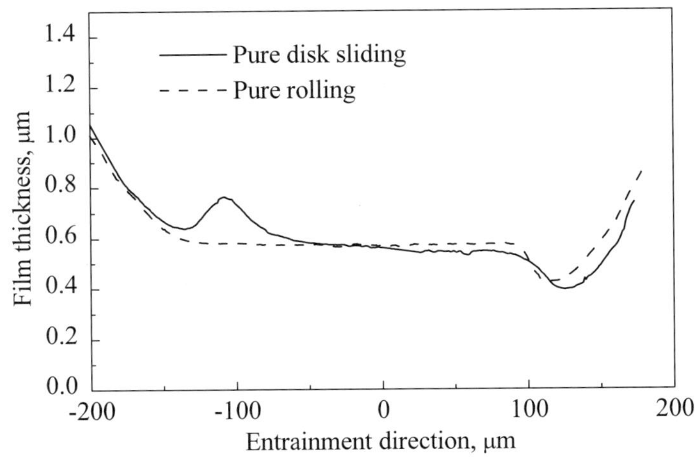

Figure 6. Different film profiles under pure-rolling and simple disc sliding, load = 16 N, PB1300.

with the solution of hexadecane/PB2400. As speed increases and a hydrodynamic film is generated, the 2 nm immobile film appears to continue to augment the film thickness of the fluid.

Conventional optical EHL measurement using a monochromatic light source enables the measurement of a thicker film. The enhancement in resolution of thickness measurement by using the fringe intensity facilitates the detection of local, very small film thickness variations in typical EHL films of thickness in the micron-scale. It is practical for the study of anomalous EHL film shapes such as the one shown in Figure 6. The film profile deviates from the conventional EHL film shape. There is evidence of a tiny inlet dimple. More similar examples can be found in our accompanying paper[11].

6. CONCLUSION

The relation between intensity and film thickness for a typical optical EHL contact was obtained through a full optical analysis using multi-beam interferometry. It was demonstrated that the film thickness can be accurately determined from the measured intensity using the known relation. The minimum detectable film thickness is no longer limited to that of the first interference fringe but can be down to nanometer scale. It is thus concluded that the conventional optical EHL interferometry is now practical for ultra-thin film lubrication studies.

ACKNOWLEDGEMENT

The authors would like to express gratitude to the City University of Hong Kong for financial support for the work (SRG grants: 7001563 and 7001432).

REFERENCES

1. F. J. Westlake, *An Interferometric Study of Ultra Thin Fluid Films.* Ph.D. Thesis, University of London (1970).
2. H. A. Spikes and P. M. Cann, The development and application of the spacer layer imaging method for measuring lubricant film thickness, *Journal of Engineering Tribology* **215**(J3), 261–277 (2001).
3. A. D. Roberts and F. R. S. D. Tabor, The extrusion of liquids between highly elastic solids, *Proc. R. Soc. Lond.* **A325**, 323–345 (1971).
4. J. Luo, S. Wen and P. Huang, Thin film lubrication Part I: Study on the transition between EHL and thin film lubrication using a relative optical interference intensity technique, *Wear* **194**, 107–115 (1996).
5. P. L. Wong and F. Guo, Effects of absorbing media on relative optical interference intensity ROII technique, *Science in China, Series A*, **44** (Supp., the 3rd International Symposium on Tribology, Beijing, China), 431–435 (2001).
6. F. Guo and P. L. Wong, A multi-beam intensity-based for lubricant film measurements in non-conformal contacts, *Journal of Engineering Tribology* **216**(5), 281–291 (2002).
7. P.L. Wong, F. Guo and C. Feng, The measurement of the refractive index of a liquid lubricant at high pressure within an EHL impact dimple using a single optical interferogram, *Tribology International* **36**(7), 497–504 (2003).
8. O. S. Heavens, *Optical Properties of Thin Solid Films*, Dover Publications, New York (1955).

9. E. D. Palik, *Handbook of Optical Constants of Solids II*, Academic Press (1991).
10. G. J. Johnston, R. Wayte and H. A. Spikes, The measurement and study of very thin lubricant films in concentrated contacts, *STLE Tribology Transactions* **34**, 187–194 (1991).
11. F. Guo and P. L. Wong, Variations of an EHL film under boundary slippage, in *Proceedings of the IUTAM Symposium on Elastohydrodynamics and Microelastohydrodynamics*, Cardiff, September 2004, R.W. Snidle and H.P. Evans (eds), Springer, Dordrecht (2006) (this volume).

SESSION 7

SURFACE MODIFICATION FOR PISTON RING AND LINER

Nathan W. Bolander and Farshid Sadeghi
School of Mechanical Engineering, Purdue University, West Lafayette, Indiana, U.S.A.

Abstract The objectives of this paper are to numerically and experimentally investigate the potential for surface patterning and features to reduce friction at the piston ring cylinder liner interface. A test rig is used to experimentally measure the frictional loss throughout the stroke. A fully deterministic model is applied to provide explicit detail of the mixed lubrication phenomena near the dead points. The effectiveness of laser textured piston rings is examined both experimentally and numerically. Frictional performance of the surface-modified piston rings is shown to have improvement over OEM rings.

Keywords: friction, mixed lubrication, piston ring, surface modification.

1. INTRODUCTION

The reduction of parasitic loss in internal combustion engines has far reaching implications in terms of fuel efficiency and engine performance. A major source of friction in reciprocating engines is the piston ring assembly, which by some estimates can account for as much as 40% of the mechanical loss [1]. It is well known that the piston ring encounters the entire range of lubrication regimes through each stroke (i.e. boundary, mixed, elastohydrodynamic (EHL), and hydrodynamic). Thus, investigations into frictional losses and potential loss reduction mechanisms at the piston ring/cylinder liner (PRCL) contact must take into account the transitions from a state of full film lubrication to boundary lubrication.

A number of experimental and theoretical studies have concluded that intentionally modifying a surface with micro-pits can benefit lubricated contacts. Ronen et al. [2] presented a hydrodynamic computational investigation in reciprocating parallel surface bearings meant to simulate a piston ring. They conclude that appropriately sized dimples can significantly reduce frictional loss, an effect that was later confirmed in the experimental work of Ryk et al. [3], Zhao et al. [4] show that pressurization of trapped lubricant due to deformation of a pocket under load will result in reduced solid contact pressures and

R.W. Snidle and H.P. Evans (eds), IUTAM Symposium on Elastohydrodynamics and Microelastohydrodynamics, 271–283.

that surface roughness in mixed and boundary lubricated contacts may cause a similar effect. Dumont et al. [5] studied starved circular rolling EHL contact with dimples in one of the surfaces. They explain that the dimples act to increase the amount of lubricant carried into a starved contact. Zhao and Sadeghi [6] studied the effect of a stationary surface pocket in line contact mixed-EHL startup of smooth surfaces. They found that dimples can essentially transform the single contact to that of two micro-EHL contacts starting in parallel. The resulting reduction in startup time has potential benefits for applications with frequent start-stop events such as reciprocating motion.

Consideration of elastic deformation is critical in determining the lubrication of piston rings, as first discussed in the investigation of Dowson et al. [7]. Here, a survey of several European diesels found that a majority were operating under EHL conditions during a portion of the stroke. Deformation of the surfaces along with piezoviscous behavior of the lubricant was found to substantially increase the predicted film thickness. Yang and Keith [8] in their series of papers extended the one dimensional EHL analysis to consider flow in the circumferential direction and cavitation.

Accurate prediction of the frictional loss requires consideration of the roughness of the mating surfaces, as the portion of the load that is carried by asperities in contact is directly proportional to the dominant source of frictional loss near TDC and BDC. Rohde et al. [9], Hu et al. [10], and Bolander et al. [11] are a few examples of researchers who have successfully applied the Greenwood and Tripp [12] model for asperity contact pressure. However, the statistical models typically assume a Gaussian distribution of asperity heights unless specially formulated, while the surfaces of interest are usually far from Gaussian. In addition, the specific details of the contact are not calculated explicitly, thus analysis of the behavior of surface modifications under mixed lubrication conditions is not possible.

A fully deterministic approach to mixed lubrication describes the process by which the fluid and solid contact regions are solved simultaneously to provide explicit detail of the fluid flow throughout the contact. Jiang et al. [13] presented a deterministic model for mixed lubrication wherein the solid contact pressure was calculated using an inverse-FFT method. Hu and Zhu [14] developed a model in which a reduced Reynolds equation is used to calculate pressure in areas where the film thickness approaches zero. Zhao and Sadeghi [15] presented a mixed lubrication model in their investigation of smooth surface startup phenomena. In their model an iterative scheme between lubricated and solid contact regions is used to achieve the solution to the mixed lubrication problem.

In this paper experimental measurements of friction for modified piston rings are presented. The mixed lubrication model of Zhao and Sadeghi [15] has been extended to include rough surfaces in a transient elliptical contact

Figure 1. Piston ring reciprocating liner test rig.

typical of the current test rig. Details of the mixed contact region are used to predict the frictional losses throughout the stroke for correlation with experimental findings. The contacts are examined in detail for the portion of the stroke near bottom dead center (BDC). Surface modifications are imposed upon the rough surface and their effects are discussed.

2. EXPERIMENTAL MEASUREMENT OF FRICTION FOR MODIFIED SURFACES

2.1 Piston Ring Reciprocating Liner Apparatus

Figure 1 illustrates the bench-scale Piston Ring Reciprocating Liner (PRRL) test rig designed, developed and constructed for this study. The test rig was designed to have the cylinder liner reciprocate while keeping the piston ring stationary and can accommodate a wide range of loads, speeds, and lubricant conditions at the PRCL interface. Details of the apparatus and experimental procedure are available in Bolander et al. [11]. The piston rings considered in this study are symmetrically crowned and typically have a negatively skewed, pitted surface. The OEM cylinder liner surfaces were finished with the standard cross-hatch pattern with a surface roughness much larger than that of the piston ring. Surface modifications in the form of circular dimples were explored as a means to reduce the losses throughout the mixed and boundary lubrication regimes. Figure 2 depicts an example laser dimpled piston ring surface. Due to the challenges associated with laser machining, dimensions of surface features on the samples received from the supplier had a fair amount of variance from those specified. In addition, post treatment of the ring surface was required to remove the rim of recast material deposited during the laser ablation process. Table 1 presents the dimensions of the surface dimples studied in this paper.

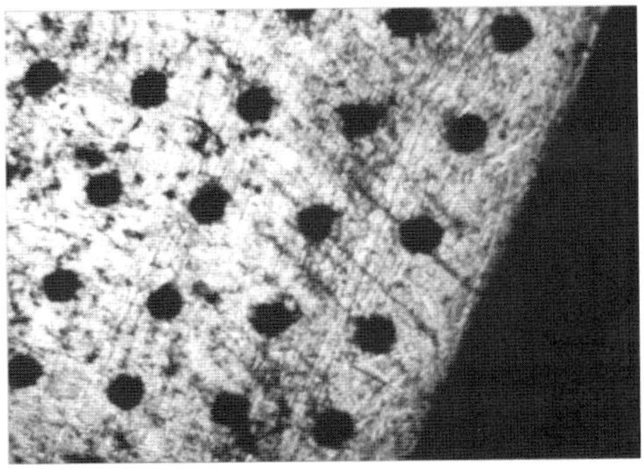

Figure 2. Photograph of laser dimpled ring surface.

Table 1. Dimensions of circular dimples.

	Dimple Diameter [mm]	Dimple Depth [mm]
Ring 'O'	150	1.5
Ring 'P'	75	5
Ring 'Q'	123	4
Ring 'R'	122	11

Table 2. Measured piston ring and liner surface characteristics.

	Rq [mm]	Sk	Ku
OEM Piston Ring 'A'	0.26	−3.04	44.7
OEM Cylinder Liner	1.21	−0.65	3.2
Polished Cylinder Liner	0.007	−1.04	5.1

The ring and liner surfaces were characterized using an optical surface profilometer. Table 2 presents the measured root-mean-square roughness Rq, skewness Sk, and kurtosis Ku for the unmodified ring and liner samples. Table 3 contains the nominal contact pressure between the piston ring and liner, which can be thought of as the sum of ring elastic pressure and cylinder gas pressures acting on the backside of a ring in an operating engine. This nominal pressure is calculated by dividing the test rig applied load by the ring width (3.78 mm) and the true Hertzian contact length in the circumferential direction. Hertzian contact ellipse dimensions are calculated from the applied load and principal radii of curvature of the installed ring and liner as measured using a coordinate measuring machine. A typical installation pressure for the OEM ring used in this investigation is 0.24 MPa, a typical peak engine com-

Table 3. Relationship of nominal load to equivalent pressure over the ring back.

Normal Force [kgf]	Contact Length (Circumferential) [μm]	(Axial) Contact Width [μm]	Nominal Pressure [MPa]
3	1843	293	4.2
6	2322	369	6.7
9	2658	422	8.8

pression pressure is 1.0 MPa and a typical peak combustion pressure is 2.0 to 4.0 MPa.

2.2 Experimental Results

Piston ring frictional behavior was first studied using the OEM liner segment finished with the typical cross-hatch pattern. It was found that the friction results were very near one another throughout the entirety of the stroke, with friction actually increasing slightly in the mixed lubrication regions. This was not the desired effect of applying the dimples, as no noticeable benefit was shown with modified piston rings run against the OEM liner segment. It was hypothesized that the dimples were not proving beneficial due to the OEM liner surface topography, as the cross-hatch pattern is of the same order of magnitude as the dimples in the piston ring. In order for any surface modification to reduce friction in mixed lubrication, the modification must cause a greater fraction of the load to be carried by the lubricant thereby reducing direct surface-to-surface contact. The dimple geometry is meant to generate higher pressure in the lubricant allowing the lubricant to carry a higher fraction of the load either from a squeezing motion or from additional hydrodynamic action. The magnitude of the OEM liner cross-hatch patterning is apparently reducing the effectiveness of the supplementary pressure generation mechanisms.

In order to remove the variability of the OEM surface finish, a liner segment was fabricated from A2 tool steel, hardened to 57 Rockwell C and polished to a final surface roughness of 7 nm (Rq). Figure 3 illustrates the friction results of unmodified ring 'A' and four modified rings on the polished liner. Test conditions were 16 kgf with SAE 30 oil at a speed of 15 rpm. The slow crank speed used for these tests ensures that mixed lubrication is present over a significant portion of the stroke. In this case, friction of the modified rings was consistently lower than the unmodified ring with a few exceptions. The frictional loss of ring 'R' for example, was consistently greater than that of the unmodified ring. It is interesting to note that ring 'R' possessed some of the deepest dimples studied. This reinforces the idea that many factors such as

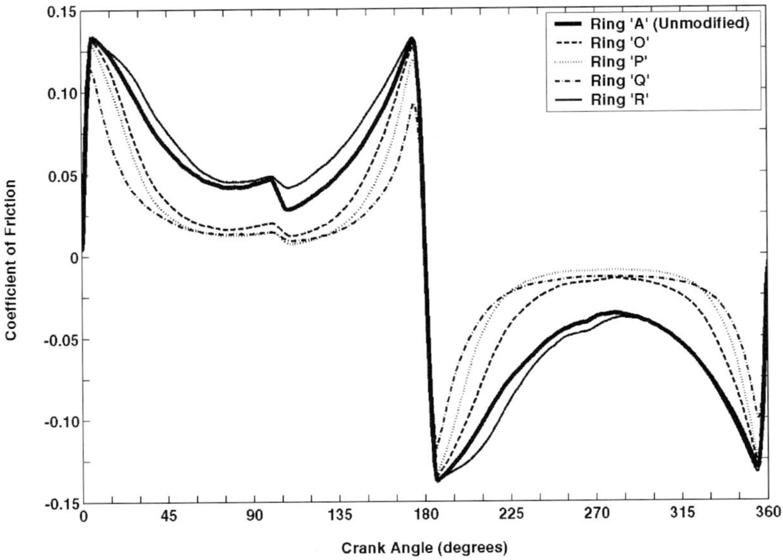

Figure 3. Measured friction traces for various modified piston rings.

dimple geometry, placement, and surface roughness play an important role in determining the overall performance of the modified surface.

In summary, modified piston rings showed no benefit over OEM piston rings on the OEM liner. However, modified rings that had not shown benefit on the OEM liner did exhibit lower friction on a polished liner segment as compared to OEM ring 'A'. While this apparent performance increase is encouraging, it is possible that this may be attributed to OEM ring variability. The apparent reduction in friction attributed to the modified surface was within the range of ring-to-ring variability found among additional OEM rings, leaving the comparison between modified and OEM rings inconclusive. Thus, while piston rings modified by laser dimpling have shown the potential to reduce friction at the piston ring/cylinder liner interface, further investigation is necessary to confirm the effect and to eliminate the influence of OEM ring variability.

3. DETERMINISTIC MODELING OF MIXED LUBRICATION AT THE PRRL CONTACT

The PRRL test rig described in the previous section was specifically designed to reduce the number of variables associated with piston ring lubrication under normal operating conditions. In doing so, the contact area experienced by the piston ring is modified from the conformal contact typical of installation to an elongated elliptical contact. Elliptical contacts have been investigated extensively and thus provide a good foundation for extension of existing mixed lubrication models.

3.1 Governing Equations

The non-dimensional, isothermal, time dependent Reynolds equation for elliptical contact is given by:

$$\frac{\partial}{\partial X}\left(\frac{\bar{\rho}H^3}{12\bar{\eta}}\frac{\partial P}{\partial X}\right)+\frac{\partial}{\partial Y}\left(\frac{\bar{\rho}H^3}{12\bar{\eta}}\frac{\partial P}{\partial Y}\right)=\lambda\frac{\partial(\bar{\rho}H)}{\partial X}+\sigma\frac{\partial(\bar{\rho}H)}{\partial\theta}. \quad (1)$$

It is assumed in this study that both solid contact and liquid lubricated regions may exist simultaneously within the nominal contact area. If the film thickness drops below a critical value, $\varepsilon < 5$–10 nm, it is assumed that the film fails and jump-to-contact behavior ensues. The existence of both solid and lubricated regions within the contact requires that appropriate boundary conditions be applied to the Reynolds equation at each interface. To ensure mass conservation throughout the domain, the fluid flow entering any given control volume must be exactly balanced by fluid flow through the remaining open cell boundaries. The condition of zero flowrate through an internal solid boundary is used to derive a Nuemann type boundary condition for pressure. Equation (2) describes the pressure gradient that must exist at a fluid/solid boundary under the assumption of zero mass flow through this face.

$$\left.\frac{\partial P}{\partial X}\right|_e=(\lambda_1+\lambda_2)\left(\frac{6\bar{\eta}}{H^2}\right). \quad (2)$$

The friction force acting between the piston ring segment and the cylinder wall consists of a viscous shear force in the lubricant film and friction between the asperities in contact. An experimentally measured value of 0.14 is used for the dynamic asperity coefficient of friction μ_{solid}. The friction contributions from viscous shear solid and solid contact may be expressed as:

$$F_{f,\text{fluid}}=\iint_{\Omega}(\tau_{zx})_{z=0}da, \quad F_{f,\text{solid}}=\iint_{\Omega}P_s dA. \quad (3)$$

To maintain generality it is beneficial to characterize the surfaces of interest using appropriate statistical measures and then obtain a numerical equivalent. In this way as many surfaces as may be required can be generated and studied while maintaining the characteristics of the original surface. In this procedure, the Johnson translator system [16] is used to transform a Gaussian random sequence to a sequence with prescribed non-Gaussian moments (Rq, Sk, Ku). This non-Gaussian system is then transformed by a linear filter to the output surface having the desired characteristics [17]. Additional details of the numerical model and iterative procedure used to solve the mixed lubrication contact are available in the work of Zhao and Sadeghi [15].

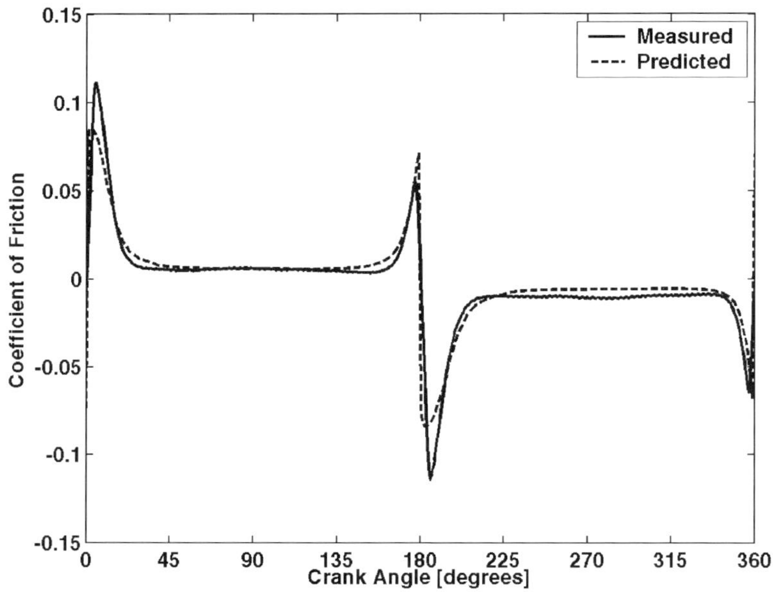

Figure 4. Comparison of measured and predicted friction.

3.2 Numerical Results

Surface modifications in the form of micro scale dimples were investigated as a means of reducing parasitic loss. Due to uncertainty in dimple dimensions and positioning in the experimental test samples it was determined that a direct comparison between experiment and analytics was not the best alternative. Previous studies of micro dimples [3] have recommended dimples in the range of 100 μm in diameter and 10 μm deep with area coverage of about 13%. Using these recommendations, a 6×14 array of dimples was arranged on a generated rough surface to simulate contact of a modified piston ring with the polished liner segment. Recall the results presented in Figure 3, where the experiments showed that ring 'R' (the ring with the deepest dimples) performed worse than the unmodified surface whereas the shallower dimples (<5 μm) tended to perform better. Applying this information, dimple depths at both extremes (1 and 10 μm) were analyzed using the numerical model.

Figure 4 presents a comparison of friction traces measured experimentally and obtained using the deterministic model under operating conditions of 60 rpm and 6 kgf load. Good agreement exists through a large portion of the stroke. It is noted that slight discrepancies exist in the value of the peak friction, however the model was able to predict the existence of the friction spikes with reasonable accuracy.

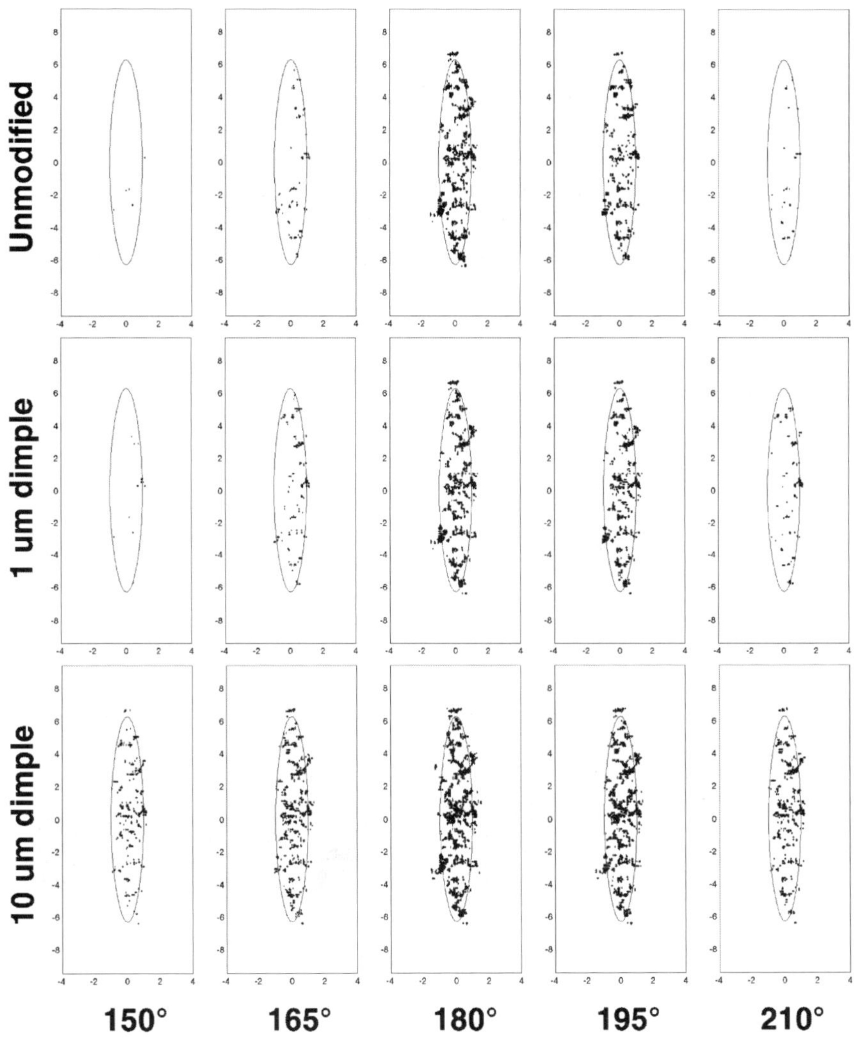

Figure 5. Contact near BDC for modified and unmodified surfaces.

Figure 5 depicts the asperity contact regions for modified and unmodified rings in the portion of the stroke near BDC. As expected, contact intensity increases as the sliding velocity decreases approaching BDC. The amount of contact then begins to drop off as the velocity increases on the return stroke. It is noted here that reversal of entrainment direction has not been considered in the current work, i.e. the sliding direction remains constant. Reversal of entrainment will be considered in future analysis. It is apparent that the addition of 10 μm dimples has greatly increased the amount of solid contact present

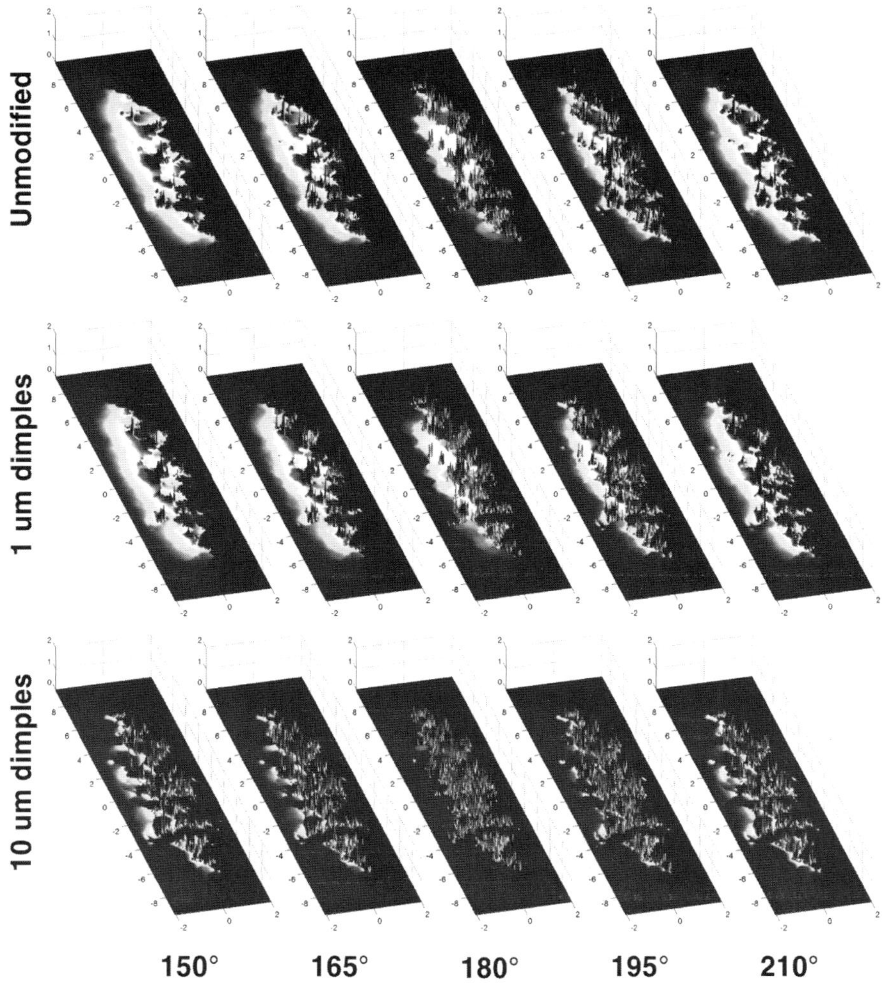

Figure 6. The effect of dimple depth on pressure buildup near BDC.

through the portion of the stroke near BDC, while the contact regions for the 1 μm dimple case remain similar to the unmodified ring.

Figure 6 depicts the pressure buildup in the contact, which is a combination of both solid contact pressure and fluid pressure. Solid contact pressures are generally much higher than the fluid pressures, especially in regions such as BDC where sliding velocity is low. Note in this figure the lighter regions illustrating pressure buildup in the lubricant in the converging portion of the ring face. In the 10 μm dimple case the converging portion of the ring face no longer develops the same hydrodynamic pressure that is seen in the unmodified or 1 μm dimple cases. It seems that dimples of this depth have a disruptive

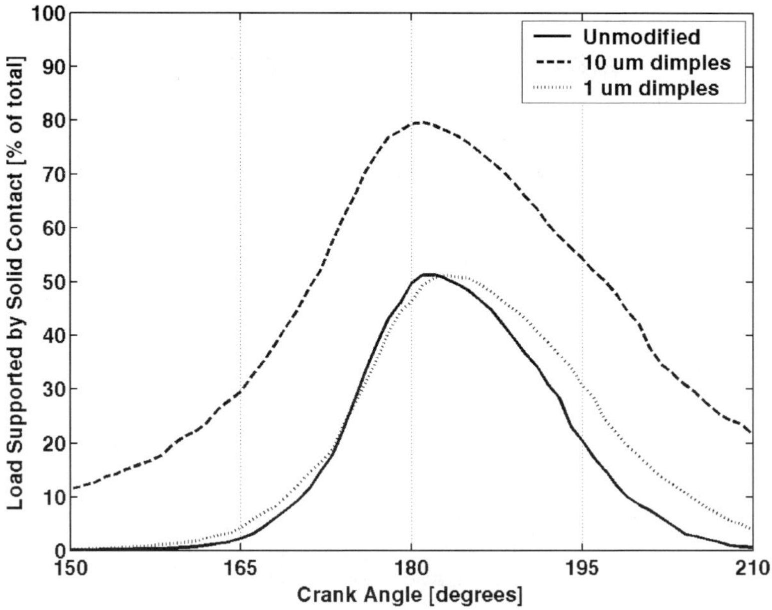

Figure 7. Comparison of load supported by asperity contact near BDC.

effect on the fluid flow and act to stifle the pressure buildup mechanism in the converging section.

Figure 7 presents a comparison of the portion of the load carried by asperity contact near BDC. As expected, hydrodynamic action is decreased as the ring (liner) decelerates, which leads to a larger portion of the load being carried by areas in solid contact. It is noted in this figure that due to squeeze action the point of minimum film thickness (and therefore maximum contact intensity) is offset slightly from BDC due to squeeze effects. Consistent with the greatly increased contact area for the 10 μm dimple case indicated previously, a much larger portion of the load in the BDC region is supported by areas in solid contact as compared to the unmodified ring. Of course, an increase in contact intensity leads to increased friction. This is in contrast to the 1 μm dimple case, where the asperity contact load support has been slightly decreased, meaning that slightly less of the load is being carried by the regions in solid contact. However, it is also apparent that the solid load support contribution curve has widened somewhat.

Figure 8 depicts a comparison of frictional losses through an entire stroke. Not only has the friction for the 10 μm dimple case increased in the area near bottom dead center, but dramatic increases are seen throughout the entirety of the stroke. In comparison, the addition of 1 μm dimples has blunted the friction spikes at the ends of stroke. However, the width of the friction spike

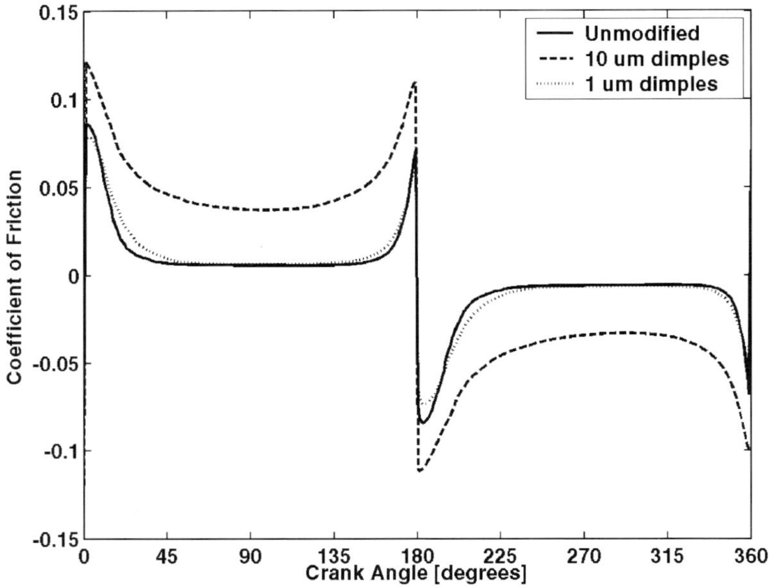

Figure 8. Comparison of friction for modified and unmodified surfaces.

has increased slightly. In other words, the maximum intensity of the solid contact has been reduced by addition of shallow (1 μm) dimples, while the duration of contact has been slightly increased.

4. CONCLUSION

Experimental results indicate that piston rings modified by laser dimpling have the potential to reduce friction at the piston ring/cylinder liner interface. However, further investigation is necessary to confirm the effect and to eliminate the influence of OEM ring variability. A deterministic mixed lubrication model has been applied to provide explicit detail of the contact found in the PRRL test rig. Friction traces obtained using this model show good comparison to the experimental measurements. In addition, trends observed in the experimental work have also been found in the numerical predictions. Specifically the depth of the dimples has been found to play a major role in the contact behavior, with deep dimples (10 μm) being detrimental while shallower dimples (1 μm) provide some benefit. Future work, both experimental and analytical, will be focused on shallow dimples while also expanding the results to include the effects of dimple diameter and area coverage.

REFERENCES

[1] Y. Jeng, Theoretical Analysis of Piston-Ring Lubrication Part 1 – Fully Flooded Lubrication, *STLE Tribology Trans.* **35**, 696–706 (1992).

[2] A. Ronen, I. Etsion and Y. Kligerman, Friction-Reducing Surface-Texturing in Reciprocating Automotive Components, *STLE Tribology Trans.* **44**, 359–366 (2001).

[3] G. Ryk, Y. Kligerman and I. Etsion, Experimental Investigation of Laser Surface Texturing for Reciprocating Automotive Components, *STLE Tribology Trans.* **45**, 444–449 (2002).

[4] J. Zhao, F. Sadeghi and H.M. Nixon, A Finite Element Analysis of Surface Pocket Effects in Hertzian Line Contact, *ASME J. Tribology* **122**, 47–54 (2000).

[5] M.L. Dumont, P.M. Lugt and J.H. Tripp, Surface Feature Effects in Starved Circular EHL Contacts, *ASME J. Tribology* **124**, 358–366 (2002).

[6] J. Zhao and F. Sadeghi, The Effects of a Stationary Surface Pocket on EHL Line Contact Start Up, *ASME J. Tribology* **126**, 672–680 (2005).

[7] D. Dowson, B.L. Ruddy and P.N. Economou, The Elastohydrodynamic Lubrication of Piston Rings, *Proc. Roy. Soc. Lond.* **A386**, 409–430 (1983).

[8] Q. Yang and T.G. Keith, An Elasto-Hydrodynamic Cavitation Algorithm for Piston Ring Lubrication, *STLE Tribology Trans.* **38**, 97–107 (1995).

[9] S.M. Rohde, K.W. Whitaker and G.T. McAllister, A Mixed Friction Model for Dynamically Loaded Contacts with Application to Piston Ring Lubrication, Surface Roughness Effects in Hydrodynamic and Mixed Lubrication, *ASME Winter Annual Meeting*, pp. 19–50 (1980).

[10] Y. Hu, H.S. Cheng, T. Arai, Y. Kobayashi and S. Aoyama, Numerical Simulation of Piston Ring in Mixed Lubrication – A Non-Axisymmetrical Analysis, *ASME J. Tribology* **116**, 470–478 (1994).

[11] N.W. Bolander, B.D. Steenwyk and F. Sadeghi, Lubrication Regime Transitions at the Piston Ring Cylinder Liner Interface, *Proc. Instn. Mech. Engrs., Part J: J. Engineering Tribology* **219**, 19–31 (2005).

[12] J.A. Greenwood and J.H. Tripp, The Contact of Two Nominally Flat Rough Surfaces, *Proc. Instn. Mech. Engrs.* **185**, 625–633 (1971).

[13] X. Jiang, D.Y. Jua, H.S. Cheng, X. Ai and S.C. Lee, A Mixed Elastohydrodynamic Lubrication Model with Asperity Contact, *ASME J. Tribology* **121**, 481–491 (1999).

[14] Y. Hu and D. Zhu, A Full Numerical Solution to the Mixed Lubrication in Point Contacts, *ASME J. Tribology* **122**, 1–9 (2000).

[15] J. Zhao, F. Sadeghi and M.H. Hoeprich, Analysis of EHL Circular Contact Start Up: Part I – Mixed Contact Model with Pressure and Film Thickness Results, *ASME J. Tribology* **123**, 67–74 (2001).

[16] I.D. Hill, R. Hill and R.L. Holder, Fitting Johnson Curves by Moments, *Applied Statistics* **25**, 180–189 (1976).

[17] V. Bakolas, Numerical Generation of Arbitrarily Oriented Non-Gaussian Three-Dimensional Rough Surfaces, *Wear* **254**, 546–554 (2003).

VARIATIONS OF AN EHL FILM UNDER BOUNDARY SLIPPAGE

F. Guo[1] and P.L. Wong[2]
[1]*School of Mechatronic Engineering, Qingdao Polytechnic University, Qingdao, P.R. China*
[2]*Department of Manufacturing Engineering and Engineering Management, City University of Hong Kong*

Abstract The typical film shape in the classical EHL theory, featuring a parallel flat contact region and a constriction at the outlet, has been readily demonstrated by both numerical analyses and laboratory tests. Recently, the lubricant limiting-shear-stress property and boundary slippage have been incorporated into EHL numerical computations. Instead of the parallel flat part in the conventional EHL film, a dimple (film thickness increasing) occurs at the inlet, and a bump (film thickness decreasing) appears in the downstream contact region. In a previous paper by the authors, some purposely devised tests were carried out and a film shape with an inlet dimple, qualitatively the same as the numerical results, was obtained. More recently, tests have been carried out over an extensive range of slide-roll ratios and interesting variations of film shape are found. The present paper reports some of the results. A viscosity wedge, referred to as the drastic drop of the effective viscosity of lubricant adjacent to the bounding surfaces, is proposed tentatively to explain the film deviation.

Keywords: elastohydrodynamic lubrication, boundary slippage, limiting shear stress.

1. INTRODUCTION

In the past decades full numerical solutions have played a key role in EHL studies and impressive progress has been made. The classical characteristics of EHL films, including the parallel flat contact region, the film constriction at the outlet and the pressure spike, were first obtained by numerical analyses, for example, the inverse-solution results by Dowson et al. in the 1950s–1960s[1]. These earlier numerical results were thereafter confirmed successfully by laboratory experiments[1,2].

R.W. Snidle and H.P. Evans (eds), IUTAM Symposium on Elastohydrodynamics and Micro-elastohydrodynamics, 285–296.

Earlier numerical EHL analyses were focused on the Newtonian isothermal model. However, it cannot correlate the experimental traction curves, and therefore thermal and non-Newtonian effects were taken into account. Most of the presented non-Newtonian EHL analyses generally employed the Ree–Eyring model for its apparent theoretical background in the thermal activation theory. In the numerical results with the thermal and non-Newtonian effects included, the film shape is still of the classical type[3-5]. However, due to the effects of shear heating and shear thinning, numerical film thickness is reduced. These numerical results also support the classical inlet-region-dependence theory that EHL film thickness mainly depends upon the conditions at the inlet region. Along with these analyses regarding the thermal effect and non-Newtonian property of lubricants, some model experiments were also presented and good correlation has been achieved.

However, the Ree–Eyring model has been criticized in that it allows unlimited increase in shear stress with shear rate, which is contrary to the experimental observations of lubricants' limiting shear stress, such as those reported by Bair and Winer[6]. Due to the limiting shear stress, lubricant slippage may occur at the bounding surfaces where the maximum shear stress occurs. Studies on solid-liquid interfaces have also revealed another kind of wall slippage, which is attributed to the poor affinity between the liquid and the solid surface. This wall slippage has been experimentally displayed for both non-Newtonian polymer melts and simple Newtonian liquids[7,8]. The wall slippage, occurring when the shear stress reaches the limiting shear stress of lubricants or the critical sustainable shear stress of the liquid/solid interface, undoubtedly affect the lubrication behaviour and further studies are needed.

Various rheological models with limiting shear stress have been proposed, and are used in numerical EHL. With the circular model, Hamrock and his co-workers showed that under large slide-roll ratios the film would not preserve its classical shape, instead a dimple at the inlet and a bump in the downstream region of the contact might occur[9,10]. Similar changes in film shape were also produced using other limiting-shear-stress models, for example, the linear model[11], the Bair & Winer model[6], and the ideal viscoplastic model[13,14]. Among these models, the linear and the ideal viscoplastic models represent the two extreme rheological behaviours of liquids with limiting shear stress, and the characteristic of a real liquid lubricant may behave somewhere between these two.

In the 1970s, Chiu and Sibley[15] reported an abnormal dimple film in optical EHL experiments under ultraslow motion conditions, and they attributed the dimple to the Weissenberg effect. Due to the limited resolution of the measuring technique, the dimple film profiles could not be reconstructed accurately from the interferograms. For further understanding of this film shape deviation, similar experiments were repeated with better quality optical parts by the authors. By advanced optical measurement technique, the authors could trace the film shape precisely and demonstrated an anomalous EHL film, in which there is a dimple in the inlet region followed by a distinct wedge gap. Variations of the inlet dimple film under different conditions were presented. Unlike Chiu and Sibley, the authors tentatively explained the dimple formation by the limiting shear stress and wall slippage, and considered that it largely confirmed the numerical findings above. The previous tests by Chiu and Sibley[15] and the authors[16] were only conducted under simple disc sliding (steel ball stationary) conditions, and no appreciable bump was produced. According to the numerical results, a typical downstream bump can occur when higher levels of sliding are applied. For the purpose of confirming the downstream bump, larger slide-roll ratios and polybutenes of high molecular weight are employed in the present experiments. Furthermore, it is found that a ZEV (zero entrainment velocity) pre-running before tests can largely modify the resultant film shape, which is possibly attributed to the modification of interface affinity through this pre-running.

2. EXPERIMENTAL CONDITIONS

The details of the test rig used in the present experiments were documented by the authors[16,19]. The experiments were performed at a temperature of 20 ± 0.5 °C. The diameter of the steel ball was 25.4 mm. Two loads, 16N and 22N, were applied. The entrainment speed varied from 10 to 800 μm/s.

In the experiments, slow entrainment speeds were adopted to avoid any influential shear heating[16]. Two polybutenes of high molecular weight, PB1300 and PB2400, were chosen as the lubricants. At the experiment temperature, PB1300 and PB2400 are polymer melts, and their very high viscosities can ensure EHL film formation at very slow entrainment speeds. No detailed rheological data for them can be found in the literature. However, high-molecular-weight polymers are usually non-Newtonian, and would have an obvious decrease in the shear viscosity even with a relatively low shear rate. Furthermore, polymer liquid/solid interfaces are prone to wall slippage, which has been extensively demonstrated by experiments and theoretical analyses[7,17,18]. Therefore, it is speculated that the chosen lubricants can behave in a non-Newtonian fashion and slippage can occur easily at the interface.

3. EXPERIMENTAL RESULTS AND DISCUSSION

3.1 Zero-Entrainment-Velocity Pre-Running (ZEVPR) and Its Effect on Film Shapes

In the experiments under ultraslow motion conditions by Chiu and Sibley[15] and the authors[16], only simple disc sliding was studied. Since shear heating is excluded, at the same entrainment speed the simple disc sliding (ball stationary) and the simple ball sliding (disc stationary), which are two kinematic equivalents, were expected to produce the same film according to the EHL theory. However, in the current experiments, it was revealed that the simple ball sliding and simple disc sliding produced two different film profiles. By accident, it was found that the resultant film shape in simple sliding tests would be largely different if a zero-entrainment-velocity (ZEV) operation, in which the ball surface and the disc surface ran at equal ultraslow speeds but in opposite directions, was applied prior to the experiment. The preceding zero-entrainment-velocity operation is hereby referred to as ZEV pre-running or ZEVPR. The effect of ZEVPR on the consequent film shape varies with the pre-running conditions, for example the load, the sliding speed and the running time. In the present experiment, the ZEV pre-running condition is fixed at load = 22 N and sliding speed u_s = 66 μm/s. It was shown that for PB1300 under such a ZEVPR condition, very much the same film shape could be obtained for both simple ball sliding and disc sliding. Figure 1 illustrates some representative interferograms.

Figure 2 depicts the film profile reconstruction along the central entrainment direction for the entrainment speed u_e = 465.5 μm/s in Figure 1. For better characterization of this kind of film profile, four quantities are used, h_{minI} – the inlet minimum film thickness, h_{dim} – the maximum film thickness at the dimple, h_{cen} – the central film thickness and h_{minO} – the outlet minimum film thickness. For simple disc sliding, it can be seen that ZEV pre-running has a significant influence on the film. The wedge after the inlet dimple is largely suppressed by ZEV pre-running, and the inlet dimple has a depth (h_{dim}–h_{minI}) of 72 nm comparing to 147 nm for the case without ZEVPR. At the outlet, the constriction exists in both cases, but less pronounced in the case without ZEV pre-runnnig. Due to the larger inclination of the wedge, the simple disc sliding without ZEVPR has a higher h_{minI} and h_{dim} than those with ZEVPR. Measured data from Figure 1 show that for simple disc sliding when the entrainment speed u_e is larger than about 300 μm/s, h_{minO} and h_{cen} with ZEV pre-running is larger than those without ZEV pre-running. For simple ball sliding, the difference in film profiles produced by ZEV pre-running are much less pronounced. In Figure 2 both the h_{minI} and h_{dim} (722nm and 798 nm) with ZEVPR are larger than those (670nm and 776 nm) without ZEVPR. However, according to all the curves from the measurement data, it cannot be concluded that ZEVPR has an influential effect on the film thickness for simple ball sliding in the present case. In fact, for simple ball sliding h_{cen} and h_{minO} are 574 nm and 407 nm with ZEVPR, and 571 nm and 412 nm without ZEVPR, nearly the same.

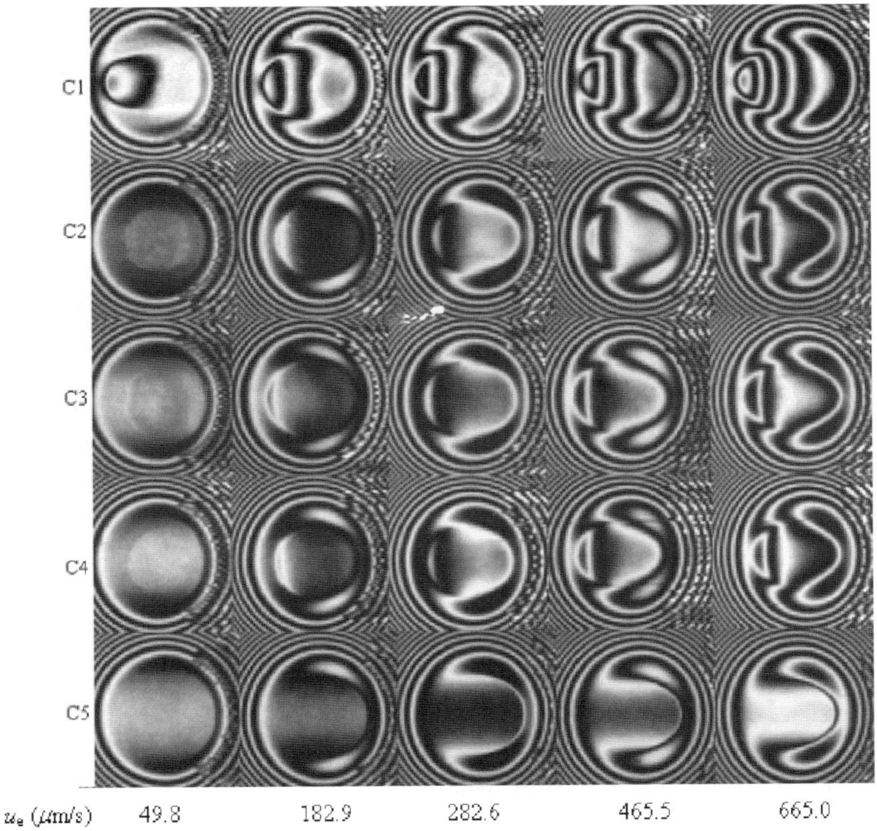

Figure 1. Effect of ZEV pre-running on the film formation, PB1300, load = 16N, condition definition: C1- simple disc sliding without ZEV pre-running; C2 - simple ball sliding without ZEV pre-running; C3 - simple disc sliding with ZEV pre-running; C4 - simple ball sliding with ZEV pre-running; C5 - Pure rolling.

For polymer melt/solid interfaces, it has been shown[7] that the critical sustainable shear stress and the degree of slippage are influenced by the interface properties. In the present set up, the tested lubricant was bounded

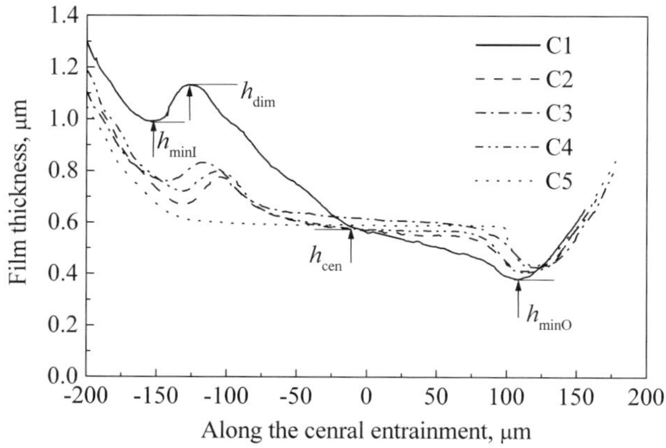

Figure 2. Film profiles along the central entrainment direction under different conditions, load = 16N, entrainment speed u_e = 465.5 μm/s, PB1300.

by surfaces of different materials i.e. two different interfaces oil/steel and oil/Cr. Hence, even under the same entrainment speed, simple disc sliding and simple ball sliding may induce different film shapes due to different slippage levels at the interface. Perhaps the effect of ZEV pre-running is due to the enhanced adsorption at the disc/lubricant interface under highly stressed conditions. No convincing theoretical explanation has yet been found to account for the interface modification effect of ZEVPR. Hitherto, the effect of ZEVPR has only been empirically confirmed.

3.2 Film Profiles under Different Slide-Roll Ratios

Since the film shape deviation described above is closely related to the boundary slippage, experiments were conducted under different slippage levels for further understanding of the film formation. Figure 3 gives the interferograms for PB1300 under different slide-roll ratios, in which no ZEV pre-running is applied. When changing the slide-roll ratios, the glass disc keeps running in the same direction. At pure rolling (slide-roll ratio $S = 0$), the film shape is that of conventional EHL. As the slide-roll ratio increases, a shallow dimple is gradually generated just before the outlet constriction and becomes enlarged, which is illustrated by interferograms at $S = 0.6$ and $S = 1.2$ in Figure 3(a). These dimples are unsteady and seem to be floating in the contact region, however, the minimum film thickness at the outlet remains stable and decreases with slide-roll ratio. With further increase of the slide-roll ratio, the dimple becomes steady and moves toward the inlet quickly.

Around S = 1.6, the depth of the dimple reaches the maximum of 630nm. At much higher slide-roll ratios, the depth of the dimple decreases. A further interesting observation is the change of the film in the downstream contact and the outlet constriction. With increasing sliding, the outlet constriction degrades continuously. As shown from S = 2.0 to S = 5.4 in Figure 3(a), a small bump occurs steadily in the region before the constriction. With the increase of the slide-roll ratio, this bump and the constriction merge into each other. The film profiles along the central entrainment are plotted in Figure 3(b). In the experiments, when increasing the sliding further, film collapse at the downstream region can be observed.

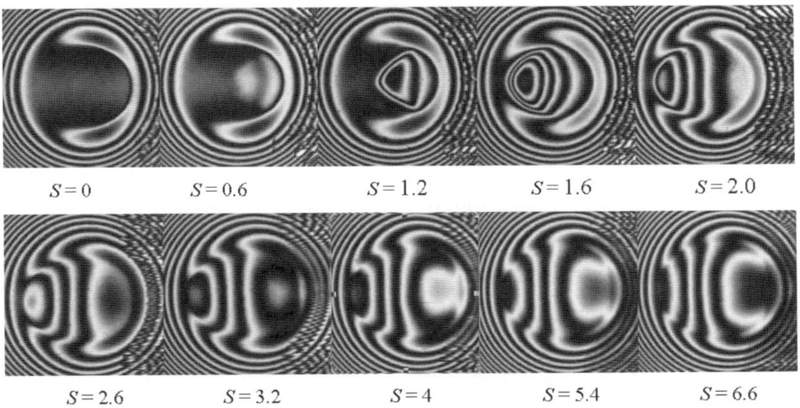

(a)

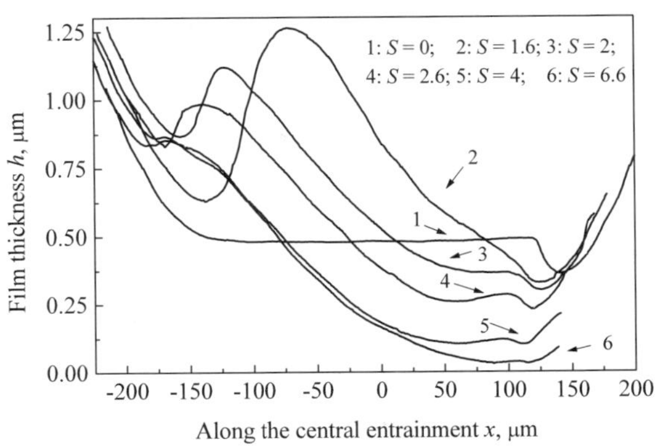

(b)

Figure 3. Film variation with slide-roll ratio, load = 22N, u_e = 332 μm/s, PB1300, no ZEV pre-running: (a) interferograms at different slide-roll ratios; (b) film profiles along the central entrainment at different slide-roll ratios.

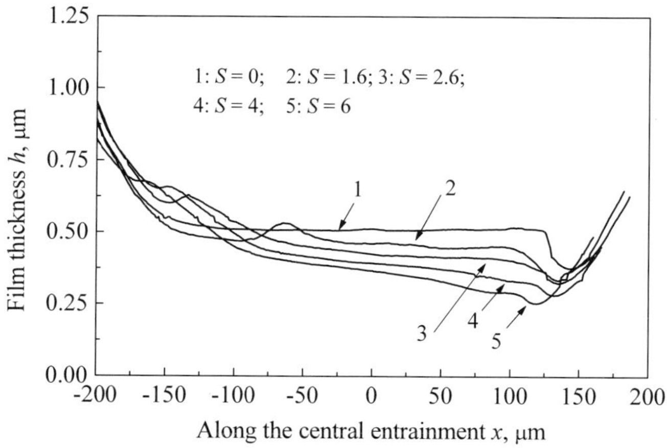

Figure 4. Film variation with slide-roll ratios along the central entrainment, load = 22N, u_e = 332 µm/s, PB1300, with ZEV pre-running.

For comparison, the same experiments have also been carried out with ZEV pre-running, the results are given in Figure 4. No obvious downstream bump is observed, which could be attributed to the enhanced interface affinity and then the decrease of slippage level by the ZEV pre-running.

PB2400 also was used in the experiments for obtaining higher levels of slippage due to its higher molecular weight. Similar phenomena can be observed for increasing slide-roll ratios. The dimple appears unsteady for slide-roll ratios lower than 1.0, and after that become stable and moves upstream. A downstream bump also occurs and the outlet constriction gradually becomes less pronounced. Most interesting is that with increasing sliding, the minimum film thickness changes its location from the outlet constriction to the bump, as shown from $S = 2.5$ (curve 4) to $S = 7$ (curve 7) in Figure 5(b). This change can also be found in some numerical work[10,13]. Compared to the result in Figure 3, the wedge part is much steeper, which is shown by the increasing number of strips in the interferograms. In Figure 5, when the slide-roll ratios are greater than 4, the dimple nearly moves out of the contact and disappears. With further increase of the slide-roll ratio, the film thickness decreases and the formed bump becomes flattened, and after that film rupture occurs.

It is thought that the observed film variations are closely linked to a change in boundary slippage. With increasing slide-roll ratios, the slippage is enhanced in two aspects. One is the increased relative sliding at the interface. The other is the extension of the slippage zone within the contact. These two aspects can generate the observed film variation. For more confirmation of the proposition, numerical analysis is needed.

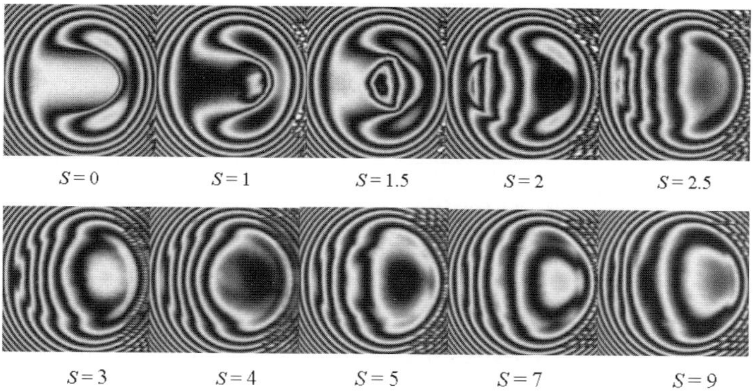

(a)

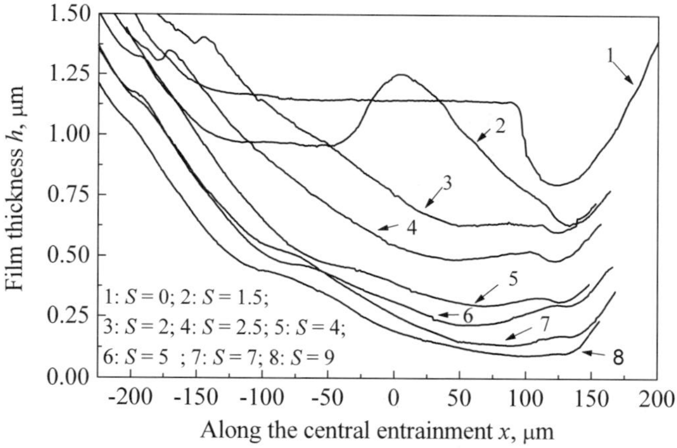

(b)

Figure 5. Film variation with slide-roll ratio, load = 22N, u_e = 199.5 μm/s, PB2400, with ZEV pre-running: (a) interferograms at different slide-roll ratios; (b) film profiles along the central entrainment at different slide-roll ratios.

4. DISCUSSION

The formation of the observed EHL film can be explained alternatively by an effective viscosity wedge which refers to the drastic drop of the effective viscosity adjacent to the bounding surfaces. The effective viscosity η_{eff} is defined by

$$\eta_{\text{eff}} = \frac{\tau}{\dot{\gamma}} \tag{1}$$

where τ is the shear stress and $\dot{\gamma}$ is the shear rate. The fall of the effective viscosity may come from different physical origins. One is closely related to the limiting shear stress of the lubricant. For the asymptotic limiting shear stress models proposed, for example, the Bair & Winer model and the circular model, there is a sharp decrease in the effective viscosity when the shear stress approaches to its limiting value. For the ideal viscoplastic model, if the shear stress reaches its limiting value, the effective viscosity will change discontinuously from the Newtonian viscosity to an infinitesimal value.

Another effective viscosity decrease may be generated by the wall slippage, which has been shown experimentally and theoretically[7,8]. The wall slippage can occur when the lubricant's shear stress exceeds the critical sustainable shear stress of the liquid/solid interface. Several physical models have been proposed and the slippage effect is linked with a decrease in the effective viscosity of a boundary layer immediately close to the solid surfaces[20].

In the present experiments, since thermal effects are neglected, the effective viscosity of lubricants is determined exclusively by the shear stress. In the present case, within the contact region near the inlet, usually the shear stress reaches its maximum at the moving glass surface. When the shear stress is close to (or attains) its limiting value (or the critical shear stress), the effective viscosity of the lubricant near the moving surface is larger than that away from the surface. Figure. 6 gives a schematic illustration of the viscosity wedge of two different models, it is assumed that the shear stress across the film increases from $0.6\tau_L$ at $Z = 0$ to $0.9999\tau_L$ at $Z = 1$, and the critical shear stress of the interface is also $0.9999\tau_L$ for illustration purposes, where τ_L is the limiting shear stress of the lubricant. The film thickness is one unit. For the Bair & Winer model, in the vicinity of the moving surface, there is a rapid drop of the effective viscosity. For the wall-slippage model, when the shear stress reaches the critical value, wall slip occurs and within the thin boundary layer of thickness δ, the viscosity is very low compared to the bulk viscosity. Due to this effective viscosity distribution, the lubricant flux across the film is smaller than that under Newtonian conditions, hence the local film thickness is forced to increase to maintain lubricant flow continuity. In a similar way, the wedge and the downstream bump can be explained. For more insight of this effective viscosity mechanism, deterministic numerical simulations are needed.

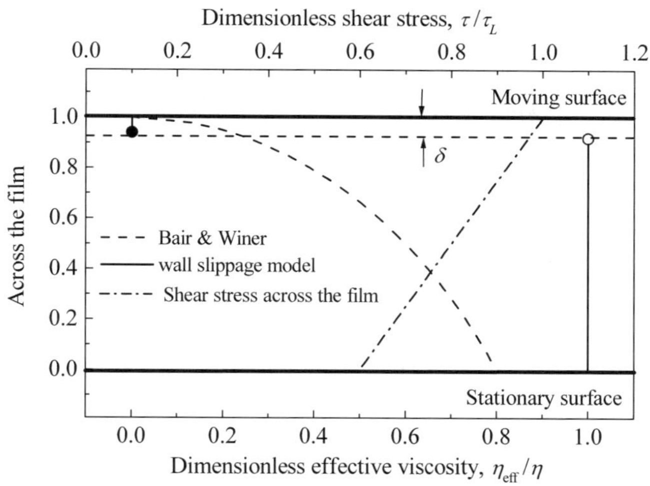

Figure 6. Schematic illustration of non-Newtonian viscosity wedge.

5. CONCLUSIONS

Using purpose-designed experiments, changes in film profiles under various slide-roll ratios have been studied to investigate the effect of limiting shear stress or wall slippage on EHL film formation. The observed film deviations can be explained by an effective viscosity wedge.

At the same entrainment speed, the simple disc sliding and the simple ball sliding do not generate the same film profiles. A ZEV pre-running was found to affect the film profile in the subsequent simple disc sliding test.

At low slide-roll ratios, an unstable dimple is formed. With increasing slippage levels, an inlet dimple, a wedged film gap in the upstream region and a downstream bump may occur. These film changes are related to different levels of the viscosity wedge and its extension within the contact region. In order to gain further understanding of the formation mechanism, deterministic numerical analysis is needed.

ACKNOWLEDGEMENTS

The authors would like to express their thanks to City University of Hong Kong for the provision of financial support for this work (strategic research grant number: 7001563). Dr. F. Guo also expresses his thanks to Qingdao Polytechnic University for partial support of this work by a Science Research Funding (C2004116).

REFERENCES

1. D. Dowson, G. Higginson and J. Archard, *Elastohydrodynamic Lubrication*, SI Edition, Pergamon Press, Oxford (1977).
2. R. Gohar, *Elastohydrodynamics*, Second Edition, Imperial College Press, London (2001).
3. K. H. Kim and F. Sadeghi, Non-Newtonian elastohydrodynamic lubrication of point contact, *Journal of Tribology* **113**, 703–711 (1991).
4. S. Wang, C. Cusano and T. F. Conry, Thermal analysis of elastohydrodynamic lubrication of line contact using the Ree–Eyring fluid model, *Journal of Tribology* **109,** 232–244 (1987).
5. H. J. Kim, P. Ehret, D. Dowson and C. M. Taylor, Thermal elastohydrodynamic analysis of circular contacts, Part 2: Non-Newtonian model, *Journal of Engineering Tribology* **215**, 353–362 (2001).
6. S. Bair and W. O. Winer, A rheological model for elastohydrodynamic contacts based on primary laboratory data, *Journal of Lubrication Technology* **101**, 258–265 (1979).
7. L. Leger, H. Hervet, G. Massey and E. Durliat, Wall slip in polymer melts, *Journal of Phys: Condens. Matter* **9**, 7719–7740 (1997).
8. Y. Zhu and S. Granick, Rate-dependent slip of Newtonian liquid at smooth surfaces, *Phys. Rev. Lett.* **87**, Paper 096105 (2001).
9. R. T. Lee and B. J. Hamrock, A circular non-Newtonian fluid model: Part I – Used in elastohydrodynamic lubrication, *Journal of Tribology* **112**, 486–496 (1990).
10. J. Shieh and B. J. Hamrock, Film collapse in EHL and micro-EHL, *Journal of Tribology* **113**, 372–377 (1991).
11. H. Iivonen and B. J. Hamrock, A new non-Newtonian fluid model for elastohydrodynamic lubrication of rectangular contacts, *Wear* **143**, 297–305 (1991).
12. P. Ehret, D. Dowson and C. M. Taylor, On lubricant transport conditions in elastohydrodynamic conjunctions, *Proc. Roy. Soc. Lond. A* **454**, 763–787 (1998).
13. Y. B. Zhang and S. Wen, An analysis of elastohydrodynamic lubrication with limiting shear stress: Part I – Theory and solutions, *Tribology Transactions* **45**, 135–144 (2002).
14. J. Stahl and Bo O. Jacobson, A lubricant model considering wall-slip in EHL line contacts, *Journal of Tribology* **125**, 523–532 (2003).
15. Y. P. Chiu and L. B. Sibley, Contact shape and non-Newtonian effects in elasto-hydrodynamic point contacts, *ASLE Lubrication Engineering* **28**, 48–60 (1972).
16. F. Guo and P. L. Wong, Experimental observation of a dimple-wedge elasto-hydrodynamic lubricating film, *Tribology International* **37**, 119–127 (2004).
17. D. W. Van.Krevelen and N. V. Arnhem, *Properties of Polymers*, 3rd Edition, Elsevier, Amsterdam (1997).
18. R. I. Tanner, *Engineering Rheology*, 2nd Edition, Oxford University Press, Oxford (2000).
19. F. Guo and P. L. Wong, A multiple-beam intensity-based approach for thin lubricant film measurement in non-conformal contacts, *Journal of Engineering Tribology* **216**, 281–291 (2002).
20. O. I. Vinogradova, Slippage of water over hydrophobic surfaces, *Int. J. Miner. Process* **56**, 31–60 (1999).

ELASTOHYDRODYNAMIC LUBRICATION IN '*SOFT-ON-SOFT*' NATURAL SYNOVIAL JOINTS; '*HARD-ON-SOFT*' CUSHION AND '*HARD-ON-HARD*' METAL-ON-METAL TOTAL JOINT REPLACEMENTS

Duncan Dowson
School of Mechanical Engineering, The University of Leeds, Leeds LS2 9JT, U.K.

Abstract The modes of lubrication in natural, compliant and metal-on-metal hip replacements are examined. While mixed lubrication appears to dominate the performance of natural and replacement joints, elastohydrodynamic action can make substantial and important contributions to load support in both. It is therefore important to analyse EHL and micro EHL to ascertain and optimize the contributions that these actions can make to mixed lubrication of human joints. A simple EHL design guide is offered for metal-on-metal total joint replacements.

Keywords: synovial joints, compliant/cushion joints, metal-on-metal total joint replacements, boundary lubrication, elastohydrodynamic lubrication (EHL), mixed lubrication.

1. INTRODUCTION

It has emerged over the past forty or fifty years that elastohydrodynamic action plays a role in the lubrication of both natural and certain forms of total replacement human joints.

1.1 '*Soft-on-Soft*' Natural Synovial Joints

While it has long been recognized that major synovial joints such as the hip and the knee are remarkable plain bearings a full understanding of their basic mechanism of lubrication has been elusive. Some experimental findings and some theoretical predictions have supported the concept of boundary lubrication, while others have suggested fluid-film action. It is unfortunate that much effort has been spent seeking support exclusively for one or other of these

R.W. Snidle and H.P. Evans (eds), IUTAM Symposium on Elastohydrodynamics and Microelastohydrodynamics, 297–308.

polarized views, while in reality both seem to contribute to the impressive tribological behavior of joints through mixed lubrication.

If fluid-film lubrication does play a role in the normal functioning of natural synovial joints, it must be elastohydrodynamic (EHL) in nature. This concept was first investigated in the 1960s, but although the predicted film thicknesses of about 1 μm were two or three orders of magnitude greater than the values derived from conventional hydrodynamic analysis, they provided little support for the concept of fluid-film action since the cartilage bearing surfaces each exhibited surface roughnesses in the range 1–5 μm.

The position changed dramatically in the 1980s (Dowson and Jin, 1986) when micro-elastohydrodynamic action was shown to result in much smoother effective bearing surfaces. It then became clear that effective elasto-hydrodynamic lubrication could be achieved under typical activity levels in highly loaded synovial joints in the body.

The development of understanding of beneficial EHL action in synovial joints is outlined in Section 2.

1.2 '*Hard-on-Soft*' Cushion Implants

During the past two decades tribologists have considered the possibility of mimicking more effectively nature's approach to joint lubrication. In particular, the recognition of EHL action in natural synovial joints prompted the suggestion that low elastic modulus bearing layers on a hard backing, instead of the increasingly hard materials now being used, might improve the development of fluid-film lubrication. These proposals are considered in Section 3, under the title of compliant or cushion bearings.

1.3 '*Hard-on-Hard*' Metal-on-Metal Total Joint Replacements

The use of polymeric acetabular cups in association with metallic or ceramic femoral heads has provided the basis of implant development during the past fifty years. The life of such bearings is limited primarily by osteolysis following the generation of large quantities of polymeric wear debris and this has resulted in a renaissance of ceramic-on-ceramic and metal-on-metal implants; the so called '*hard-on-hard*' solutions.

Remarkably, this enhances the possibility of effective EHL action throughout part of each cycle of articulation. This finding has resulted in the full analysis of lubrication in such joints and the development of useful design guides. These developments are outlined in Section 4.

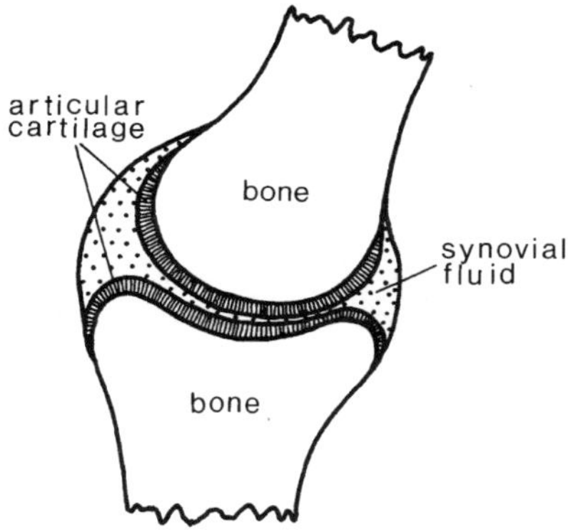

Figure 1. Representation of synovial joint.

2. '*SOFT-ON-SOFT*' ELASTO-HYDRODYNAMIC LUBRICATION OF SYNOVIAL JOINTS

It has long been recognised that synovial joints are simple plain bearings with lubricant contained within the joint capsule (Figure 1).

Two hundred and sixty years ago Sir William Hunter (1743) noted that the bony ends of articulating joints were:

> *enclosed in a bag that contains a proper fluid deposited there, for lubricating the two contiguous surfaces.*

In his classical paper establishing the fundamental principles of fluid-film lubrication, Osborne Reynolds (1886) perceived that:

> *The only other self-acting system of lubrication is that of reciprocating joints with alternate pressure on and separation (drawing the oil back or a fresh supply) of the surfaces ... and is as fundamental to animal mechanics as the lubricating action of the journal is to mechanical contrivances.*

These early suggestions that synovial joints were lubricated by hydrodynamic action prevailed throughout much of the early half of the 20th Century and were endorsed by MacConaill (1932) from anatomical considerations and Jones (1936) on the basis of his direct measurement of friction in a freshly amputated human finger joint.

Charnley (1979) was of the view that boundary lubrication, so clearly revealed by Hardy and Doubleday (1936) in 1922, was the most likely mode of lubrication, since it was:

> *ideally suited to the lubrication of slow-moving, heavily loaded animal joints and especially since these were exposed to oscillating motion.*

McCutchen (1959), proposed the concept of weeping lubrication in a joint acting as a self-pressurised hydrostatic bearing. Maroudas (1967) drew attention to the existence of gels having thicknesses in the range 50 to 200 Å on the surface of cartilage after loading, while Walker et al. (1968) introduced the concept of boosted lubrication with enhanced lubricant viscosity resulting from increased hyaluronic acid concentration as the porous surfaces approached each other in squeeze-film motion.

The case for fluid-film action in synovial joints received renewed support from friction experiments on a new form of pendulum machine introduced by Unsworth et al. (1975) in the mid 1970s. When dynamic loads were applied to a hip joint mounted on hydrostatic bearings (O'Kelly et al., 1978), it became clear that the deduced coefficients of friction and their variation with speed or load were entirely consistent with the occurrence of fluid-film rather than boundary lubrication.

2.1 Elasto-Hydrodynamic Lubrication

The principal load bearing joints in the lower limb are the hip, the knee and the ankle. The former is modeled as a sphere in a near hemispherical cup, the latter as a cylindrical shaft in a partial arc bearing and the knee as ellipsoidal condyles in two ellipsoidal partial arc support bearings on the tibia. All have been subjected to elastohydrodynamic analysis for the appropriate elastic-isoviscous conditions. The joints experience dynamic loads and varying entraining speeds as they are subjected to reciprocating motion in normal articulation. Studies have ascertained the magnitudes and directions of these quantities with reasonable accuracy, but uncertainty still exists about the effective viscosity of the synovial fluid at the high shear rates encountered, the elastic characteristics of the layers of cartilage and the effective radius of the various geometrical configurations encountered in the joints of the lower limb. Full and rather lengthy dynamic analyses of the elastohydrodynamic action in these joints have been completed, but important findings by Medley and Dowson (1984) and Dowson and Jin (1987) for the hip, ankle and knee respectively indicated that the minimum film thicknesses remained remarkably constant throughout each severe cycle of dynamic operation as shown in Figure 2. These findings were attributable to the resilience of squeeze film action, which maintained a near constant film thickness notwithstanding the severe cyclic variations of load and speed.

This important finding provided a convenient basis for the estimation of EHL film thicknesses in synovial joints. It demonstrated that if a representative set of data could be introduced into the appropriate steady-state film thickness equation the quasi-static film thicknesses representative of the full dynamic solutions could be predicted with fair accuracy. The ranges of values obtained for the hip, knee and ankle are shown in Table 1.

The three most important features of the results displayed in Table 1 are that:

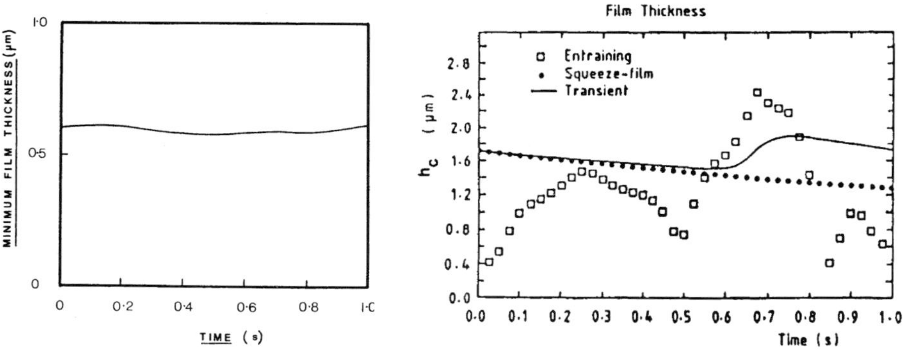

(a) Human ankle joint,
Medley and Dowson (1984)

(b) Human hip joint,
Jin et al. (1993)

Figure 2. Cyclic variation of predicted EHL minimum film thickness.

Table 1. Calculated EHL minimum film thicknesses in synovial joints in the lower limb.

	Hip	Knee	Ankle
Minimum EHL film thickness (μm)	(4–6)[1]	(2.0)[1]	(0.6)[4]
Minimum EHL film thickness (μm)		(0.1)[2]	
Minimum EHL film thickness (μm)	(1.3)[3]	(1.25)[3]	
Central EHL film thickness (μm)	(0.5–1.9)[5]		
Minimum hydrodynamic film thickness (μm)		(0.000125)[1] (0.00049)[3]	

[1]Dowson (1967); [2]Higginson (1977); [3]Dowson (1981); [4]Medley and Dowson (1984); [5]Jin et al. (1993).

- the film thicknesses in each of the load bearing joints in the lower limb are similar;
- the predicted hydrodynamic film thicknesses are at least three orders of magnitude smaller than the calculated EHL film thicknesses;
- the EHL film thicknesses have an order of magnitude of about 1 μm.

An indication of the mode of lubrication in bearings can often be obtained by relating the predicted film thicknesses to the composite surface roughnesses. In the early 1960s (Davies et al., 1962) it was thought that cartilage was very smooth with surface irregularities in the range (0.02–0.2) μm. However, later in the decade Jones and Walker (1968) recorded R_a values of 1 μm for foetal cartilage, 2.7 μm for healthy adult cartilage and 5.3 μm for osteoarthritic cartilage. These values were at least 2000 times greater than the hydrodynamic

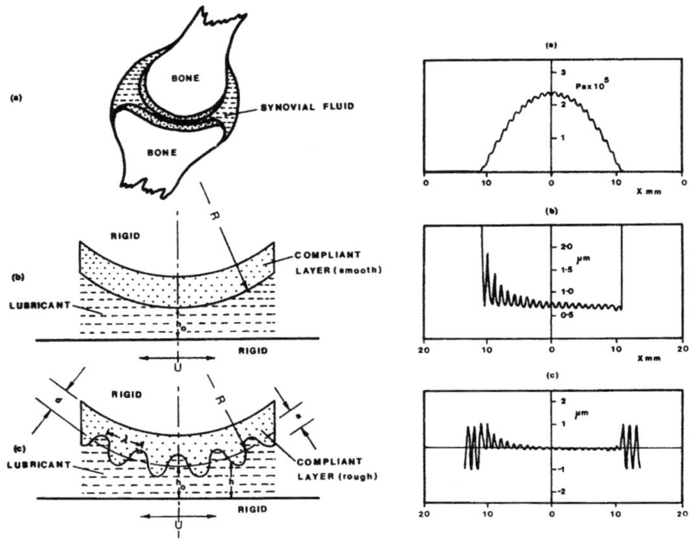

Figure 3. Micro-EHL solution for synovial joint.

film thickness, but similar to the predicted elasto-hydrodynamic film thicknesses. The mode of lubrication in synovial joints thus appeared to be *mixed*, but with EHL action well able to support part of the applied load and to relieve the severity of asperity contact.

2.2 Micro-Elastohydrodynamic Lubrication

Even stronger support favouring elasto-hydrodynamic action in synovial joints emerged in the late 1980s with the development of micro-elastohydrodynamic analysis of rough cartilage surfaces. Dowson and Jin (1986) completed a micro-elastohydrodynamic analysis of cartilage of thickness 2.4 mm, with surface roughness modeled by a sinusoidal profile of amplitude 1 μm and wavelength 1 mm. The results were remarkable, as depicted in Figure 3.

The pressure profiles predicted for smooth surfaces were little disturbed in the rough surface solutions and the small ripples of pressure generated by the asperities reduced their amplitudes enormously. This resulted in micro-EHL film thickness predictions of surface separations of about 0.6 μm (hip) and 0.5 μm (knee). The smooth surface prediction for the latter case was 0.58 μm.

The case for EHL, or rather micro-EHL, contributing to effective synovial joint lubrication during those parts of the cycle conducive to micro-EHL action was thus greatly strengthened. The near constancy of the predicted smooth surface film thickness demonstrated earlier must still be related to the unloaded roughness of cartilage during the cycle of articulation when the sliding velocities are low and the potential for micro-EHL is limited.

Table 2. Elastic modulus of natural and prosthetic materials.

Material	Modulus of Elasticity (GPa)	Material	Modulus of Elasticity (GPa)
natural rubber	0.0025–0.1	UHMWPE	0.5
articular cartilage	0.001–0.17	bone cement (PMMA)	3.0
bone	10–30	Co-Cr-Mo alloy	230
		alumina	350

The overall assessment of the lubrication mechanism in synovial joints thus favours mixed lubrication, just as it did in the 1960s. The feature that has changed is that the contribution of EHL can now be determined much more clearly.

3. *'HARD-ON-SOFT'* CUSHION IMPLANTS

While current total replacement hip joints perform well enough to bring great relief to many patients, a brief examination of their form shows that they are quite different from natural (*'soft-on-soft'*) joints in some important respects. The most widely used prostheses during the past fifty years have used ultra-high molecular weight polyethylene (UHMWPE) acetabular cups with metallic or ceramic femoral heads; a *'hard-on-soft'* configuration. The properties of the bearing materials are thus quite different from those of the elastic porous layers of cartilage on relatively hard foundations of bone in the natural joint. Furthermore, they yield substantial quantities of UHMWPE wear debris which promote osteolysis and perhaps loosening and failure of the implant.

In recent times there has been an increasing tendency to use ceramic-on-ceramic or metal-on-metal pairs of materials; the so called *'hard-on-hard'* material combinations. These forms of prostheses are becoming increasingly popular, primarily because they avoid the issue of UHMWPE wear debris generation, but also because they yield much reduced volumes of ceramic or metallic wear debris, although the wear particles are smaller and more numerous. The drift towards harder and harder material combinations has moved ever further away from the natural plain bearing arrangement as shown in Table 2.

Can tribologists design joint replacements that more closely reflect the bearing solution deemed to be most appropriate by nature? One intriguing possibility would be to use a compliant or cushion layer of relatively soft and maybe porous bearing material in the acetabular cup articulating against a metallic or ceramic femoral head. This biomimetic concept is illustrated in Figure 4.

A considerable amount of experimental and analytical work has been carried out on this simple bearing configuration (Cudworth and Higginson, 1976; Unsworth et al., 1987; Dowson, 1989; Auger et al., 1990; Gladstone and Med-

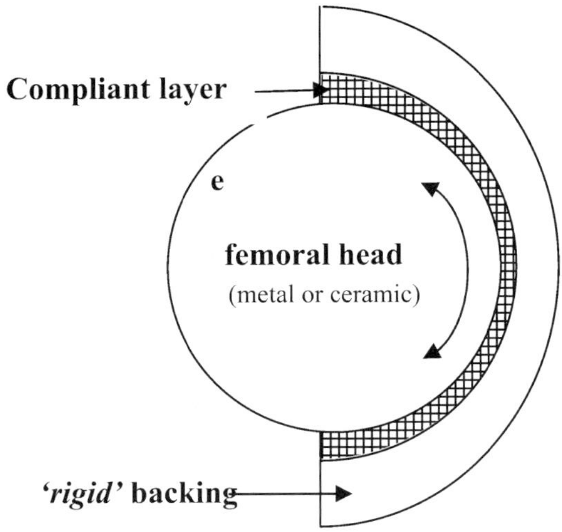

Figure 4. Compliant/cushion hip joint concept.

ley, 1990; O'Carroll et al., 1990; Blamey et al., 1991; Dowson et al., 1991; Jin et al., 1991; Caravia et al., 1993; Auger et al., 1995a, b; Smith et al., 1997; Bigsby et al., 1998). It has been demonstrated in laboratory tests that it is possible to design the bearing and to select the materials, both hard and soft, to achieve satisfactory fluid-film lubrication (EHL), but some concerns about long term stability of the soft, elastic liner and the integrity of its attachment to the hard backing persist.

It is evident however that this is a clear example of EHL action in potential joint replacements, for if the compliant or cushion bearing proves to be successful it will need to call upon elastohydrodynamic action to lubricate and protect the relatively soft bearing surfaces.

4. '*HARD-ON-HARD*' METAL-ON-METAL TOTAL JOINT REPLACEMENTS

The dominant Charnley type metal or ceramic on UHMWPE hip replacement (*hard-on-soft*) is being challenged by new generations of ceramic-on-ceramic or metal-on-metal (*hard-on-hard*) implants. The background has been outlined elsewhere (Dowson, 2001), but the major driver behind this development has been concern for the limitations imposed upon function and life of implants by excessive generation of UHMWPE wear debris.

The metal or ceramic on UHMWPE joints appear to operate in the boundary or at best severely taxed mixed lubrication regime, with little opportunity for fluid film lubrication to relieve substantially the severity of contact between the articulating surfaces. Indeed, the low friction arthroplasty concept introduced by Charnley (1961) relied upon the use of a low friction polymer sliding on a relatively hard counterface. The materials chosen were widely used in polymeric rubbing bearings in engineering applications.

The development of ceramic and metallic acetabular cups offering much smoother bearing surfaces at once opened up the possibility of developing fluid-lubricating films during at least part of the operating cycle. It can readily be shown that the mode of lubrication attainable is elastohydrodynamic. The appropriate minimum film thickness equation for the dominant isoviscous-elastic regime has been presented by Hamrock and Dowson (1987) in the form:

$$H_{\min} = \frac{h_{\min}}{R_x} = 2.8 \left\langle \frac{\eta u}{E' R_x} \right\rangle^{0.65} \left\langle \frac{w}{E' R_x^2} \right\rangle^{-0.21}. \tag{1}$$

The entraining velocity is proportional to the head diameter and it is clear that:

$$\frac{1}{R_x} = \frac{1}{R_h} - \frac{1}{R_c} = \frac{c_r}{R_h R_c} \quad \text{or} \quad \frac{2c_d}{d^2}. \tag{2}$$

The minimum film thickness for fixed values of load, rotational speed, viscosity and elastic constants is related to the joint geometry by

$$h_{\min} \propto (\text{head diameter})^{2.19} \times (\text{clearance})^{-0.76}. \tag{3}$$

The most beneficial film thickness will thus be secured by making the head diameter as large as possible while keeping the clearance as small as practicable. This result provides a good tribological (EHL) basis for the design of hard-on-hard implants, whether they be of monolithic or surface (cartilage) replacement form.

The effects of head diameter and clearance upon wear of metal-on-metal hip replacements has been demonstrated experimentally on a ten station hip joint simulator by Smith et al. (2001a, b) and Dowson (2003).

As the EHL film thickness increases it bears an increasing proportion of the applied load, thus reducing the wear rate associated with metal-to-metal contact. This is well demonstrated by Figure 5 which shows the steady state wear rate of different designs of metal-on-metal joints measured on joint simulators in several laboratories around the world. It appears that theoretical film thicknesses in excess of about (20–30) nm ensure modest steady-state wear rates of 0.1 $mm^3/10^6$ cycles or less. This observation is quite consistent with the surface finish of well manufactured metal-on-metal hip joints, since most current femoral heads and acetabular cups exhibit roughnesses in the range 5 to 10 nm.

5. CLOSURE

Elastohydrodynamic lubrication studies of natural synovial joints and total joint replacements have:

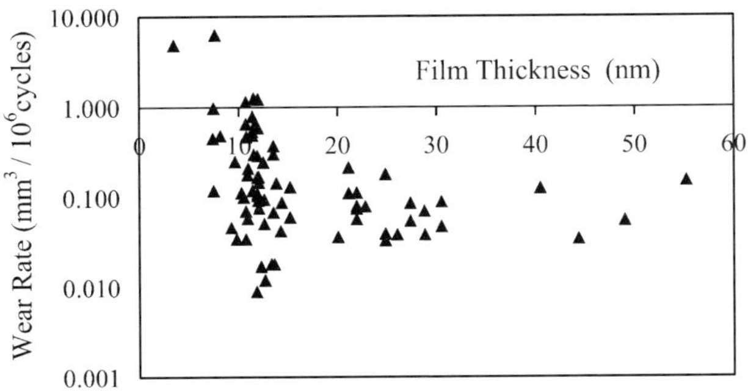

Figure 5. Metal-on-metal hip joint; steady-state wear rate vs EHL film thickness.

- enlightened understanding of the mechanism of lubrication in natural synovial joints;
- guided the design and manufacture of the new forms of ceramic-on-ceramic and metal-on-metal hip replacements;
- offered guidance for the development of compliant/cushion joints which more closely reflect natures design solution.

Elastohydrodynamic action alone does not appear to describe fully the mode of lubrication in natural synovial joints or total joint replacements. This is not at all unexpected for such highly taxed reciprocating bearings. If it applied throughout the activity cycles of daily life, no wear would occur and this is clearly not the case in either natural or total replacement joints.

Mixed lubrication undoubtedly applies in the joints considered, with full load support by fluid-film lubrication being possible during part of the cycle and reliance upon boundary lubrication under the most severe conditions, as it does in many engineering situations.

The importance of EHL studies in this field is that they can more completely define the contribution that fluid-film lubrication undoubtedly makes to the very effective total ('boundary' plus 'EHL') load support system in mixed lubrication.

ACKNOWLEDGEMENTS

The author is pleased to acknowledge permission to include in this paper the two parts of Figure 2 which were originally published by (a) *ASLE Transactions*, **27**, (1984) and (b) Elsevier (*Proceedings of the 19th Leeds–Lyon Symposium on Tribology*, 1993), respectively.

REFERENCES

Auger, D.D., Dowson, D. and Fisher, J., 1995(a), Cushion form bearings for total knee replacement. Part 1: Design, friction and lubrication, *Proc. Instn. Mech. Engrs., Part H*, **209**(H2), 72–81.

Auger, D.D., Dowson, D. and Fisher, J., 1995(b), Cushion form bearings for total knee replacement. Part 2: Wear and durability, *Proc. Instn. Mech. Engrs., Part H*, **209**(H2), 83–91.

Auger, D.D., Medley, J.B., Fisher, J. and Dowson, D., 1990, A preliminary investigation of the 'cushion bearing' concept for joint replacement implants, in *Mechanics of Coatings*, Tribology Series 17, D. Dowson, C.M. Taylor and M. Godet (eds), Elsevier, Amsterdam, pp. 251–260.

Auger, D.D., Dowson, D., Fisher, J. and Jin, Z.-M., 1993, Friction and lubrication in cushion form bearings for artificial hip joints, *Proc. Instn. Mech. Engrs., Part H*, **207**, 25–33.

Bigsby, R.J.A., Auger, D.D., Jin, Z.-M., Dowson, D., Hardaker, C.S. and Fisher, J., 1998, A comparative tribological study of the wear of composite cushion cups in a physiological hip joint simulator, *J. Biomechanics*, **31**, 363–369.

Blamey, J., Rajan, S., Unsworth, A. and Dawber, R., 1991, Soft layered prostheses for arthritic hip joints: A study of materials degradation, *J. Biomed. Engng.* **13**, 180–184.

Caravia, L., Dowson, D. and Fisher, J., 1993, Startup and steady state friction of thin polyurethane layers, *Wear*, **160**, 191–197.

Charnley, J., 1961, Arthroplasty of the hip. A new operation, *Lancet*, **1**, 1129–1132.

Charnley, J., 1979, *Low Friction Arthroplasty of the Hip*, Springer-Verlag, Berlin/Heidelberg, 376 pp.

Cudworth, C.J. and Higginson, G.R., 1976, Friction of lubricated soft surface layers, *Wear*, **37**, 299–312.

Davies, D.V., Barnett, C.H., Cochrane, W. and Palfrey, A.J., 1962, Electron microscopy of articular cartilage in the young rabbit, *Ann. Rheum. Dis.*, **21**, 11.

Dowson, D., 1967, Modes of lubrication in human joints, 'Lubrication and wear in living and artificial human joints', in *Instn. Mech. Engrs. Proceedings 1966–1967*, Volume 181, Part 3J, London, pp. 45–54.

Dowson, D., 1981, Lubrication of joints; A natural joints, in *An Introduction to the Biomechanics of Joints and Joint Replacement*, D. Dowson and V. Wright (eds), Mechanical Engineering Publications Ltd., London, pp. 120–133.

Dowson, D., 1989, Are our joint replacements adequate?, in *The Changing Role of Engineering in Orthopaedics*, Mechanical Engineering Publications, London, pp. 1–5.

Dowson, D., 2001, New joints for the millennium: Wear control in total replacement hip joints, James Clayton Memorial Lecture, *Proc. Instn. Mech. Engrs., Part H*, **215**, 335–358.

Dowson, D., 2003, The Relationship between steady-state wear rate and theoretical film thickness in metal-on-metal total replacement hip joints, in *Proceedings of the 29th Leeds–Lyon Symposium on Tribology*, D. Dowson, M. Priest, G. Dalmaz A.A. and Lubrecht (eds), Tribology Series 41, Elsevier, Amsterdam, pp. 273–280.

Dowson, D. Fisher, J., Jin, Z.M., Auger, D.D. and Jobbins, B., 1991, Design considerations for cushion form bearings in artificial hip joints, *Proc. Instn. Mech. Engrs., Part H*, **205**(H2), 59–68.,

Dowson, D. and Jin, Z.-M., 1986, Micro-elastohydrodynamic lubrication of synovial joints, *Engrg. Med.*, **15**(2), 63–65.

Dowson, D. and Jin, Z.M., 1987, An analysis of micro-elasto-hydrodynamic lubrication in synovial joints considering cyclic loading and entraining velocities, in *Proceedings of the 13th Leeds–Lyon Symposium on Tribology*, 8–12 September 1986, Elsevier, Amsterdam, pp. 375–386.

Gladstone, J.R. and Medley, J.B., 1990, Comparison of theoretical and experimental values for friction of lubricated elastomeric surface layers under transient conditions, in *Mechanics of Coatings*, D. Dowson, C.M. Taylor and M. Godet (eds), Tribology Series 21, Elsevier, Amsterdam, pp. 241–250.

Hardy, W.B. and Doubleday, I., 1922, Boundary lubrication-the paraffin series, *Proc. Roy. Soc.*, **A100**, 550–574.

Higginson, G.R., 1977, Elastohydrodynamic lubrication in human joints, *Proc. Instn. Mech. Engrs.*, **131**, 34–77.

Hunter, W., 1743, Of the structure and diseases of articulating cartilages, *Phil. Trans.*, **42**, 514.

Jin, Z.M., Dowson, D. and Fisher, J., 1991, Stress analysis of cushion form bearings for total hip replacements, *Proc. Instn. Mech. Engrs., Part H*, **205**(H4), 219–226.

Jin, Z.M., Dowson, D. and Fisher, J., 1993, Fluid film lubrication in natural hip joints, in *Proceedings of the 19th Leeds–Lyon Symposium on Tribology*, 8–11 September 1992, D. Dowson, C.M. Taylor, T.H.C. Childs, M. Godet and G. Dalmaz (eds), Tribology Series 25, Elsevier, Amsterdam, pp. 545–555.

Jones, E.S., 1936, Joint lubrication, *Lancet*, **1**, 1426.

Jones, H.P. and Walker, P.S., 1968, Casting techniques applied to the study of human joints, *J. Inst. Sci. Technol.*, **14**(3), 57–58.

MacConaill, M.A., 1932, Function of intra-articular fibrocartilage with special reference to the knee and inferior radio-ulnar joints, *J. Anat.*, 66, 210.

Maroudas, A., 1967, Hyaluronic acid films, in *Lubrication and Wear in Living and Artificial Human Joints, Proc. Instn. Mech. Engrs.*, **181**(3J), 122–124.

McCutchen, C.W., 1959, Sponge-hydrostatic and weeping bearings, *Nature*, Lond., 184, 1284.

Medley, J.B. and Dowson, D., 1984, Lubrication of elastic-isoviscous line contacts subject to cyclic time-varying loads and entrainment velocities, *ASLE Trans.*, **27**(3), 243–251.

O'Carroll, S., Jin, Z.-M., Dowson, D., Fisher, J. and Jobbins, B., 1990, Determination of contact area in 'cushion form' bearings for artificial hip joints, *Proc. Instn. Mech. Engrs., Part H*, **204**(H4), 217–223.

O'Kelly, J.O., Unsworth, A., Dowson, D., Hall, D.A. and Wright, V., 1978, A study of the role of synovial fluid and its constituents in the friction and lubrication of human joints, *Engrg. Med.* **7**(2), 73–83.

Reynolds, O., 1886, On the theory of lubrication and its application to Mr. Beauchamp Tower's experiments, including an experimental determination of the viscosity of olive oil, *Phil. Trans. Roy. Soc.*, **177**, 157–234.

Smith, S.C., Burgess, I.C. and Unsworth, A.U., 1997, Friction and wear testing of compliant layer hip joints on a hip simulator, *Br. Orthop. Res. Soc.*, **40**.

Smith, S.L., Dowson, D. and Goldsmith, A.A.J., 2001(a), The effect of femoral head diameter upon the lubrication and wear of metal-on-metal total hip replacements, *Proc. Instn. Mech. Engrs., Part H*, **215**(H2), 161–170.

Smith, S.L., Dowson, D. and Goldsmith, A.A.J., 2001(b), The effect of diametral clearance, motion and loading cycles upon lubrication of metal-on-metal total hip replacements, *Proc. Instn. Mech. Engrs., Part C*, **215**(C1), 1–5,

Unsworth, A., Dowson, D. and Wright, V., 1975, The frictional behaviors of human synovial joints – Part I: Natural joints, *Trans. ASME, F, J. Lubrication Technol.*, **97**(3), 369–376.

Unsworth, A.U., Pearcy, M.J., White, E.F.T. and White, G., 1987, Soft layer lubrication of artificial hip joints, in *Proceedings of IMechE Int. Conf. on Fifty Years On*, pp. 715–724.

Walker, P.S., Dowson, D., Longfield and Wright, V., 1968, Boosted lubrication in synovial joints by fluid entrapment and enrichment, *Ann. Rheum. Dis.*, **6**, 512–520.

SESSION 8

STRIBECK CURVE FOR STARVED CONCENTRATED CONTACTS

I.C. Faraon and D.J. Schipper
Laboratory of Surface Technology and Tribology, Faculty of Engineering Technology, University of Twente, Enschede, The Netherlands; E-mail: i.c. faraon@ctw.utwente.nl

Abstract This paper discusses a mixed lubrication model in order to predict the Stribeck curve for starved lubricated line contacts. The model is an extension of the mixed lubrication model of Gelinck and Schipper [1]. In order to build the starved Stribeck curve model, the contact model of Greenwood and Williamson [2] and the EHL film thickness for starved line contacts making use of the assumption of Johnson et al. [3] is combined. The starved solution to be implemented in the EHL component is obtained by fitting numerical data of Wolveridge et al. [4] who computed the starved film thickness for smooth line contacts. Calculations are presented for different oil supply layer thickness over roughness values. For values of oil layer thickness over roughness ratio (h_{oil}/σ_s) larger than approximately 6, the Stribeck curve and separation do not change. If the oil layer thickness over roughness ratio is in the range of 6 to 0.7 friction starts to increase and the film thickness decreases. When the oil layer thickness over roughness ratio is less than approximately 0.7 the Stribeck curve tends to transform into a straight line and separation stays on the same value as in the BL regime.

Keywords: Stribeck curve, starved lubrication, mixed lubrication, film thickness, oil layer thickness over roughness ratio.

1. INTRODUCTION

One of the developments in design is to reduce the size of the components in constructions while transmitting the same or even higher loads, resulting in severe operational contact conditions. This means that the contact between the different components do not operate anymore under (elasto-) hydrodynamic lubrication (EHL) conditions but mixed lubrication (ML) or even boundary lubrication (BL) is more likely to occur. Therefore, the Stribeck curve is an important tool to determine in which lubrication regime a contact operates. In Gelinck and Schipper [1] a model is presented in order to predict the Stribeck curve for line contacts. This model is based on the combination of the Greenwood and Williamson [2] contact model and the full film theory using the

R.W. Snidle and H.P. Evans (eds), IUTAM Symposium on Elastohydrodynamics and Microelastohydrodynamics, 311–320.

mixed lubrication model of Johnson et al. [3]. With this model one is able to predict friction and determine the transitions between the different lubrication regimes i.e. (Elasto-) Hydrodynamic Lubrication (EHL), Mixed Lubrication (ML) and Boundary Lubrication (BL). This model is based on the assumption that enough lubricant is supplied to the contact, e.g. fully flooded conditions. However, in many applications the amount of lubricant available is not sufficient and the contacts operate under starved lubrication conditions. For a designer it is very important to know the friction and the transitions between the different lubrication regimes (EHL, ML and BL) under starved lubrication conditions.

In the literature much attention is paid to starvation, however, the investigations were conducted for the full-film situation (no contact takes place between the opposing surfaces) with the emphasis on the inlet boundary conditions and cavitation. Most of these investigations were applied to the circular contact situation. For the starved line contact situation, a few were published, for instance Wolveridge et al. [4].

In the work of Wolveridge et al. [4] a correction on the film thickness formula for line contacts due to starvation is presented. Combining this modified film thickness relation for starved line contacts with the model of Gelinck and Schipper [1] will result in a mixed lubrication model for starved line contacts.

In this article the consequences of starvation on friction depicted in the generalised Stribeck curve due to change in film formation are discussed.

2. MIXED LUBRICATION MODEL

The total normal load F_T is shared by the hydrodynamic action and the interacting asperities:

$$F_T = F_C + F_H, \tag{1}$$

where F_C is the load carried by the asperities and F_H the load carried by the hydrodynamic component. Therefore, friction in mixed lubrication is a summation of the friction due to asperity interaction (boundary lubrication) and friction due to shearing the lubricant.

2.1 Load Carried by Asperity Interaction

Greenwood and Williamson [2] showed that the load carried by the asperities is given by:

$$p_a(x) = \frac{2}{3} n \beta \sigma_s \sqrt{\frac{\sigma_s}{\beta}} E' F_{3/2}\left(\frac{h(x)}{\sigma_s}\right) \tag{2}$$

with h the separation between two surfaces, n the density of the asperities, β the average radius of the asperities, and σ_s the standard deviation of the height distribution of the asperities. $F_{3/2}(h)$ is defined as:

$$F_{3/2}(h) = \int_h^{\infty} (s-h)^{3/2}\, \phi(s)\, \mathrm{d}s, \tag{3}$$

where $\phi(s)$ is the height distribution of the asperity summits. In this paper a Gaussian height distribution of the asperities has been assumed:

$$\phi(s) = \frac{1}{\sqrt{2\pi}}\, \mathrm{e}^{-1/2s^2}. \tag{4}$$

The elasticity modulus used in Equation (2) is defined as:

$$\frac{2}{E'} = \frac{1-\nu_1^2}{E_1} + \frac{1-\nu_2^2}{E_2} \tag{5}$$

with E_i the elasticity modulus of the surfaces and ν_i the Poisson's ratio.

In Gelinck and Schipper [1] the pressure distribution was calculated on the basis of Equation (2), making the nominal pressure $p[h(x)]$ due to the deformed asperities consistent with the gap $h[p(x)]$ resulting from the bulk deformation. A relation for the central (maximum) pressure p_C compared to its smooth value as characterising the complete pressure distribution is derived:

$$\frac{p_C}{p_{\mathrm{Hertz}}} = \left[1 + \left(a_1 n'^{\,a_2} \sigma_{s'}^{\,a_3} W^{a_2-a_3}\right)^{a/4}\right]^{1/a_4} \tag{6}$$

in which p_{Hertz} is the maximum Hertzian pressure (for the other parameters the reader is referred to the Appendix).

2.2 EHL Component

Based on EHL calculations Moes [7] derived a fit in dimensionless form for the central film thickness in an EHL line contact:

$$H_C = \left[\left(H_{RI}^{7/3} + H_{EI}^{7/3}\right)^{3/7s} + \left(H_{RP}^{7/2} + H_{EP}^{7/2}\right)^{2/7s}\right]^{s^{-1}} \tag{7}$$

with

$$s = \frac{1}{5}\left(7 + 8\exp\left[-2\frac{H_{EI}}{H_{RI}}\right]\right). \tag{8}$$

H_{RI}, H_{EI}, H_{RP} and H_{EP} are the dimensionless film thickness asymptotes for respectively the Rigid-Isoviscous, Elastic-Isoviscous, Rigid-Piezoviscous and Elastic-Piezoviscous situation as described in the Moes-diagram. The reader is referred to the Appendix, Gelinck and Schipper [1] or to Moes [7] for further details of the Moes diagram and the asymptotes.

2.3 The Model

According to Johnson et al. [3] the total pressure in an ML contact can be divided into an EHL component and a BL component.

Two coefficients γ_1 and γ_2 according to Johnson et al. [3] are defined:

$$\gamma_1 = \frac{F_T}{F_H}, \quad \gamma_2 = \frac{F_T}{F_C}. \tag{9}$$

It is shown by Johnson et al. [3] and Gelinck and Schipper [1] that the results of the EHL and BL calculations can be combined in the ML regime, if E' is replaced by E'/γ_1, F_T by F_T/γ_1 and n by $n \cdot \gamma_2$. For the EHL component this results in:

$$H_C = \left[\gamma_1^S \left(H_{RI}^{7/3} + \gamma_1^{14/15} H_{EI}^{7/3}\right)^{3/7S} + \left(H_{RP}^{7/2} + H_{EP}^{7/2}\right)^{2/7S}\right]^{1/S} \tag{10}$$

with

$$s = \frac{1}{5}\left(7 + 8\exp\left[-2\frac{H_{EI}}{H_{RI}} \cdot \gamma_1^{-2/5}\right]\right) \tag{11}$$

and for the BL component is:

$$\frac{p_C}{p_{\text{Herz}}} = \left[1 + (a_1 n'^{\,a_2}\, \sigma_{s'}{}^{a_3} W^{a_2 - a_3} \gamma_2^{a_2})^{a_4}\right]^{1/a_4} \frac{1}{\gamma_2}. \tag{12}$$

As the load increases or the surface becomes smoother, p_C approaches the Hertzian pressure corresponding to the fraction of load carried by the asperities.

With the equations given so far the fractions of load of the BL component and the EHL component can be calculated.

2.4 Friction Force

The total friction force F_f for a certain velocity is the sum of the balanced friction forces of the interacting asperities and the shear force of the hydrodynamic component:

$$F_f = \sum_{i=1}^{N} \iint_{A_{C_i}} \tau_{C_i}\, \mathrm{d}A_{C_i} + \iint_{A_H} \tau_H\, \mathrm{d}A_H \tag{13}$$

with N the number of asperities in contact, A_{Ci} the area of contact of a single asperity, i; τC_i the shear stress at the asperity contact i; A_H the contact area of the hydrodynamic component; and τ_H the shear stress of the hydrodynamic component.

The coefficient of friction f_{C_i} of a single asperity can be written as:

$$f_{C_i} = \frac{\tau C_i}{p_{C_i}} \tag{14}$$

with p_{C_i} the normal pressure of a single asperity. According to Briscoe et al. [8] the ratio of the shear strength and the local contact pressure is nearly constant and therefore the coefficient of friction is approximately constant for all asperity contacts. Thus the first term in Equation (13) can also be written as:

$$\sum_{i=1}^{N} \iint_{A_{C_i}} \tau C_i \, \mathrm{d}A_{C_i} = f_C F_C, \tag{15}$$

where F_C is the total normal load carried by the asperities. The value of f_C is determined from experiments.

For the shear force in the lubricant film the Eyring model is used:

$$\tau_H(\dot{\gamma}) = \tau_0 \cdot \operatorname{arcsinh}\left(\frac{\eta\dot{\gamma}}{\tau_0}\right), \tag{16}$$

where η is calculated according to the Roelands equation, the pressure used to calculate the viscosity is the average pressure of the hydrodynamic component, $\dot{\gamma}$ is the shear rate (ν^{dif}/h, ν^{dif} is the difference in velocity of the two surfaces or sliding velocity) and τ_0 is the Eyring shear stress.

The coefficient of friction can now be written as:

$$f = \frac{F_f}{F_T} = \frac{f_C F_C + \iint_{A_H} \tau_H(\dot{\gamma}) \, \mathrm{d}A_H}{F_T}. \tag{17}$$

Combining this result with Equation (16) results in the coefficient of friction for Stribeck curves:

$$f = \frac{F_C F_C + \tau_0 \, A_H \operatorname{arcsinh}\left(\frac{\eta\nu^{\mathrm{dif}}}{h\tau_0}\right)}{F_N}. \tag{18}$$

3. THE STARVED MODEL

In Figure 1 a two dimensional representation of the starved contact situation is presented in which h_{oil} is the supplied oil layer thickness and h'_{oil} is the oil layer thickness when the surface tension is taken into account; x_i and x'_i are the lubricant inlet lengths belonging to the aforementioned lubricant layer thicknesses; h^*, b and V are the starved film thickness, half contact width and the velocity respectively.

The surface tensions depend on the properties of the lubricant and the surface, Tian and Bhushan [10]. In this paper h_{oil} and h'_{oil} are taken equal. According to Crook [11] the oil layer thickness in the converging entry is given

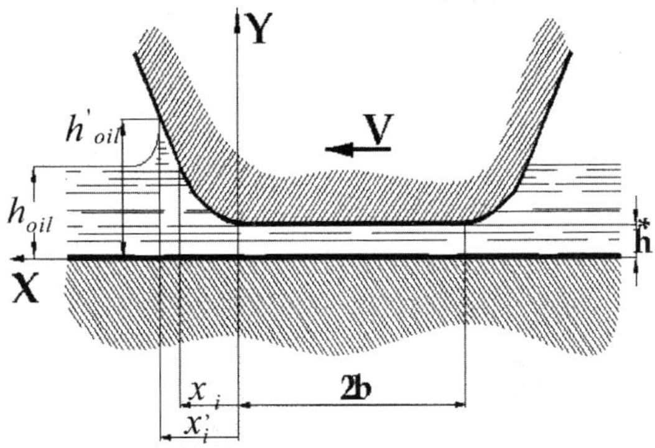

Figure 1. A schematic representation of starved situation.

with good approximation by:

$$h_{\rm oil} = h^* \left[1 + \frac{4\sqrt{2}}{3} \phi^{3/2} \right], \tag{19}$$

where ϕ is a nondimensionless coordinate:

$$\phi = b^{1/3} x_i \big/ (2Rh^*)^{2/3} \tag{20}$$

in which R is the contact radius.

The ratio β between the film thickness for the starved (h^*) and fully flooded condition (h) is derived by Wolveridge et al. [4], based on computational solutions of Orcutt and Cheng [9], as a function of the dimensionless lubricant inlet length ψ_i:

$$\frac{h^*}{h} = \frac{H_C^*}{H_C} = \beta = f(\psi_i), \tag{21}$$

where

$$\psi(i) = b^1 3 x_i \big/ (2Rh)^{2/3}. \tag{22}$$

In order to implement the numerical solution of Wolveridge et al. in Gelinck and Schipper's model an analitical solution is needed.

The numerical solutions are well fitted by the following equation:

$$\beta = \frac{h^*}{h} = \frac{2}{\pi} \arctan\,(2.7 \cdot \psi_i) \tag{23}$$

as may be seen in Figure 2. If in the EHL component of the mixed lubrication model, the film thickness formula is replaced by the fitted starved film thickness formula (23), the starved mixed lubrication model is completed.

In the next section the results of the presented starved Stribeck model for two different sets of roughness parameters are presented.

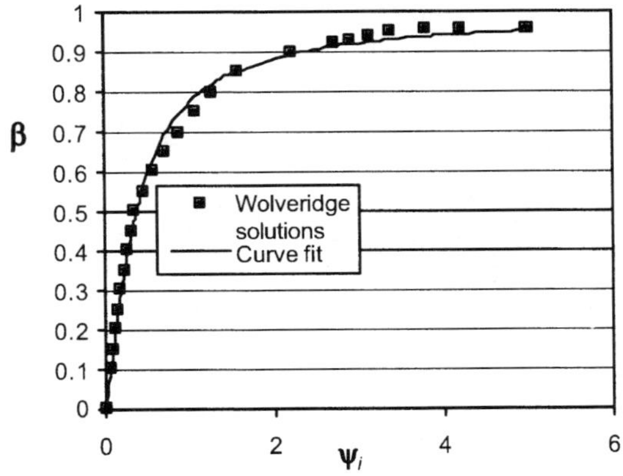

Figure 2. The approximation of numerical solutions of Wolveridge et al. [4].

Table 1. Parameters used for calculations.

Property	Value Reference Case 1	Value Reference Case 2
n	$1 \cdot 10^{11}$ m^{-2}	$13.0 \cdot 10^{9}$ m^{-2}
β	$1 \cdot 10^{-5}$ m	$2.6 \cdot 10^{-6}$ m
σ_s	$0.05 \cdot 10^{-6}$ m	$0.487 \cdot 10^{-6}$ m
R_a	$0.0565 \cdot 10^{-6}$ m	$0.55 \cdot 10^{-6}$ m
B	$10.0 \cdot 10^{-3}$ m	$12.7 \cdot 10^{-3}$ m
E'	$231 \cdot 10^{9}$ Pa	$231 \cdot 10^{9}$ Pa
R	$20.0 \cdot 10^{-3}$ m	$19.0 \cdot 10^{-3}$ m
η_0	0.0202 Pa·s	0.0374 Pa·s
α	$2.0 \cdot 10^{-8}$ Pa^{-1}	$2.0 \cdot 10^{-8}$ Pa^{-1}
τ_0	$2.5 \cdot 10^{6}$ Pa	$2.5 \cdot 10^{6}$ Pa
f_c	0.13	0.13
F_N	1000 N	1000 N

4. RESULTS OF STARVED STRIBECK CURVE CALCULATIONS

The Stribeck curve for starved lubrication conditions can now be calculated by varying the velocity. Two sets of surface parameters in combination with the supplied oil layer thickness are used to show their influence on friction under starved conditions. The parameters used are presented in Table 1.

The surface parameters of reference case 1 are of a spur gear transmission and the parameters of reference case 2 are are taken from Johnson and Spence [12].

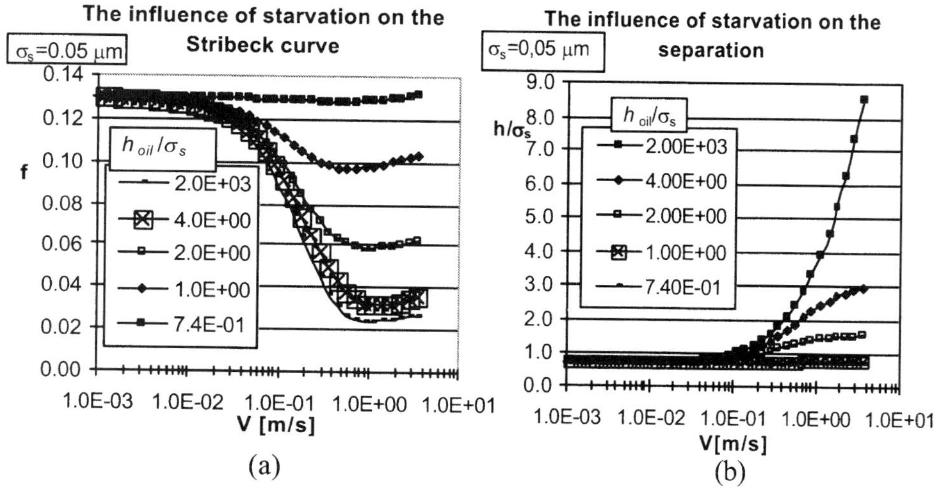

Figure 3. Stribeck curve (a) and separation (b) for reference case 1.

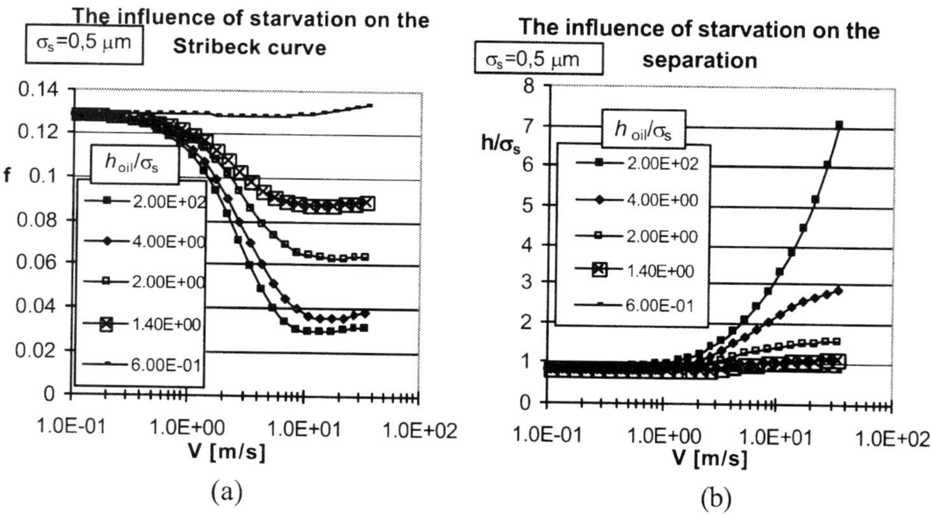

Figure 4. Stribeck curve (a) and separation (b) for reference case 2.

The results of the calculation of the Stribeck curve and the separation for the two reference cases performed for different oil layer thickness over rougness ratios, h_{oil}/σ_s are given in Figures 3 and 4.

The tendency of the starved Stribeck curve and the corresponding separation as a function of oil layer thickness ratio h_{oil}/σ_s can be described as follows:

- when the oil layer thickness (h_{oil}) is larger than, say, 6 times the standard deviation of summits heights (σ_s) the Stribeck curve and separation do not change compared to the fully flooded condition.

- if the oil layer thickness ratio (h_{oil}/σ_s) is between 6 and, say, 0.7 the friction level in the EHL and ML regimes starts to increase and the separation decreases, Figures 3b and 4b. The mixed lubrication regime becomes less steep in this range of oil layer thickness over roughness ratio. The transition from BL to ML and from ML to EHL respectively occurs approximately at the same transition velocity only the friction level changes (Figures 3a and 4a). The Stribeck curve for starved lubricated contact "pivots" at the transition ML/BL as the ratio of h_{oil}/σ_s decreases.
- for values of oil layer thickness over roughness ratio's (h_{oil}/σ_s) less than ~ 0.7 the Stribeck curve tends to transform into a straight line, Figures 3a and 4a and the separation stays on the same level as in the BL regime, Figures 3b and 4b.

Similar results have been found for point contacts by Liu [5].

5. CONCLUSIONS

A model has been developed in order to calculate the starved Stribeck curve and separations in the mixed lubrication regime. The model is a combination of the Gelinck and Schipper mixed lubrication model and numerical calculations for starved smooth lubricated line contacts of Wolveridge et al. [4]. The surface tensions which give a thicker film and heat development which give lower film are neglected, however in real contacts these two main effects may compensate each other.

The calculations have been performed for two sets of roughness parameters. In terms of oil layer thickness over roughness ratio (h_{oil}/σ_s) the results show the same trend for the two reference cases. For values of oil layer thickness over roughness ratio (h_{oil}/σ_s) larger than approximately 6 the Stribeck curve and separation do not change. If oil layer thickness over roughness ratio is in the range of 6 to 0.7 the friction starts to increase and the film thickness decreases. When the oil layer thickness over roughness ratio is less than approximately 0.7 the Stribeck curve tends to transform into a straight line (constant friction level) and separation stays at the same value as in BL regime.

APPENDIX

In Equation (6) the dimensionless numbers n', σ_s' and W, as well as the values of the constants a_i were used. These numbers are defined as:

$$n' = \frac{342}{\pi} n R \sqrt{\beta R}, \quad \sigma_s' = \frac{\pi}{8}\frac{\sigma_s}{R} \quad \text{and} \quad W = \frac{F_T}{BE'R}.$$

The values of the parameters a_i in eq. (6) are $a_1 = 0.953$, $a_2 = 0.0337$, $a_3 = -0.442$ and $a_4 = -1.70$, respectively.

In Equation (7) the function fit of the Moes-diagram is given. The asymptotes in the Moes-diagram are given as:

$$H_{RI} = 3M^{-1}, \quad H_{RP} = 1.287L^{2/3}$$
$$H_{EI} = 2.621M^{-1/5} \quad \text{and} \quad H_{EP} = 1.31M^{-1/8}\, L^{-3/4}$$

in which

$$H_C = \frac{h}{R}\left(\frac{E'R}{\eta_0 \nu^+}\right)^{1/2}, \quad M = \frac{F_T}{BE'R}\left(\frac{E'R}{\eta_0 \nu^+}\right)^{1/2}$$
$$L = \alpha E'\left(\frac{E'R}{\eta_0 \nu^+}\right)^{-1/4},$$

where h is the film thickness and ν^+ is the sum velocity.

REFERENCES

[1] Gelinck E.R.M. and Schipper D.J. (2000), "Calculation of Stribeck Curves for line contacts", *Tribology International* **33**, 175–181.

[2] Greenwood J.A. and Williamson J.B.P. (1966), "Contact of nominally flat surfaces", *Phil. Trans. Roy. Soc. London, Series A* **295**, 300–319.

[3] Johnson, K.L., Greenoowd, J.A. and Poon, S.Y. (1972), "A simple theory of asperity contact in elasto-hydrodynamic lubrication", *Wear* **19**, 91–108.

[4] Wolveridge, P.E., Baglin, K.P. and Archard, J.F. (1971), "The starved lubrication of cylinders in line contact", *Proc. Instn. Mech. Engrs.* **181**, 1159–1169.

[5] Liu, Q. (2002), "Friction in mixed and elasto-hydrodynamic lubricated contacts including thermal effects", Ph.D. Thesis, University of Twente, Enschede, The Netherlands.

[6] Moes, H. (2000), "Lubrication and beyond", Lecture Notes, University of Twente, Enschede, The Netherlands.

[7] Briscoe, B.J. (1981), "Friction and wear of organic solids and the adhesion model of friction", *Phil. Mag. A* **43**(3), 511–527.

[8] Orcutt, F.K. and Cheng, H.S. (1966), "Lubrication of rolling contact instrument bearings, giro spin-axis", in *Proceedings of the Hydrodynamic Bearing Symposium*, Vol. 2, MIT Instrument Lab., Cambridge, MA, December 1966.

[9] Tian, X. and Bhushan, B. (1996), "The micro-meniscus effect of a thin liquid film on the static friction of rough surface contact", *J. Phys. D: Appl. Phys.* **29**, 163–178.

[10] Crook, A.W. (1958), "The lubrication of rollers-I", *Phil. Trans. Roy. Soc. A* **250**, 387.

[11] Johnson, K.L. and Spence, D.I. (1991), "Determination of gear tooth friction by disc machine", *Tribology International* **24**(5), 269–275.

INVESTIGATION ON THERMAL DISTRESS AND SCUFFING FAILURE UNDER MICRO EHL CONDITIONS

A. Polacco[1], G. Pugliese[1], E. Ciulli[1], G.M. Bragallini[2] and M. Facchini[2]
[1]*Dipartimento di Ingegneria Meccanica, Nucleare e della Produzione, Pisa University, Italy,*
E-mail: alessandro.polacco@ing.unipi.it, g.pugliese@ing.unipi.it
[2]*Avio Propulsione Aerospaziale S.p.A., Rivalta (TO), Italy*

Abstract Scuffing is a degenerative form of adhesive wear caused by local welding between mating surfaces. Classical methods for scuffing prediction indicate the surface temperature rise (flash temperature) due to friction generated heat as the most important index to assess the risk of scuffing. Nowadays numerical analyses constitute a useful tool to provide a good knowledge in mixed lubrication. However, the difficulties of taking into account the many parameters involved in wear, such as mechanical and chemical changes, do not allow us to use them to simulate a wear process or as a practical design tool. Moreover the model used to represent the surface roughness is of great importance since scuffing is associated with asperity interactions. Some roughness models are discussed. Aramaki and co-worker's roughness model has been chosen and implemented in a numerical code for the assessment of contact temperature and pressure in lubricated rough contacts. The new simplified approach proposed seems to be promising as a tool for a practical and easy to apply scuffing criterion.

Keywords: scuffing, flash temperature, roughness characterization, numerical analysis, micro-EHL.

1. INTRODUCTION

It is well known that micro-elastohydrodynamic lubrication (micro-EHL) occurs in highly loaded machine elements such as roller bearings, cams and gears. During the few last years the increasing performance demanded, in terms of load and speed, operating life and required maintenance, means that mechanical components commonly work under severe conditions.

R.W. Snidle and H.P. Evans (eds), IUTAM Symposium on Elastohydrodynamics and Micro-elastohydrodynamics, 321–332.

Consequently, the fluid film is not able to keep the bodies completely apart, so direct contacts can occur between the asperities present on the mating surfaces (the lubrication regime is mixed or boundary). In these conditions one of the most important roles of the lubricant is the interposition of a thin film of a few molecules that can prevent local welding between asperities.

Micro-EHL numerical analysis is a useful tool to assess the severity of asperity interaction allowing the evaluation of the pressure field and the film thickness between the mating surfaces (Tao et al., 2002 and Wang et al., 2004). Lubricated rough contacts can also be investigated with micro-EHL codes in order to predict the occurrence of wear (Kato, 2000).

Scuffing, pitting and micropitting are common wear modes in several applications specifically in aeronautical transmissions where high power and torque are commonly employed. Scuffing is a severe form of adhesive wear due to local welding caused by friction generated heat. Its occurrence is sudden and various theories have been proposed on its starting mechanisms. Pitting and micropitting are due to fatigue and their occurrence depends on contact conditions and the number of loading cycles. These phenomena are linked because some tribological parameters are shared by their failure modes. In particular, several studies have been carried out for the prediction of scuffing, but, despite the progress made in this field, some questions are still open and need further researches.

Commonly the criteria employed for failure prediction assume simplified conditions in order to obtain methods (usually single formulas) that are easy to apply to practical design (Grosberg and Lee, 1977 and Castro and Seabra, 1998). The main problem is that one needs a deep and reliable knowledge of the real tribological conditions to evaluate how these can lead to failure and to include every relevant parameter into the simplified models. It is well known that the risk of scuffing is high under critical situations such as heavily loaded contacts or inadequate lubrication that can lead to asperity interactions with significant heat generation. The common scuffing criteria (e.g. total contact temperature or frictional power intensity methods) consider the temperature values over the contact region as an important parameter in estimating the probability of failure. The use of micro-EHL analysis can furnish a more detailed simulation of the real phenomena than the aforementioned methods, but some problems are still present. For example, it is not easy to model correctly every aspect of the real contacts, such as the lubricant behavior and the asperity interactions. In particular surface roughness can evolve significantly during the life cycle when wear occurs. It is difficult to take into account the many parameters involved in wear of lubricated couples, where the nature and severity of asperity contact depends on mechanical (e.g. hardness), chemical (e.g. reactivity, tribological layers, lubricant adsorption) and geometrical characteristics (e.g. thermal expansion), all of which are influenced by the friction generated heat. The difficulties involved in extrapolating the results obtained for specific conditions to different ones and to fit the needs of an easy to apply qualitative analysis are also to be considered. The development of a methodology that combines the power of a micro-EHL analysis and the simplicity of the classical criteria would be very useful for practical scuffing prevention.

In this paper the development of an investigation of thermal distress and scuffing failure based on a simplified micro-EHL simulation is described. Due to the big influence of roughness on the results, fundamental research work related to roughness characterization and thermal distress is firstly analyzed and selected results are reported. The numerical code developed, which is able to assess the contact temperature by coupling a detailed roughness model with the EHL solution, is briefly described. Some initial results are reported. The developed methodology could provide a helpful tool to estimate the risk of scuffing.

This work has been developed as a result of a joint research programme of the University of Pisa and Avio.

2. ROUGHNESS CHARACTERIZATION AND FLASH TEMPERATURE

A thermal investigation in sliding contacts cannot ignore surface roughness. A wide bibliographical search has been conducted and the various approaches studied. A selection of the works mainly used for the numerical analysis developed by the authors is presented below.

For many years the heavy influence of roughness on contact conditions and lubrication has been assessed (Johnson et al., 1972 and Cheng and Dyson, 1978) and the problem is still the subject of current research (see Jang and Khonsari, 2004, Siripuram and Stephens, 2004 and van der Heide and Schipper, 2004).

To assess the influence of roughness on lubrication and contact pressure a roughness model is necessary. There are two ways to represent roughness: the deterministic or the statistical approach. The deterministic characterization of surface roughness is generally adopted to develop accurate, numerical solutions of dry or lubricated contacts; these solutions give a useful and accurate picture of the contact, but the picture is limited to the studied roughness and it is difficult to extrapolate to the general case. The statistical approach allows us to obtain approximated methods more easily but without the same accuracy. In this work both approaches are investigated.

Several approximate methods for the study of contact of rough surfaces start from the statistical approach proposed by Greenwood and Williamson, 1966 (the GW model). The GW model has been regularly applied, studied, validated (Handzel-Powierza et al., 1992) and extended to different roughness, including plastic contacts and anisotropic surfaces (e.g. McCool, 1986). Some uncertainties have been found in the application of the model. The main hypotheses of the GW method are:

- the asperities are represented as hemispheres with constant radius
- rough surfaces contact can be studied as the contact of a rigid smooth plane and a rough surface with equivalent roughness
- the contacts are considered elastic
- there are no interactions between adjacent asperities
- the asperities heights follow a Gaussian distribution[1]

The Gaussian probability density function *f*(*z*) is given by:

$$f(z) = \frac{1}{\sigma\sqrt{2\pi}}\exp(-\frac{z^2}{2\sigma^2}) \tag{1}$$

z being the distance from the mean plane and σ the standard deviation of the asperities heights. Defining a function

$$F(n,y) = \int_y^\infty (z-y)^n f(z)\,dz \tag{2}$$

the probability that an asperity is higher than z is given by:

$$p(z > z') = F(0,z') \tag{3}$$

Let the equivalent rough surface have an equivalent elasticity modulus E' and a composite roughness with a standard deviation σ. The pressure for each asperity in contact with the plane can be found from the Hertz theory as a function of its height. The mean pressure on the contact area, caused by asperity deformation, can be found by coupling these results with statistical considerations:

$$P_{asp} = \frac{4}{3}\eta E'\sqrt{\beta}\sigma'^{3/2}\,F(3/2,d') \tag{4}$$

[1] The Gaussian distribution is suitable for machined surfaces and allows an easy composition of the two rough surfaces into an equivalent one

$d' = d/\sigma$ is the dimensionless separation, d the distance of the smooth rigid plane from the mean line of the equivalent rough surface, β the radius of the spheres, η the number of asperities per unit area. In a similar way the real area of contact and the number of contacts can be obtained.

Commonly (e.g. Tomanik et al., 2003) it is supposed that σ, β and η are functions of the surface spectral moments of order 0, 2 and 4 (m_0, m_2 and m_4):

$$\sigma' = \sqrt{\left(1 - \frac{0.8968}{\psi}\right) m_0} \quad (5); \qquad \beta = \frac{3\sqrt{\pi}}{8\sqrt{m_4}} \quad (6); \qquad \eta = \frac{m_4 / m_2}{6\pi\sqrt{3}} \quad (7)$$

where the bandwidth parameter ψ and the spectral moments are defined as:

$$m_0 = E\left\{(z(x))^2\right\} \quad (8); \qquad m_2 = E\left\{\frac{d^2(z(x))}{dx^2}\right\} \quad (9)$$

$$m_4 = E\left\{\left(\frac{d^2(z(x))}{dx^2}\right)^2\right\} \quad (10); \qquad \psi = \frac{m_0 m_4}{m_0^2} \quad (11)$$

$z(x)$ is the measured profile and $E\{\ \}$ is the expectancy function. Other functions for σ', β and η have been proposed to take into account different roughness (e.g. anisotropic or asymmetric, Yu and Polycarpou, 2004). The moment m_4 cannot be measured by a roughness measurement device and various methods have been proposed in the literature for its evaluation. Nevertheless different results are obtained for a given profile. The problem is that m_4 is related to the measuring instruments and to the signal processing method applied to measured data. The GW method depends greatly on these inputs and the effects of this uncertainty are noticeable and should be assessed with care.

Several other methods have been proposed in the last years for the characterization and the analysis of rough surface contacts: Iida and Ono, 2003, Lin and Ovaert, 2004a and Lin and Ovaert, 2004b to cite a few. Various trials have also been made with fractal geometry (Majumdar and Bhushan, 1991, Ganti and Bhushan, 1995, Lopez et al., 1995 and Willner, 2004). A comprehensive survey of contact models has been presented by Bhushan, 1998.

An interesting method for roughness characterization is proposed in Aramaki et al., 1993a and Aramaki et al., 1993b. This method allows a more refined approach and overcomes the difficulties arising from the application of the GW model to actual surfaces. Moreover, because it is not formulated from a statistical approach, it can be introduced into a numerical code for lubricated rough contact problems. According to this method, rough surfaces can be modelled using three parameters: root mean square roughness, autocorrelation length and bearing area. In this model, one asperity is defined as "one which makes one contact spot" while its width is defined as the distance between the two points at which the profile crosses the roughness mean plane. An equivalent parabolic asperity is defined as having the same width on the mean line and a height which gives the same cross-sectional area above the mean line as the original profile. An exponential asperity width distribution vs. asperity heights has been experimentally found by the authors of this model for typical engineering surfaces thereby validating the assumption of this model. The relationship between asperity height ξ and asperity width L can be expressed as:

$$\langle \xi(L^*) \rangle = \sqrt{\frac{2\pi}{L^{*2}+1}} L^* \sigma \tag{12}$$

where $\xi(L^*)$ is the height of an asperity of width L^*, $L^* = \varphi L/\pi$, where φ is the coefficient of the exponential autocorrelation function, and $\langle \; \rangle$ the mean operator. The value of the mean contact pressure P_{mean} acting on a single asperity is given by the following expression:

$$P_{mean} = \frac{2}{2-1.39(a/S)^2}\left(\frac{a}{L}\right)\left(\frac{\pi E'}{4}\right)\frac{\xi}{L} \tag{13}$$

where S is the distance between two asperities and a is the diameter of the asperity contact area.

Whatever roughness model is used the flash temperature is commonly assessed assuming the frictional heating process to be stationary. The first expression for the flash temperature evaluation has been proposed by Blok, 1937 under the following assumptions:

- all the frictional work is assumed to be transformed into heat
- all the heat flux flows into the bodies
- the bodies exchange heat only through the contact area
- the temperature of the bodies on the contact area is the same
- thermal properties of the materials are constant

According to these hypotheses the heat flux per unit area q, due to frictional work, is:

$$q = \mu v_s p \tag{14}$$

where μ is the friction coefficient, v_s the sliding velocity and p the contact pressure.

Some approximate solutions were also proposed by Jaeger, 1942, and extended by Kuhlman-Wilsdorf, 1987a and Kuhlman-Wilsdorf, 1987b. Starting from Blok's hypotheses other expressions were proposed by Cameron et al., 1965 and Tian and Kennedy, 1994.

The local values of temperature T can be obtained by substituting q in Kuhlman-Wilsdorf's steady state temperature equation:

$$T = \frac{CRq}{\dfrac{K_1}{Z_1 Sh_1} + \dfrac{K_2}{Z_2 Sh_2}} \tag{15}$$

where C is a function of speed and shape, R the equivalent radius, K the thermal conductivity, Z and Sh the speed and the shape factors respectively. Alternative approaches are available by using FFT methods such as that of Gao et al., 2000.

3. NUMERICAL ANALYSIS

In an attempt to develop a simplified approach for the design of lubricated rough contacts a numerical investigation has been carried out coupling roughness, lubrication and thermal analyses. A brief discussion of roughness characterization and the lubrication problem solution technique adopted for the analysis are presented below.

3.1 Roughness

Although the GW contact model allows a relatively easy numerical implementation, as mentioned above, the effects of the uncertainty in the evaluation of the inputs to the model could be significant and should be assessed with care. A sensitivity study of these effects has been conducted and the results will be discussed later. Moreover, the statistical approach on which the GW method is based does not allow its easy incorporation into numerical methods for the EHL problem solution.

As an alternative approach Aramaki and co-worker's model has been adopted. Using this approach a numerical code has been developed in which an actual surface roughness profile is represented as a series of parabolas. In Figures 1a and 1b are shown an example of acquired roughness and that assumed in the model

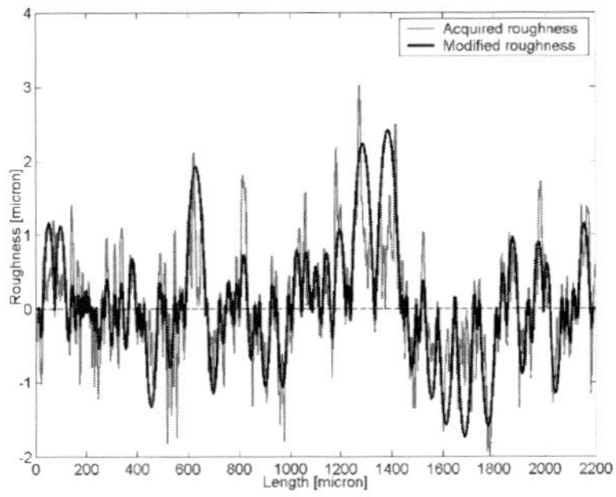

Figure 1a. Acquired and calculated roughness.

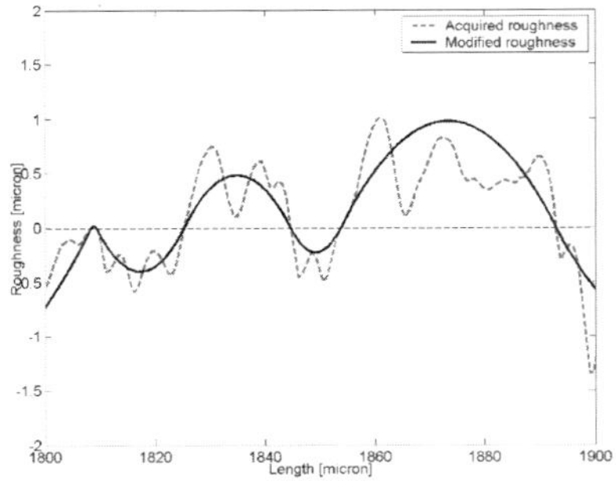

Figure 1b. Magnification of Figure 1a.

3.2 Lubrication

A numerical code has been developed coupling the roughness and thermal models as described above with the EHL line contact problem. The code evaluates the values of the local contact pressure and temperature. For the EHL analysis the lubricant is considered as Newtonian, and conditions are assumed isothermal and stationary. The solution is obtained using finite difference and Newton-Raphson methods with an under-relaxation factor.

Taking average values of the lubricant properties across the film, the Reynolds equation for the EHL line contact, in dimensionless form, can be expressed as:

$$\frac{\partial}{\partial X}\left(\frac{\overline{\rho}}{\overline{\eta}} H^3 \frac{\partial P}{\partial X}\right) - \lambda \frac{\partial\left(\overline{\rho} H\right)}{\partial X} = 0 \tag{16}$$

where X is the dimensionless coordinate, H and $\overline{P}$ the dimensionless film thickness and pressure, λ a dimensionless velocity parameter, η and ρ the dimensionless viscosity and density, respectively. A finite difference form of eq.16 has been used with a non-uniform grid of n points to avoid excessive fineness in the computation near the start of solution domain,

As suggested in Venner et al., 1990 the film shape formulation for the generic grid point i, in dimensionless form, is:

$$H_i = H_0 + \frac{X_i^2}{2} - \frac{1}{\pi} \sum_{j=1}^{n} K_{ij} P_j \tag{17}$$

where H_0 is the dimensionless film constant and K_{ij} is defined as:

$$K_{ij} = (i-j+1/2)\Delta X_i\left[\ln(|i-j+1/2|\Delta X_i)-1\right]+ \\ -(i-j-1/2)\Delta X_i\left[\ln(|i-j-1/2|\Delta X_i)-1\right] \quad (18)$$

where ΔX is the meshing spacing in coordinate direction X.

Mass density and viscosity of the lubricant are assumed to depend on pressure. The expressions proposed in Dowson and Higginson, 1966 for density (eq.19) and by Barus (eq.20) for viscosity have been adopted:

$$\bar{\rho} = 1 + \frac{0.6 \times 10^9 p_H\ P}{1 + 1.7 \times 10^9 p_H\ P} \quad (19); \qquad \bar{\eta} = \exp(\alpha p_H P) \quad (20)$$

where p_H is the maximum Hertzian pressure and α is the pressure-viscosity coefficient.

3.3 Numerical Code and Results

The flow chart for the iteration process is shown in Figure 2. During each iteration the fluid film is checked and any calculated negative film thickness value is set to zero. This is interpreted as dry contact by the code and in the points with fluid film equal to zero a contact pressure is calculated and imposed. The calculation of these contact pressures is made according to Aramaki's model. In the next iteration the load value is updated to take into account the load sharing between fluid film and asperity contact.

When convergence is reached the contact temperature is calculated for each asperity in contact. Surface temperature rise is evaluated as discussed in the previous sections. Figs. 3 and 4 show an example of the dimensionless contact pressure and temperature distribution evaluated by the code.

4. CONCLUSIONS

A wide bibliographical research on topics concerning rough lubricated and dry contacts has been undertaken. Simplified methods (GW, Aramaki et al.) for the analysis of rough contact problems have been analyzed. The difficulty of obtaining a model of scuffing able to take into account all the parameters related to scuffing has led us to adopt a simplified approach for scuffing prediction. A methodology has been developed based on a

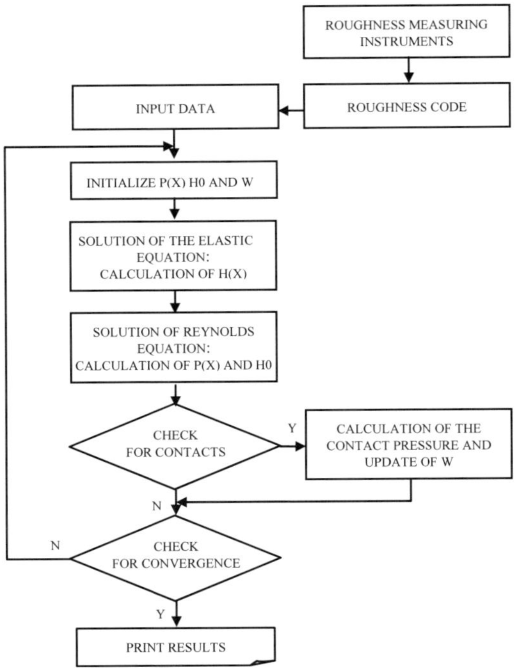

Figure 2. Flow chart for computation.

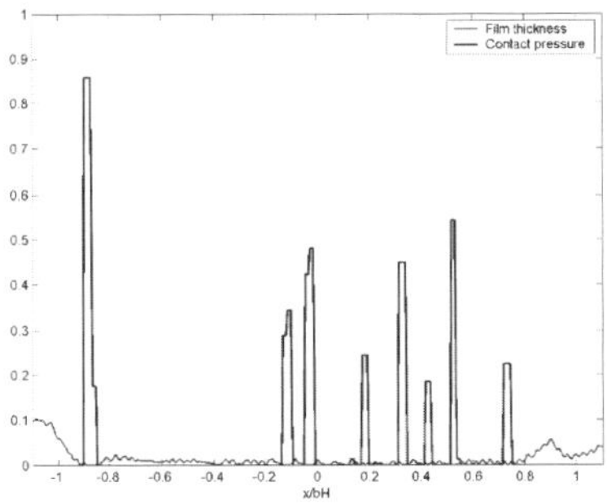

Figure 3. Contact pressure distribution.

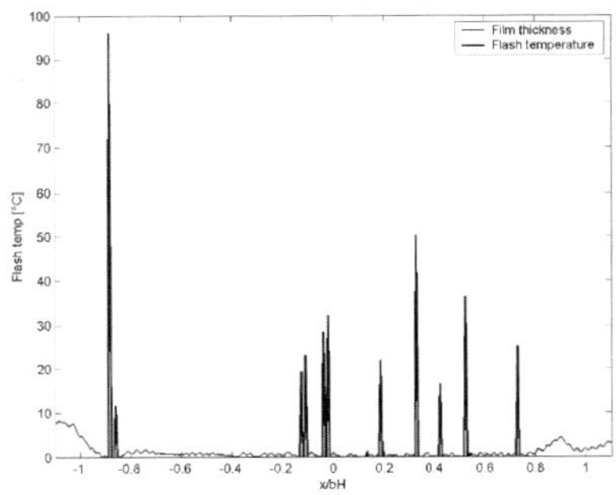

Figure 4. Flash temperature due to asperity contacts.

numerical code for lubricated rough line contacts coupled with roughness and temperature models. This methodology seems to be promising as a tool for the development of a practical and easy to apply scuffing criterion. Comparisons of this simplified approach with more complex ones and experimental validation of the results are programmed and will be proposed in future developments.

REFERENCES

Aramaki, H., Cheng, H. S., and Chung, Y. (1993a). The contact between rough surfaces with longitudinal texture part I: Average contact pressure and real contact area. *ASME J. of Tribology*, 115:419–424.

Aramaki, H., Cheng, H. S., and Chung, Y. (1993b). The contact between rough surfaces with longitudinal texture part ii: Flash temperature. *ASME J. of Tribology*, 115:425–431.

Bhushan, B. (1998). Contact mechanics of rough surfaces in tribology: multiple asperity contacts. *Tribology Letters*, 4:1–35.

Blok, H. (1937). General discussion on lubrication. *Institution Mechanical Engineers*, 2:222.

Cameron, A., Gordon, A. N., and Symm, G. T. (1965). Contact temperatures in rolling/sliding surfaces. *Proc. R. Soc.*, A286:45–61.

Castro, J. and Seabra, J. (1998). Scuffing and lubricant film breakdown in FZG gears part I. analytical and experimental approach. *Wear*, 215(12):104–113.

Cheng, H. S. and Dyson, A. (1978). Elastohydrodynamic lubrication of circumferentially ground rough disks. *ASLE Trans.*, 21(1):25–40.

Dowson, D. and Higginson, G. R. (1966). *Elastohydrodynamic Lubrication*. Pergamon Press, Oxford.

Ganti, S. and Bhushan, B. (1995). Generalized fractal analysis and its applications to engineering surfaces. *Wear*, 180(12 SU):17–34. TY JOUR.

Gao, J., Lee, S. C., Ai, X., and Nixon, H. (2000). An FFT based transient flash temperature model for general three-dimensional rough surface contact. *ASME J. of Tribology*, 122(51923).

Greenwood, J. A. and Williamson, J. P. B. (1966). Contact of nominally flat surfaces. *Proc. R. Soc. London*, A295:300–319.

Grosberg, J. and Lee, C. S. (1977). A critical review of gear scoring criteria. *Wear*, 43(1):237– 242.

HandzelPowierza, Z., Klimczak, T., and Polijaniuk, A. (1992). On the experimental verification of the Greenwood-Williamson model for the contact of rough surfaces. *Wear*, 154(1):115– 124.

Iida, K. and Ono, K. (2003). Design consideration of contact/near contact sliders based on a rough surface contact model. *Transactions of the ASME F Journal of Tribology*, 125(3):562– 571.

Jaeger, J. C. (1942). Moving sources of heat and the temperature at sliding contacts. *Proc. R. Soc. New South Wales*, 56:203–24.

Jang, J. Y. and Khonsari, M. M. (2004). On the growth rate of thermoelastic instability. *Transactions of the ASME F Journal of Tribology*, 126(1):50–56.

Johnson, K. L., Greenwood, J. A., and Poon, S. Y. (1972). A simple theory of asperity contact in elastohydrodynamic lubrication. *Wear*, 19(1):91–108.

Kato, K. (2000). Wear in boundary or mixed lubrication regimes. In *Leeds-Lyon Symposium on Tribology*, Volume 40, Elsevier, Amsterdam, pp. 3–17.

Kuhlman-Wilsdorf, D. (1987a). Demystifying flash temperatures, Part I: analytical expression based on a simple model. *Materials Science and Engineering*, 93:119.

Kuhlman-Wilsdorf, D. (1987b). Demystifying flash temperatures, Part II: first order approximation for plastic contact. *Materials Science and Engineering*, 93:119.

Lin, Y. and Ovaert, T. C. (2004a). A rough surface contact model for general anisotropic materials. *Transactions of the ASME F Journal of Tribology*, 126(1):41–50.

Lin, Y. and Ovaert, T. C. (2004b). A two-dimensional thermoelastic rough surface contact model. *Transactions of the ASME F Journal of Tribology*, 126(3):430–436.

Lopez, J., Hansali, G., Zahouani, H., Le Bosse, J. C., and Mathia, T. (1995). 3d fractal-based characterisation for engineered surface topography. *International Journal of Machine Tools and Manufacture*, 35:211–217.

Majumdar, A. and Bhushan, B. (1991). Fractal model of elastic-plastic contact between rough surfaces. *ASME Journal of Tribology*, 113.

McCool, J. I. (1986). Comparison of models for the contact of rough surfaces. *Wear*, 107:37–60.

Siripuram, R. B. and Stephens, L. S. (2004). Effect of deterministic asperity geometry on hydrodynamic lubrication. *Transactions of the ASME F Journal of Tribology*, 126(3):527–535.

Tao, J., Hughes, T. G., Evans, H. P., and Snidle, R. W. (2002). Elastohydrodynamic response of transverse ground gear teeth. In *Leeds-Lyon Symposium on Tribology*, Volume 40, Elsevier, Amsterdam , pp. 447–458.

Tian, X. and Kennedy, F. E. (1994). Maximum and average flash temperatures in sliding contacts. *ASME J. of Tribology*, 116:167–74.

Tomanik, E., Chacon, H., and Teixeira, G. (2003). A simple numerical procedure to calculate the input data of Greenwood-Williamson model of asperity contact. In *LeedsLyon Symposium on Tribology*, Elsevier, Amsterdam, pp. 205–215.

van der Heide, E. and Schipper, D. J. (2004). On the frictional heating in single summit contacts: Towards failure at asperity level in lubricated systems. *Transactions of the ASME F - Journal of Tribology*, 126(2):275–281.

Venner, C., Napel, W., and Bosma, R. (1990). Advanced multilevel solution of the EHL line contact problem. *ASME Journal of Tribology*, 111:426–432.

Wang, W.Z., Liu, Y.C., Wang, H., and Hu, Y.Z. (2004). A computer thermal model of mixed lubrication in point contacts. *Journal of Tribology*, 126(1):162–170.

Willner, K. (2004). Elastoplastic normal contact of three-dimensional fractal surfaces using halfspace theory. *Transactions of the ASME F Journal of Tribology*, 126(1):28–34.

Yu, N. and Polycarpou, A. A. (2004). Combining and contacting of two rough surfaces with asymmetric distribution of asperity heights. *Transactions of the ASME F Journal of Tribology*, 126(2):225–233.

EFFECT OF SLIP RATIO ON ROLLING CONTACT FATIGUE OF BEARING STEEL ROLLERS LUBRICATED WITH TRACTION OIL

Akira Nakajima and Toshifumi Mawatari
Department of Mechanical Engineering, Faculty of Science and Engineering, Saga Uni-versity,1, Honjo-machi, Saga-shi, Saga 840-8502, Japan

Abstract Using a two-roller testing machine, the rolling contact fatigue life of bearing steel rollers with mirror-like smooth surfaces was examined. A pair of rollers were driven under nearly pure rolling (slip ratio $s \approx 0\%$) or rolling with sliding conditions of $s=-0.5\% \sim -3.2\%$, and a maximum Hertzian stress in the range of $P_H=3.2GPa \sim 4.2GPa$ was applied in point contact condition. A synthetic traction oil with viscosity grade of ISO VG 32 was used as the lubricant. As a result of experiments, the bearing steel rollers lubricated with traction oil showed a fairly high durability in pure rolling conditions, while the fatigue life decreased remarkably with a slight increase in slip ratio. Based on the numerical results of thermal EHL analysis, the authors discuss the mechanism which causes such a difference in the fatigue life.

Keywords: bearing steel, rolling contact fatigue, traction oil, slip ratio, thermal EHL.

1. INTRODUCTION

High carbon chromium bearing steel has been widely used as the material of rolling contact elements in traction drives, not to mention rolling element bearings. Flaking or spalling failure, which occurs on the high hardness contacting surfaces such as rolling elements and bearing races, is generally regarded as subsurface initiated rolling contact fatigue starting from minute defects like a nonmetalic inclusion, and so far the prolongation of bearing life has been steadily achieved owing to the improvement in the quality of bearing steels. On the other hand, as the fact that the bearing life depends

R.W. Snidle and H.P. Evans (eds), IUTAM Symposium on Elastohydrodynamics and Micro-elastohydrodynamics, 333–344.

strongly on the film parameter Λ[1] shows, the lubricating condition is another important factor. In addition, in recent works, it is supposed that the rolling contact fatigue is influenced by the property of solidified oil at high pressure as well as the oil film thickness[2,3].

Using a two-roller testing machine, the authors have examined the effect of the kind of lubricating oil on the occurrence of rolling contact fatigue of bearing steel rollers with mirror-like smooth surfaces. In the previous investigation, it was found that the fatigue life is longer with a low viscosity mineral oil than with a high viscosity traction oil[4,5]. In the present investigation, using a synthetic traction oil with a low viscosity grade of ISO VG 32, the effect of slip ratio on the rolling contact fatigue of bearing steel rollers was examined. Based on the experimental evidence and the numerical results of thermal EHL analysis, the authors discuss the mechanism which causes a difference in the fatigue life.

2. EXPERIMENTAL

2.1 Testing Machine and Test Rollers

Experiments were carried out using a two-roller testing machine having a center distance of 66 mm. The main part of this machine is shown in Figure 1.

A pair of cylindrical (D) roller and (FR) roller with crowning of R25 in an axial direction were mated. The outside diameter of both rollers is 66mm and the ellipticity parameter K_e is about 1.32. In nearly pure rolling conditions, FR roller was driven by mating D roller with the aid of rolling friction (slip ratio s≈0%). In rolling with sliding conditions, both rollers were driven by gears, and the slip ratio of s=-3.2% was obtained with a gear ratio of 31/32. In the case of the lower slip ratios of s=-0.5% and s=-0.7%, the designated slip

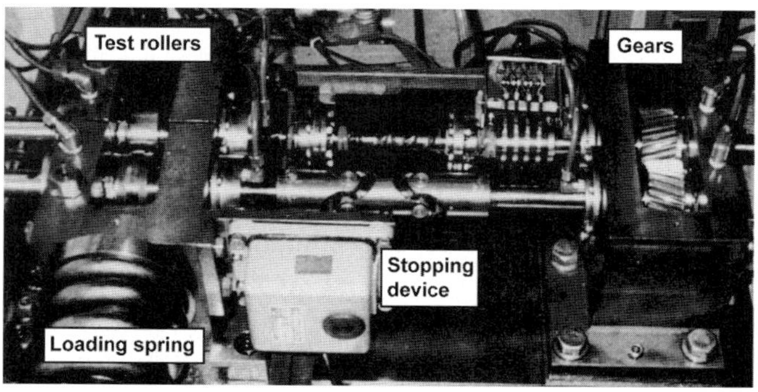

Figure 1. Two-roller testing machine.

Table 1. Summary of experiments and main results.

Test No.		Hardness[a] Hv	Hertz[b] stress P_H,GPa	Slip ratio s, %	Coeffi. of friction μ	h_{min}[c], μm	Number of cycles N,×10^4	Failure
A-1	D FR	837 835	3.4	0.0	~0.001	0.44 (0.73)	2000	D: No FR: No
A-2	D FR	807 803	3.6	0.0	~0.004	0.63 (0.72)	2000	D: No FR: No
A-3	D FR	833 834	3.8	0.0	~0.005	0.52 (0.71)	2000	D: No FR: No
A-4	D FR	874 871	4.0	0.0	~0.001	0.50 (0.70)	2000	D: No FR: No
A-5	D FR	819 819	4.2	0.0	~0.001	0.49 (0.70)	2000	D: No FR: No
B-1	D FR	787 787	3.6	-0.5	0.088 ~0.085	0.43 (0.71)	1894	D: No FR: L pit
B-2	D FR	822 825	3.6	-0.7	0.096 ~0.093	0.45 (0.72)	1197	D: L pit FR: No
C-1	D FR	776 780	3.2	-3.2	0.089 ~0.080	0.36 (0.72)	2000	D: No FR: No
C-2	D FR	778 775	3.4	-3.2	0.088 ~0.080	0.35 (0.71)	699	D: No FR: L pit
C-3	D FR	819 811	3.4	-3.2	0.087 ~0.078	0.34 (0.71)	634	D: L pit FR: No
C-4	D FR	799 791	3.4	-3.2	0.089 ~0.080	0.33 (0.71)	478	D: L pit FR: L pit
C-5	D FR	770 760	3.6	-3.2	0.086 ~0.077	0.32 (0.71)	238	D: No FR: L pit
C-6	D FR	704 702	3.6	-3.2	0.086 ~0.082	0.31 (0.71)	178	D: No FR: L pit
C-7	D FR	803 804	3.6	-3.2	0.085 ~0.077	0.32 (0.71)	122	D: No FR: L pit
C-8	D FR	786 787	3.8	-3.2	0.087 ~0.081	0.30 (0.69)	147	D: No FR: L pit
C-9	D FR	814 813	3.8	-3.2	0.088 ~0.082	0.27 (0.69)	129	D: No FR: L pit

[a] Surface hardness of D and FR roller.
[b] Maximum Hertzian stress.
[c] EHL oil film thickness calculated using the oil viscosity at roller surface temperature. ( ) is h_{min} for the inlet oil temperature.

ratio was given with the gear ratio of 31/31 and by mating rollers with slightly different diameter keeping the center distance constant. Test rollers were made of an oil hardened and tempered high carbon chromium bearing steel (SUJ 2 according to JIS G 4805-1990). The contact surfaces of D and FR rollers were finished smooth to a mirror-like condition with a roughness of R_y=0.02μm (R_y: maximum height of the profile according to ISO 4287-1997 or JIS B0601-1994, sampling length 0.25mm) by precision grinding and subsequent polishing. The surface hardness differed a little with each roller, but a pair of rollers with almost the same hardness were combined.

2.2 Lubricating Oils

As lubricant, a low viscosity synthetic traction oil (kinetic viscosity ν: 32.2 mm^2/s at 313K, 5.31mm^2/s at 373K, α: 31.2GPa^{-1} at 313K, specific gravity 288/277K: 0.962) with viscosity grade of ISO VG 32 (ISO VG32 according to ISO 3448-1992) was used. The oil was supplied on the entrance side of the rotating rollers at a flow rate of 15cm^3/s. The temperature of the oil supplied was kept at 318K during each running test.

2.3 Experimental Conditions and Procedures

A summary of the present experiments is given in Table 1. The rotational speed of the driving side roller was 3583±10 rpm and a fixed normal load which gives a maximum Hertzian stress in the range of P_H=3.2GPa~4.2GPa was applied in point contact condition. Experiments are divided into three groups according to the slip ratio as follows:

A group: slip ratio of $s \approx 0\%$

B group: slip ratios of s=-0.5% and s=-0.7%

C group: slip ratio of s=-3.2%

The theoretical EHL oil film thickness h_{min}[6] was calculated using the oil viscosity at the actual temperature of the roller surfaces, which was measured successively by the trailing thermo-couples. For reference, the film thickness for the viscosity at the inlet oil temperature was also calculated and these values are shown in Table 1.

The state of oil film formation between two rollers was continuously monitored by means of an electric resistance method[7]. The frictional force between rollers was measured using strain gauges mounted on the driving shaft (via slip rings).

3. TEST RESULTS AND DISCUSSION

3.1 Occurrence of Flaking and Fatigue Life

The main test results are summarized in Table 1. The testing machine is equipped with an automatic stopping device which operates in response to abnormal vibration induced by the occurrence of fatigue failures or any other surface damage. In the present investigation, the fatigue life N is defined as

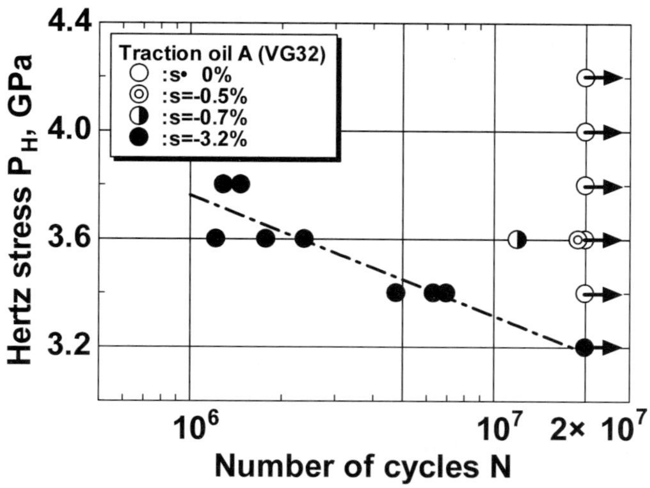

Figure 2. Effect of slip ratio on the fatigue life.

the total number of cycles of the failed roller side. In most cases, only one large sized pit (L pit) or flaking occurred on either D roller or FR roller surface was a fatal damage. When the testing machine continued to run without serious surface damage, the running was discontinued at $N=2\times10^7$ cycles.

Figure 2 shows the relation between the fatigue life and the slip ratio. The fatigue life with traction oil was strongly affected by the rolling/sliding conditions. Namely, in the case of pure rolling conditions, under a high stress of P_H=4.2GPa, no flaking occurred up to $N=2\times10^7$ cycles. On the other hand, in the case of rolling with sliding conditions, flaking was apt to occur, and the fatigue life decreased remarkably even with a slight increase in slip ratio. It was also recognized that flaking is inclined to progress on the FR roller side with slower peripheral velocity, but was restrained on the faster D roller side.

3.2 Oil Film Formation and Friction

Figure 3 shows some examples of the progress of oil film formation during operation. When the oil film is developed fully, the voltage E_{ab} recorded on a chart reaches 15mV. In the tests with the slip ratio of s≈0%, s=-0.5% and s=-0.7%, immediately after the start of running, the voltage E_{ab} showed almost 15mV at a maximum Hertz stress of P_H=3.6GPa. Especially in the case of pure rolling conditions, the full EHL oil film was formed even at a high stress of P_H=4.2GPa. On the other hand, with slip ratio of s=-3.2%, the voltage decreased slightly at the initial stage of running, but soon the voltage reached almost the full EHL condition.

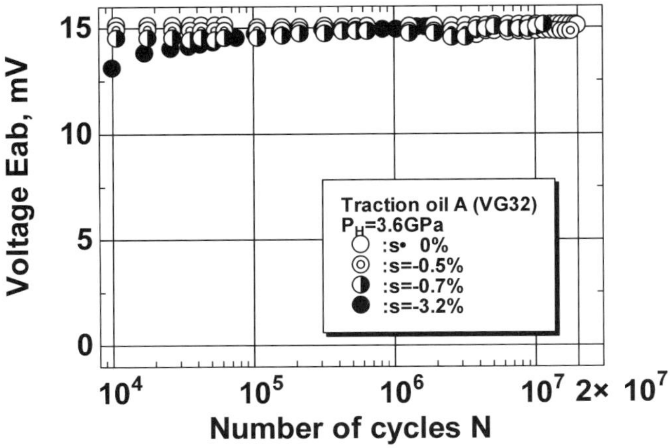

Figure 3. Changes in voltage between rollers (E_{ab}=0mV: contact, 15mV: separation).

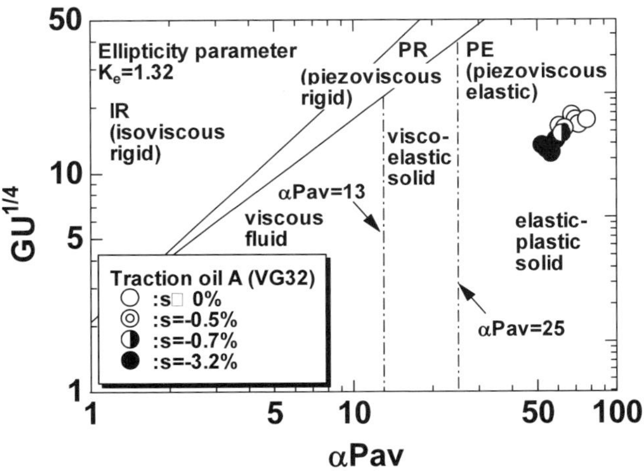

Figure 4. Map of lubrication regimes for K_e=1.32.

The results of friction measurements in each test are given in Table 1 as the coefficient of friction (initial value ~ steady value). Some differences in the coefficient of friction were recognized depending on the rolling with sliding conditions. However, the variation in friction was hardly observed and remained almost constant during each test.

3.3 Map of Lubrication Regimes

Figure 4 shows the map of lubrication regimes[2,8] for the ellipticity parameter K_e≈1.32. According to the study by Ohno et al., it was stated that the transition from viscoelastic solid to elastic-plastic solid of the lubricating oil occurs at αP_{av}=25 (α: pressure viscosity coefficient GPa^{-1}, P_{av}: average contact pressure GPa)[2,3]. As shown in the figure, the present experimental points are all included in the elastic-plastic solidifying condition $\alpha P_{av}>25$.

4. THERMAL EHL ANALYSIS

4.1 Procedure of Numerical Analysis

In view of the fact that the rolling contact fatigue life of bearing steel rollers was markedly affected by the slip ratio, the authors analyzed numerically the steady-state thermal EHL problem. In the present numerical analysis, the lubricating oil was assumed to behave as a Newtonian fluid. Although the governing equations of the thermal EHL contact problem are almost the same as those offered by Yang and Kaneta et al.[9,10], relationships of the viscosity-pressure-temperature and the density-pressure-temperature[11] based on the free volume theory[12] were adopted in the present investigation.

All equations and the boundary conditions need to be written into dimensionless form[13,14]. The numerical schemes are basically the same as that developed by Guo et al.[13]. The numerical solutions are obtained by the pressure-temperature iteration between Reynolds and energy equations. When the pressure is calculated, the temperature is considered as a known value. Multigrid method[15] was used to solve the Reynolds equation. The surface deformation was calculated by the multilevel multi-integration method[16]. A sequential line sweeping method developed by Yang and Rodkiewicz[17] was used to obtain the temperature field.

Figure 5 shows the computation domain for numerical analysis. In this figure, the following dimensionless coordinates were used:

$X=x/b$, $Y=y/b$, $Z=z/h$, $Z_a=z_a/b$ and $Z_a=z_a/b$, where b is the minor radius of the contact ellipse, h is the film thickness, z is the coordinate across the film, and z_a and z_b are the coordinates in solids A and B.

In the multigrid solver for Reynolds equation, four levels are employed. The finest grid consists of 129 nodes along the X and Y direction. Across the film, 11 equidistant nodes are used. In each solid, 6 non-equidistant nodes along Z_a or Z_b direction are adopted. The mesh size is very small at the interface, and then increases in a geometrical series to facilitate programming.

When the following three conditions are reached, the numerical calculation is stopped.

$$\sum\left|P_{i,j}^{(new)} - P_{i,j}^{(old)}\right| / \sum P_{i,j}^{(new)} < 1\times 10^{-3}$$

$$\sum\left|T_{i,j}^{(new)} - T_{i,j}^{(old)}\right| / \sum T_{i,j}^{(new)} < 1\times10^{-4}$$

$$\left|\sum \Delta X \Delta Y P_{i,j}^{(new)} - (2\pi K_e/3)\right| / (2\pi K_e/3) < 1\times10^{-3}$$

where ΔX and ΔY are the finest grid distances along the X and Y directions, respectively. Upper scripts "new" and "old" show after and before an iteration cycle.

4.2 Numerical Calculation Results

Material properties of lubricating oils and contacting solids used in the numerical analysis are shown in Table 2. The slide-roll ratios are s=0%, s= -0.6% and s=-3.2%, respectively. Other input parameters are maximum Hertzian stress P_H=1.0GPa, velocity of the faster side solid A $u_a \approx 12.4$m/s, ambient temperature t_0=318K, dimensionless load parameter $W \approx 2.66\times10^{-6}$, dimensionless material parameter $G \approx 6.75\times10^{3}$, and dimensionless velocity parameter $U \approx 9.16\times10^{-11}$(s=0%), 9.13×10^{-11}(s=-0.6%), 9.01×10^{-11}(s=-3.2%).

The dimensionless parameters used in the numerical calculation are as follows:

$P=p/P_H$, $H=hR'_x/b^2$, $X=x/b$, $Y=y/b$, $Z=z/h$, $T=t/t_0$, $U=u_e\eta_0/E'R'_x$, $G=\alpha E'$ and $W=w/E'R'^2_x$, where p is the hydrodynamic pressure, R'_x is the curvature radii in x direction, u_e is the entrainment velocity, η_0 is the ambient viscosity of oil, E' is the equivalent elastic modulus, and w is the applied load.

Figure 6 shows the dimensionless temperature distribution in the middle-layer of oil film on the plane of Y=0. As shown in the figure, a significant difference in the temperature distribution is caused depending on the slip ratio. In the case of pure rolling of s≈0%, the temperature increase is hardly observed. On the other hand, the temperature increase is very noticeable with

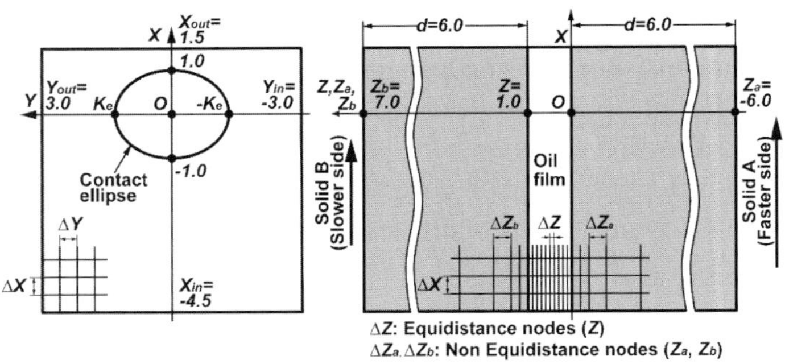

Figure 5. Computation domain for numerical analysis.

Table 2. Material properties for numerical calculation.

Elastic modulus, GPa	206
Density of steel, kg/m^3	7850
Specific heat of lubricant, J/kgK	2000
Specific heat of steel, J/kgK	470
Thermal conductivity of lubricant, W/mK	0.14
Thermal conductivity of steel, W/mK	46

an increase in the slip ratio.

Figure 7 show the film thickness on the plane of Y=0. A certain difference in the oil film thickness is observed depending on the rolling/sliding conditions. Namely, in the pure rolling condition, the oil film thickness is almost constant through the area of contact ellipse except for the outlet of oil flow. On the other hand, in the case of rolling with sliding, the flat area slightly decreases around the center part of contact ellipse, and a small variation in the contour is found at the area where pressure spike occurs. These results reveal clearly that the shear heating in the oil film increases due to an

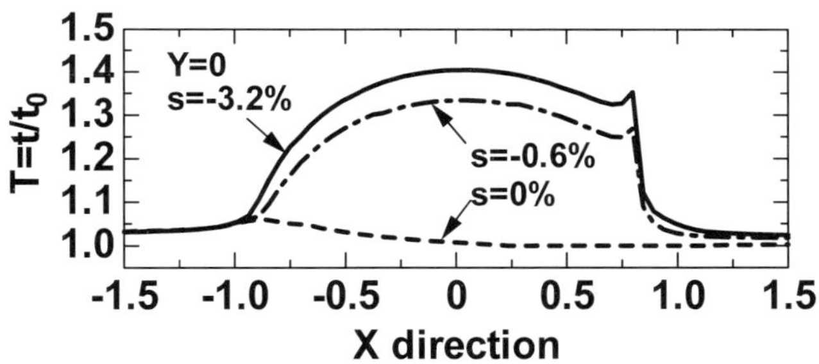

Figure 6. Effect of slip ratio on temperature distribution in middle layer of oil film (P_H=1.0GPa, W≈2.66×10^{-6}, G≈6.75×10^{3}).

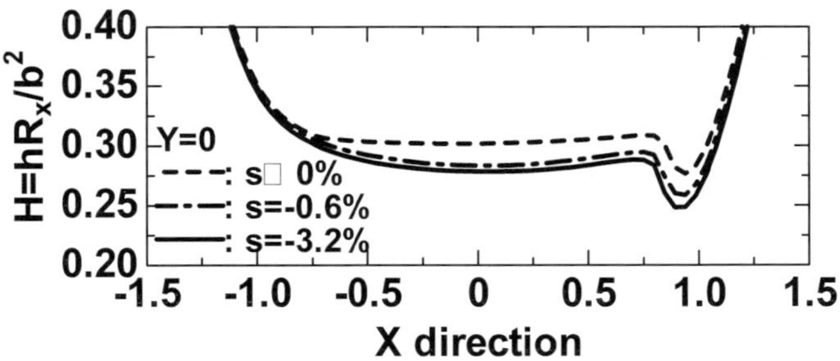

Figure 7. Film thickness on the plane of Y=0 (P_H=1.0GPa, W≈2.66×10^{-6}, G≈6.75×10^{3}).

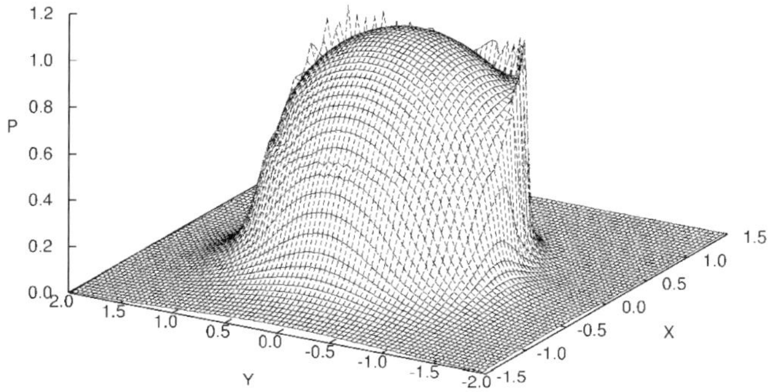

(a) Slip ratio of s=0% ($U \approx 9.16 \times 10^{-11}$)

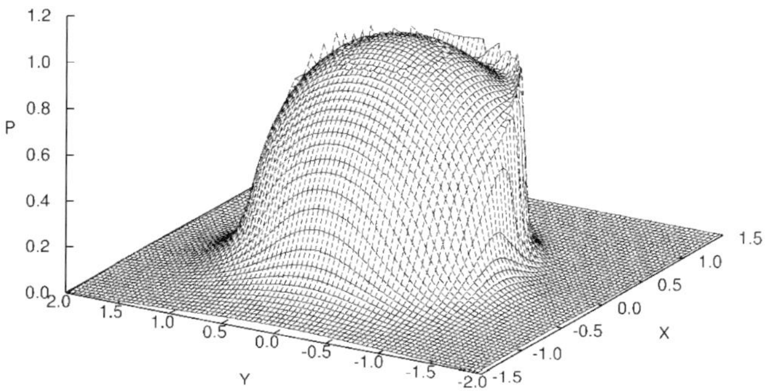

(b) Slip ratio of s=-0.6% ($U \approx 9.13 \times 10^{-11}$)

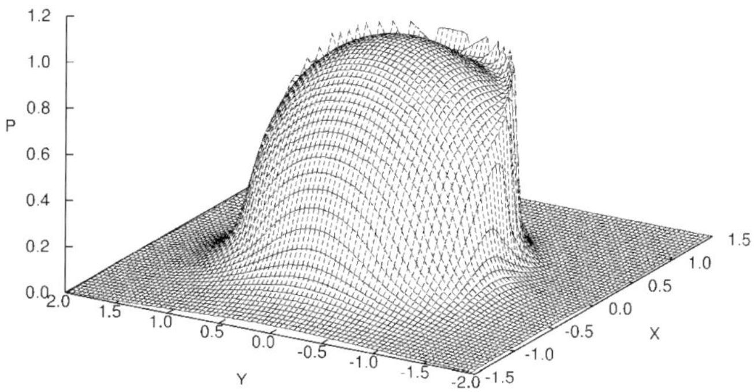

(c) Slip ratio of s=-3.2% ($U \approx 9.01 \times 10^{-11}$)

Figure 8. Pressure distribution obtained by numerical analysis (P_H=1.0GPa, $W \approx 2.66 \times 10^{-6}$, $G \approx 6.75 \times 10^{3}$).

introduction of sliding, and consequently the film thickness becomes thin.

Figure 8 shows the dimensionless pressure distributions in the contact region. Although the pressure distributions are slightly different depending on the rolling/sliding conditions, fairly high pressure spikes are observed at the outlet area of contact ellipse in any case. Sun et al. analyzed the line contact thermal EHL problem of a cylindrical roller rotated under heavy load, and also found that the oils with high viscosity and large pressure-viscosity coefficient like a traction oil can produce a high pressure spike as well as noticeable heating in the oil film, and the local stress concentration immediately beneath the contact surface of the slower side roller is very striking[18]. Furthermore, in the case of rolling with sliding conditions, the pressure spike has a tendency to increase with an increase in the slip ratio.

5. CONCLUSIONS

Using a low viscosity synthetic traction oil, the effect of slip ratio on the rolling contact fatigue of bearing steel rollers was examined. Based on the numerical results of thermal EHL analysis, the mechanism which causes a difference in the fatigue life was discussed. The main results are summarized as follows:

(1) The fatigue life with traction oil was remarkably affected by the rolling/sliding conditions. Namely, in the case of pure rolling conditions, a fairly high durability was shown.

(2) On the other hand, in the case of rolling with sliding conditions, flaking was apt to occur, and the fatigue life decreased remarkably even with a slight increase in slip ratio.

(3) From the numerical results of thermal EHL analysis, it was found that the distributions of pressure, film thickness and temperature remarkably differ depending on the rolling/sliding condition.

(4) Present theoretical results suggest that the shorter rolling contact fatigue life under rolling with sliding condition is attributed to the existence of the pressure spike and the thermal effects caused by high traction.

ACKNOWLEDGMENTS

The authors would like to express their sincere thanks to Prof. M. Kaneta of Kyushu Institute of Technology, Japan; to Prof. P. Yang and Mr. H. Y. Sun of Qingdao Institute of Architecture and Engineering, China; and to Prof. N. Ohno of Saga University, Japan, for their helpful discussions.

REFERENCES

1. T. E. Tallian, On competing failure modes in rolling contact, *ASLE Trans.*, **10**(4), 418–439 (1967).
2. N. Ohno, N. Kuwano and F. Hirano, Bulk modulus of solidified oil at high pressure as predominant factor affecting life of thrust ball bearings, *Tribology Trans.*, **38**(2), 285–292 (1995).
3. N. Ohno, N. Kuwano and F. Hirano, Diagrams for estimation of the solidified film thickness at high pressure EHD contacts, in *Dissipative Processes in Tribology*, Elsevier pp. 507–518 (1994).
4. A. Nakajima and T. Mawatari, Effect of kind of lubricating oil on rolling contact fatigue of bearing steel rollers, in *Proc. Int. Tribology Conf. Yokohama 1995, JAST*, pp. 1405–1410 (1996).
5. T. Mawatari and A. Nakajima, Rolling contact fatigue of bearing steel rollers lubricated with traction oil – Effects of slip ratio and oil temperature, in *Proc. Int. Tribology Conf. Nagasaki 2000, JAST*, pp. 1737–1742 (2001).
6. B. J. Hamrock and D. Dowson, *Ball Bearing Lubrication,* John Wiley and Sons (1981).
7. K. Ichimaru, A. Nakajima and F. Hirano, Effect of asperity interaction on pitting in rollers and gears, *Trans. ASME, J. Mech. Des.,* **103**(2), 482–491 (1981).
8. J. A. Greenwood, Film thickness in circular elastohydrodynamic contacts, *Proc. Instn. Mech. Engrs.,* **202**(C1), 11–17 (1988).
9. P. Yang, S. Qu, M. Kaneta and H. Nishikawa, Formation of steady dimples in point TEHL contacts, *ASME J. Tribol.,* **123**, 42– 49 (2001).
10. M. Kaneta and P. Yang, Effects of thermal conductivity of contacting surfaces on point EHL contacts, *ASME J. Tribol.,* **125**, 731–738 (2003).
11. N. Ohno, H. Tsutsumi and S. Yamada, High pressure rheology and tractional properties of t-CVT traction oil, *Proc. of JAST,* **2004**–5, 177–178 (2004) (in Japanese).
12. A. K. Doolittle, Studies in Newtonian flow. II. The dependence of the viscosity of liquids on free–space, *J. Appl. Phys.,* **22**(12), 1471–1475 (1951).
13. F. Guo, P. Yang and S. Qu, On the theory of thermal elastohydrodynamic lubrication at high slide-roll ratios – Circular glass-steel contact solution at opposite sliding, *ASME J. Tribol.,* **123**, 816–821 (2001).
14. M. Kaneta and P. Yang, Formation mechanism of steady multi-dimples in thermal EHL point contacts, *ASME J. Tribol.,* **125**, 241–251 (2003).
15. C. H. Venner, Multilevel solution of the EHL line and point contact problems, Ph.D. Thesis, Twente University, Enschede, the Netherlands, 1991.
16. A. Brandt and A. A. Lubrecht, Multilevel matrix multiplication and fast solution of integral equations, *J. Comput. Phys.,* **90**, 348–370 (1990).
17. P. Yang and C. M. Rodkiewicz, On the numerical analysis to the thermoelastohydrodynamic lubrication of a tilting pad inclusive of side leakage, *STLE Tribol. Trans.,* **40**, 259–266 (1992).
18. H. Y. Sun, Analysis of thermal elastohydrodynamic lubrication at high speed and extremely heavy load, Master Thesis, University of Qingdao Institute of Architecture and Engineering (2002).

PREDICTION OF FATIGUE DAMAGE IN ROUGH SURFACE EHL

H. Qiao, H.P. Evans and R.W. Snidle
Cardiff School of Engineering, Queen's Buildings, The Parade, Cardiff CF24 3TA, U.K.

Abstract The study of elastohydrodynamic lubrication (EHL) with rough surfaces results in a transient problem due to time varying geometry. In conditions where the Λ ratio (smooth surface film thickness : composite roughness) is low, significant pressure deviations from the corresponding smooth surface results occur. The relative motion of asperity features on the two surfaces leads to cycling of these pressure deviations as surface features move through the nominal contact area. The paper presents a numerical procedure for prediction of fatigue damage as a result of this load cycling. The way in which the calculated fatigue damage varies with changes in factors such as slide roll ratio, viscosity and the roughness profiles adopted for the analysis are demonstrated in the numerical examples.

Keywords: mixed lubrication, stress, fatigue, micropitting.

1. INTRODUCTION

The study of micro elastohydrodynamic lubrication (micro-EHL) of rough surfaces leads naturally to mixed lubrication where part of the load is carried by pressurised lubricant that separates the surfaces completely, and part of the load is carried by dry or boundary lubricated contact between asperity features. The motion of the roughness relative to the contact point requires that transient effects are considered. A transient analysis is thus necessary to examine the evolution of pressure and film thickness within the EHL contact. In recent years much effort has been made to address this situation and different approaches have been used in numerical analysis.

R.W. Snidle and H.P. Evans (eds), IUTAM Symposium on Elastohydrodynamics and Micro-elastohydrodynamics, 345–356.

Fatigue is a common mechanism for failure of the EHL system and is seen as the limiting factor controlling the operational life of rolling element bearings. Another possible example is micropitting which is seen in gears and may well be viewed as rolling contact fatigue on the scale of surface asperities. Although the severity of the EHL contact can be assessed qualitatively by pressure and film thickness within the contact, it is desirable to evaluate the fatigue life quantitatively using a fatigue model. Some recent developments seek to extend the evaluation of pressure variation to include their effect on fatigue life due to roughness and cyclic loading. Ai [1] employed an effective stress concept based on damage accumulation theory and introduced a stress-based fatigue equation with volume integration. Tallian [2] proposed a data fitted life prediction model for rolling contacts under variable operating conditions. Epstein et al. [3] calculated fatigue life in mixed lubrication using fatigue models developed by Ioannides and Harris [4] and Zaretsky [5]. These fatigue models grew out of the Weibull model and were originally developed for rolling element bearings. The current paper presents an alternative model for preliminary fatigue calculations based on numerical analysis of the mixed lubrication problem that uses a novel technique developed by the authors [6, 7]. Numerical simulations are performed and results are presented for several cases of line contacts of rough surfaces.

2. THEORY

2.1 Stress Cycling

Understanding stress cycling is fundamental to fatigue failure analysis. Therefore, an accurate and efficient evaluation of the stresses involved in the EHL contact is necessary in order to implement realistic fatigue prediction. For line contact analysis the solid bodies may be considered to be in plane strain so that the state of stress at any point (x, z) is determined by the three components σ_x, σ_z, τ_{xz}. These components can be evaluated at each timestep due to the EHL contact pressure $p(x, t)$ and surface shear stress loading $q(x, t)$ by direct integration of the integrals [8]

$$\begin{aligned}\sigma_x(x,z,t) &= \int p(s,t)k_1(x-s,z)ds + \int q(s,t)k_4(x-s,z)ds \\ \sigma_z(x,z,t) &= \int p(s,t)k_2(x-s,z)ds + \int q(s,t)k_3(x-s,z)ds \\ \tau_{xz}(x,z,t) &= \int p(s,t)k_3(x-s,z)ds + \int q(s,t)k_1(x-s,z)ds\end{aligned} \tag{1}$$

where the influence coefficients are defined as

$$k_1(x,z) = -\frac{2}{\pi}\frac{x^2 z}{(x^2+z^2)^2} \quad k_2(x,z) = -\frac{2}{\pi}\frac{z^3}{(x^2+z^2)^2}$$
$$k_3(x,z) = -\frac{2}{\pi}\frac{xz^2}{(x^2+z^2)^2} \quad k_4(x,z) = -\frac{2}{\pi}\frac{x^3}{(x^2+z^2)^2} \tag{2}$$

The stress components in equation (1) can be calculated efficiently by the use of Fast Fourier Transforms (FFT) and Inverse FFT (IFFT) as follows:

$$\begin{aligned}
\sigma_x(x,z,t) &\xleftarrow{\text{IFFT}} P(\omega,t)K_1(\omega,z) + Q(\omega,t)K_4(\omega,z) \\
\sigma_z(x,z,t) &\xleftarrow{\text{IFFT}} P(\omega,t)K_2(\omega,z) + Q(\omega,t)K_3(\omega,z) \\
\tau_{xz}(x,z,t) &\xleftarrow{\text{IFFT}} P(\omega,t)K_3(\omega,z) + Q(\omega,t)K_1(\omega,z)
\end{aligned} \tag{3}$$

where

$$\begin{aligned}
P(\omega,t) &\xleftarrow{\text{FFT}} p(x,t) \\
Q(\omega,t) &\xleftarrow{\text{FFT}} q(x,t) \\
K_1(\omega,z) &\xleftarrow{\text{FFT}} k_1(x,z) \\
K_2(\omega,z) &\xleftarrow{\text{FFT}} k_2(x,z) \\
K_3(\omega,z) &\xleftarrow{\text{FFT}} k_3(x,z) \\
K_4(\omega,z) &\xleftarrow{\text{FFT}} k_4(x,z)
\end{aligned} \tag{4}$$

Such a calculation establishes the stress field in terms of the x and z co-ordinates whose origin is at the nominal contact point. In order to consider the transient variation of the stress field within the material it is necessary to express these fields in terms of axes fixed in each of the two contacting surfaces.

2.2 Damage Estimation

Micro asperity interactions give rise to high localised pressures and these are applied cyclically to the surface of an asperity feature as it moves relative to prominent features on the other contacting surface. The degree of pressure cycling that an asperity experiences during its traverse of the nominal contact area can be quantified from kinematic considerations, the slide roll ratio, and the characteristic separation distance of significant surface features on the contacting surface [9]. The amplitude of the pressure cycles experienced will vary however according to the shapes and heights of the micro asperities involved in each "collision". The stress cycling experienced by the material is therefore of varying amplitude. To take this into account in a fatigue calculation the rainflow counting method [10] is used for determining the active cycles.

For each cycle the amplitude of the von-Mises equivalent stress, $\Delta\sigma_v$, is evaluated from the amplitudes of the directional stress components, $\Delta\sigma_x$, $\Delta\sigma_z$, $\Delta\tau_{xz}$, according to

$$\Delta\sigma_v = \frac{1}{2\sqrt{2}}\begin{Bmatrix}(\Delta\sigma_x - \Delta\sigma_y)^2 + (\Delta\sigma_y - \Delta\sigma_z)^2 + \\ (\Delta\sigma_z - \Delta\sigma_x)^2 + 6(\Delta\tau_{xy}^2 + \Delta\tau_{yz}^2 + \Delta\tau_{zx}^2)\end{Bmatrix}^{1/2} \tag{5}$$

after Li and de Freitas [11].

The multiaxial equivalent strain amplitude, $\Delta\varepsilon$, is obtained from $\Delta\sigma_v$ using the cyclic stress-strain correlation [12]

$$\Delta\varepsilon = \frac{\Delta\sigma_v}{E} + \varepsilon'_f\left[\frac{\Delta\sigma_v}{\sigma'_f}\right]^{1/n'} \tag{6}$$

The final step is to use a strain-life relationship [12, 13] to determine the life for the i'th strain cycle, $\Delta\varepsilon_i$,

$$\Delta\varepsilon_i = \frac{\sigma'_f}{E}(2N_i)^b + \varepsilon'_f(2N_i)^c \tag{7}$$

where N_i is the number of cycles to failure for cycles of magnitude $\Delta\varepsilon_i$. Equations (5), (6) and (7) thus allow calculation of N_i, the number of cycles to failure of the i'th loading cycle identified by the rainflow counting method. For each load cycle equation (6) determines the strain amplitude, $\Delta\varepsilon_i$, and a bisection method is used to find N_i from equation (7).

The total damage sustained from the cyclic loading is obtained using the Palmgren-Miner cumulative damage rule [13] so that the damage, D, is

$$D = \sum_{\text{all } i}\frac{1}{N_i} \tag{8}$$

and a value of D equal to unity then corresponds to fatigue failure. The value of D resulting from a single passage of an asperity through the contact area will generally be less than unity so that its reciprocal is the number of passes through the contact that will result in fatigue occurring.

2.3 Statistical Analysis

A statistical approach can be used to give an asperity damage distribution for a particular depth beneath the surface based on the damage values calculated at each "peak" of the surface profile. (A "peak" in this case is taken as a point on the surface whose local height exceeds that of its immediate neighbours.) The damage probability density function is naturally skewed towards low damage and the Weibull distribution is thus an appropriate stochastic model to use for this purpose. The cumulative density function for this distribution has the form [14]

$$F(x) = 1 - e^{-(x/\sigma)^{\beta}} \tag{9}$$

For the current application the random variable used is $\log_{10}(1/D)$. The scale and shape parameters, σ and β, are obtained from the cumulative histograms by the median rank regression method [14]. (This is used for robustness in preference to maximum likelihood estimation which requires solution of nonlinear equations in the shape parameters.)

3. RESULTS

Results are presented for damage calculations as described in Section 2 based on the transient results of two rough surfaces in line contact situation. The parameters used to specify the problem are given in Table 1. The material used for the damage modelling is BS970:En24 steel. Its nearest American equivalent is SAE4340 [15] whose material properties can be found in [12] and are also listed in Table 1. The profiles of rough surfaces used for the analyses are taken from [16] and shown in Figure 1 where they are denoted as profiles A, B and C.

Table 1. Parameters used for analysis.

a	0.335 mm	p_{hz}	1.0 GPa
b	-0.091	$\overline{u}$	25 m/s
c	-0.60	ε'_f	0.48
E	200 GPa	ν	0.3
n'	0.15	σ'_f	2.0 GPa

A series of fatigue analyses were carried out for the EHL contact between two surfaces each with profile C (C-on-C) with $\eta_0 = 0.001$ Pas. A range of ξ values between 0.25 and 1.5 was considered. Figure 2 shows contours of damage obtained for a section of the slower moving rough surface for ξ=0.5.

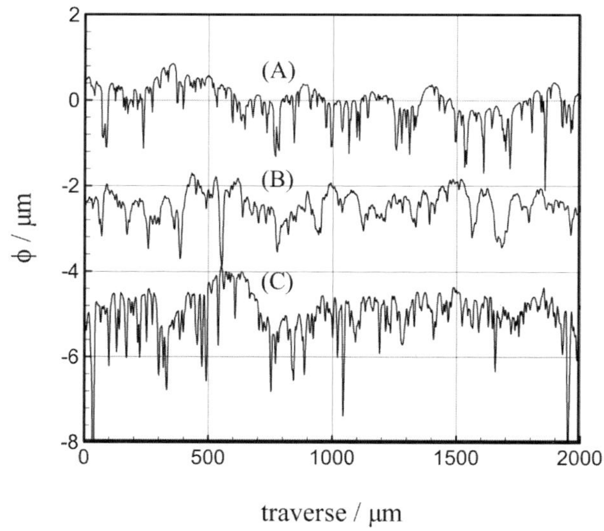

Figure 1. Surface profiles used for analyses. Ra values are (A) 0.32μm, (B) 0.22μm, (C) 0.31μ m.

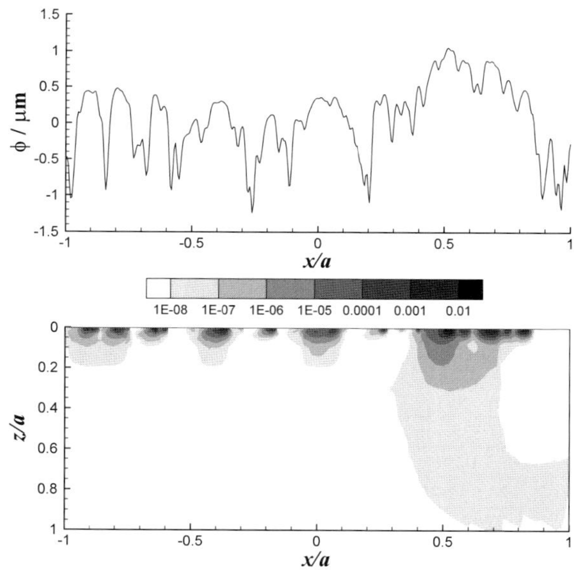

Figure 2. Contours of subsurface damage calculated for a section of the slower moving surface $2a$ long (shown in the upper figure) during its transit of the contact area with $\xi = 0.5$.

(The slower moving surface is examined because its asperities are subject to higher numbers of stress cycles in traversing the contact than are those of the faster moving surface [9].) The length of profile considered is $2a$ and its shape is also included in the figure. It is clear that accumulated damage is concentrated near the surface of particular asperity features. This is the calculated damage in one traverse of the contact area.

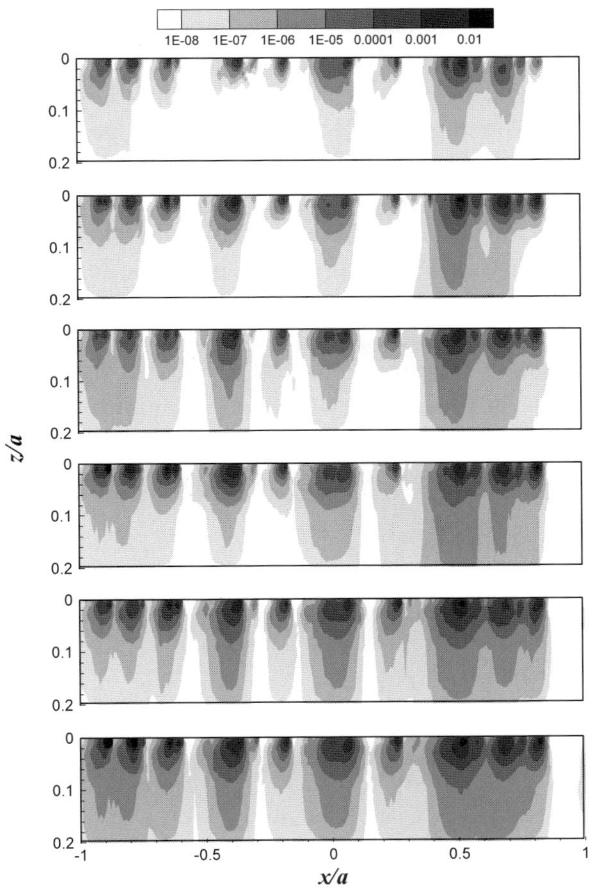

Figure 3. Contours of subsurface damage calculated for the section of the slower moving surface illustrated in figure 2 during its transit of the contact area with $\xi = 0.25$ (upper figure), 0.5, 0.75, 1.0, 1.25, 1.5 (lower figure).

Figure 3 compares the damage contours obtained with a range of values of ξ. In general increased ξ values lead to higher calculated values of D and the location of the damage hotspots is unchanged. The volume subject to a damage level of 0.001 surrounding the peak damage locations is seen to increase. The extent of the islands of high D values calculated indicates that the material subject to fatigue damage is separated by areas that are not subject to anything like the same level of damage. This may suggest that the surface distress phenomenon of micropitting could be expected in this kind of lubrication situation with micropits whose characteristic dimensions may be inferred from the typical sizes of the high D islands. For the current example which uses a rough surface profile taken from FZG gear testing this dimension would be of the order 50 μm.

Figures 2 and 3 show deterministic damage calculations and identify particular asperities that are subject to greater levels of calculated damage than others. The damage value depends on the shape of the asperity, but also on its height relative to the local surface height envelope. Prominent asperities are, not surprisingly, subject to greater damage levels than those of similar shape that are at a lower level than their neighbours.

Figure 4 shows the cumulative Weibull distributions at a series of depths beneath the surface for the case where $\xi = 0.5$. The curves are plotted so as to give the probability that the damage level is greater than the abscissa value so that for the z/a=0.03 case, for example, about 30% of asperities at that depth experience calculated damage greater than $D = 10^{-5}$ during a single pass through the contact area. The curves show that calculated damage is greatest at the z/a=0.01 level as might be expected from examination of Figure 3. At the greater depth of z/a=0.5 the level of damage has reduced by four or five orders of magnitude in comparison.

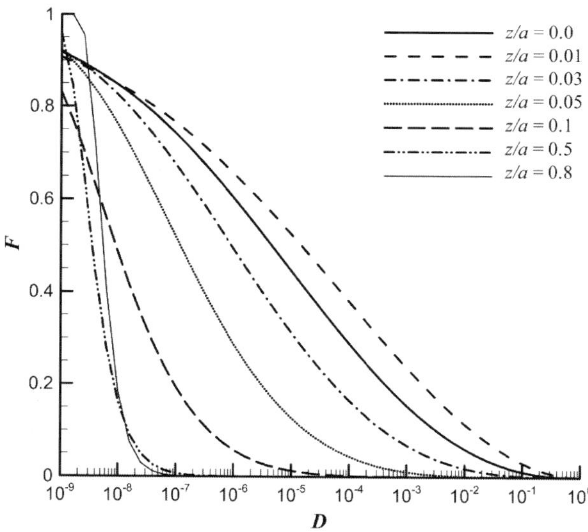

Figure 4. Cumulative asperity damage distributions for the case with $\xi = 0.5$ at a series of depths beneath the surface.

In general the highest damage probabilities occur at depths in the range $0 \leq z/a \leq 0.03$ except for the case of pure rolling (or very low sliding) where there is no stress cycling due to asperity collisions within the contact area. The effect of ξ on damage probabilities is seen in Figure 5 which shows the cumulative damage distributions at a depth of z/a=0.01 for each value of ξ considered. Here it can be seen that higher sliding generally leads to greater damage probabilities. In this comparison it is necessary to point out that the "collisions" experienced by each asperity on the surface will differ according to the specified slide roll ratio. The number of "collisions" during a traverse of the contact area will increase with slide roll ratio due to the increase in relative velocity of the surfaces [9]. Also the part of the opposing surface passing through the contact area in direct interaction with the section considered will change according to the sliding speed specified.

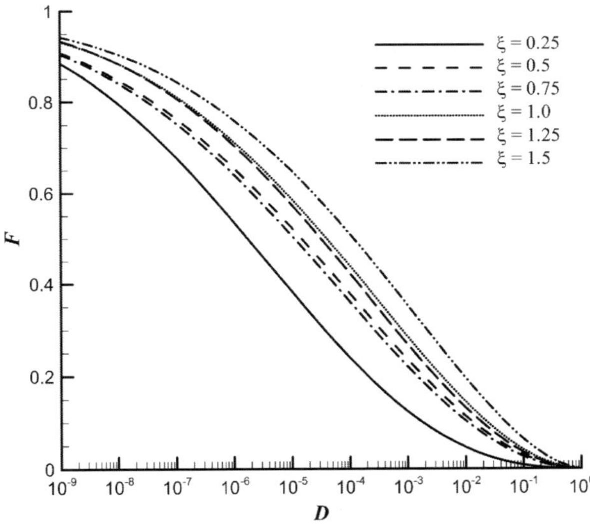

Figure 5. Cumulative asperity damage distributions at a depth of z/a = 0.01 beneath the surface for a series of ξ values.

The effect of η_0 on fatigue damage is considered for the contact of C-on-C with $\xi = 0.5$. The cumulative damage probabilities are shown in Figure 6 for five η_0 values considered at a depth of z/a=0.01. It can be seen that the fatigue damage decreases with increased η_0. This is due to the fact that decreasing viscosity generally leads to reduction in film thickness and increase in pressure. The Λ ratio values for these cases are $\Lambda = 2.448$, 1.548, 0.973, 0.611, and 0.384 in decreasing viscosity order.

The influence of roughness on the fatigue damage can be further investigated by the use of different rough profiles. Three contact conditions (A-on-A, B-on-B and C-on-C) are considered with ξ=0.5 and η_0=0.001Pas.

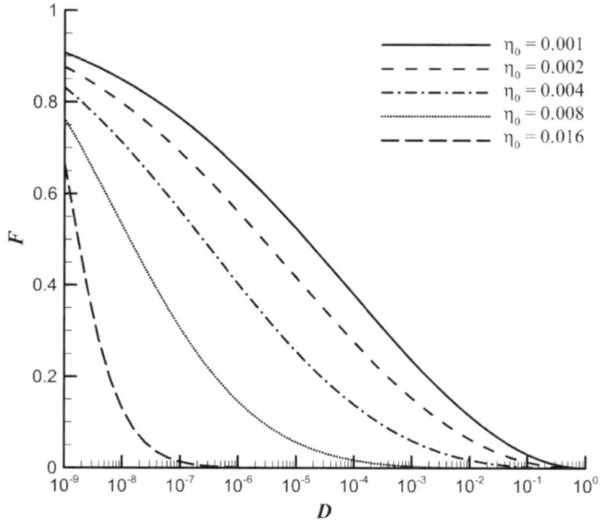

Figure 6. Cumulative asperity damage distributions at a depth of $z/a = 0.01$ beneath the surface for a series of η_0 values.

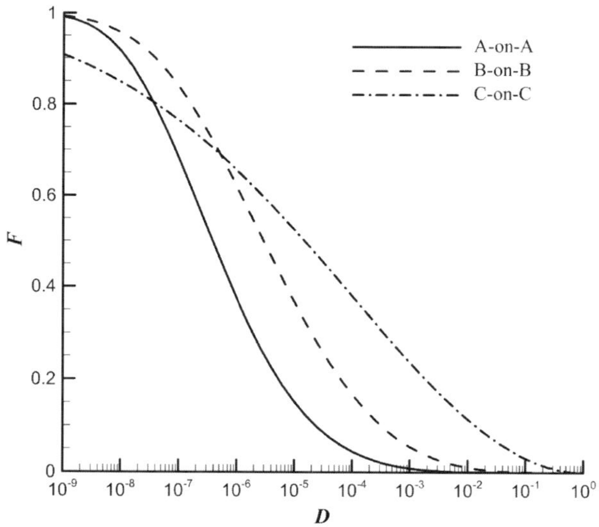

Figure 7. Cumulative asperity damage distributions at a depth of $z/a = 0.01$ beneath the surface for cases of A-on-A, B-on-B and C-on-C.

The cumulative damage probabilities are shown in Figure 7. Making comparison of results for the profiles it is clear that profile (A) is the least aggressive and profile (C) the most aggressive. This is consistent with the conclusion drawn by considering the line contact behaviour of these profiles [16] in terms of extreme pressure and film thickness behaviour. The Λ ratio values for the A-on-A and C-on-C cases are essentially the same (0.372 and

0.384) as the profiles have similar roughness values, but the calculated damage values differ by two orders of magnitude. The Λ ratio values for the B-on-B case is higher with Λ = 0.542, but the calculated damage is intermediate between the other profile results. These observations indicate that the shape of the asperities is an important property of the surface as far as fatigue calculations are concerned. Whether shape is important in terms of asperity height, wavelength, or curvature remains to be established.

4. CONCLUSION

Cycling of pressure as asperity features "pass" each other on the two surfaces leads to an oscillatory stress field being developed in the components as they traverse the contact area. Estimation of cumulative fatigue damage based on the rainflow method for non-uniform stress cycling is presented in this paper. The effects of roughness, slide roll ratio and viscosity on fatigue damage are demonstrated in numerical analyses. The results show that particular asperities can sustain damage levels that are significantly higher than that experienced by the surrounding material. The location and size of the area within the material for which high damage is calculated is consistent with the type of surface damage seen in micropitting.

5. NOTATION

a	Hertz contact dimensions (m)
b	fatigue strength exponent
c	fatigue ductility exponent
D	damage, $0 < D < 1$
E	elastic moduli of contacting bodies (Pa)
F	cumulative fatigue probability
n'	cyclic strain hardening exponent
N_i	cycles to failure of i'th cycle
p	pressure (Pa)
p_{hz}	maximum pressure in Hertzian contact (Pa)
q	surface shear stress (Pa)
t	time (s)
$\overline{u}$	mean surface velocity in x directions (m/s)
x	co-ordinates in entrainment direction (m)
z	co-ordinate normal to surface (m)
Δ	prefix indicating cycle amplitude
ε	strain

ε'_f	fatigue ductility coefficient
η_0	viscosity at ambient pressure (Pa s)
Λ	smooth surface film thickness / composite surface roughness
ν	Poisson ratios of contacting bodies
σ'_f	fatigue strength coefficient (Pa)
σ_v	von-Mises equivalent stress (Pa)
σ_x, σ_z, τ_{xz}	stress components (Pa)
ω	transformation domain for FFT
ξ	slide roll ratio
ϕ	composite roughness height profile (m)

ACKNOWLEDGEMENTS

The first author gratefully acknowledges the receipt of an ORS award that has helped to make this study possible.

REFERENCES

1. Ai, X., *ASME Journal of Tribology*, Vol. 120, pp. 159–164, 1998.
2. Tallian, T.E., *Tribology Transactions*, Vol. 42, pp. 241–249, 1999.
3. Epstein, D., Keer, L.M., Wang, Q.J., Cheng, H.S. and Zhu, D., *Tribology Transactions*, Vol. 46, pp. 506–513, 2003.
4. Ioannides, E. and Harris T.A., *ASME Journal of Tribology*, Vol. 107, pp. 367–378, 1985.
5. Zaretsky, E.V., *Journal of Propulsion and Power*, Vol. 3, pp. 76–83, 1987.
6. Elcoate, C.D., Evans, H.P., Hughes, T.G. and Snidle, R.W., *Proc. 25th Leeds-Lyon Symp. on Tribology*, Elsevier, Amsterdam, pp. 163–174, 1999.
7. Holmes, M.J.A., Evans, H.P. and Snidle, R.W., *Proc I Mech. Engrs. Part J: Journal. of Engineering Tribology*, Vol. 217, pp. 289–304 & 305–322, 2003.
8. Johnson, K.L. *Contact Mechanics*, Cambridge University Press, 1985.
9. Holmes, M.J.A., Qiao, H., Evans, H.P. and Snidle, R.W., *Proc. 30th Leeds-Lyon Symp. on Tribology*, Elsevier, Amsterdam, pp. 201–212, 2004.
10. Amzallag, C., Gerey, J.P., Robert, J.L. and Bahaud, J., *Fatigue*, Vol. 16, pp. 287–293, 1994.
11. Li, B. and de Freitas, M., *ASME J. of Mechanical Design*, Vol. 124, pp. 558–563, 2002.
12. Zahavi, E. and Torbilo, V., *Fatigue Design, Life Expectancy of Machine Parts*, CRC Press, 1996.
13. Suresh, S., *Fatigue of Materials*, Cambridge University Press, 2nd Edition, 1998.
14. Abernethy, R.B., *The New Weibull Handbook*, ISBN0-9653062-0-8, R.B. Abernethy, 2nd Edition, 1996.
15. British Iron and Steel Association (Steel User Section), compiled by J. Woolman and R.A. Mottram. The Mechanical and Physical Properties of the British Standard En Steels (B.S. 970-1955), Volume 2, En21 to En39. Pergamon Press, 1966.
16. Tao, J., Hughes, T.G., Evans, H.P. and Snidle, R.W., *Proc. 28th Leeds-Lyon Symp. on Tribology*, Elsevier, Amsterdam, pp. 447–458, 2002.

SURFACE CONTACT IN MICRO-EHL

M.J.A. Holmes, H.P. Evans and R.W. Snidle
Cardiff School of Engineering, Queen's Buildings, The Parade, Cardiff CF24 3TA, U.K.

Abstract The study of micro elastohydrodynamic lubrication (micro-EHL) of rough surfaces leads naturally to mixed lubrication where part of the load is carried by pressure in the hydrodynamic film, and part of the load by contact pressure between parts of the surfaces that make direct contact without the benefit of a hydrodynamic film. The paper is concerned with numerical analysis of this situation. Some attention is given to the details of a numerical scheme able to deal with mixed lubrication in micro-EHL in a single, integrated, solution method, and the sensitivity of direct contact calculations to mesh spacing is examined

Keywords: mixed lubrication, direct contact.

1. INTRODUCTION

The study of elastohydrodynamic lubrication (EHL) with rough surfaces has developed over the past twenty years or so and incorporates both experimental and theoretical studies. Of the two, theoretical studies using numerical analysis are more plentiful, probably due to the considerable challenge of experimental work in this field. This is not to say that numerical analyses of this type are without difficulty, quite the contrary as the developments published over the period have tended to be at the limit of what was possible with available hardware and software. Interest in the field flows from the fact that EHL lubricant films in machine elements are often thinner than the heights of asperity features on the contacting surfaces. This means that the well developed and validated theory for smooth surface EHL is of only limited applicability. The dependence of solutions on the shape of moving roughness features also increases the dimension of the

R.W. Snidle and H.P. Evans (eds), IUTAM Symposium on Elastohydrodynamics and Micro-elastohydrodynamics, 357–368.

problems to be considered to include time. As a result new transient effects, unseen in steady state situations, come into play and can dominate the outcome of the analysis.

Transient studies of micro EHL where individual roughness features produce pressure and film thickness pertubations have been presented based on sinusoidal surface features [1, 2], individual surface defects [1], and measured surface roughness profiles [3, 4].

In recent years analyses have attempted to address situations of mixed lubrication where part of the load is carried by pressurised lubricant that separates the surfaces completely, and part of the load is carried by dry or boundary lubricated contact between asperity features. Different approaches have been used to this numerical problem. These include making simplifying assumptions so as to examine conditions at micro contacts [5], partitioning the contact area into sub areas where lubricant is and is not present [6, 7], and modifying the Reynolds equation for conditions of very low film thickness in order to develop a unified approach [8, 9]. Some of the results produced by the latter approach have been found to be questionable [10–12]. The most recent developments seek to extend the evaluation of pressure variation within the contact to include their effect on the contacting materials due to cyclic loading [13].

Mixed lubrication is a particularly challenging and important regime to understand as, in truth, almost all failures of surfaces in rolling/sliding contact with an EHL lubricant film occur in this regime. The current paper presents an alternative model for mixed lubrication analysis where an unified approach to dry and lubricated areas of the contact is made possible by use of a differential approach to the deflection equation. The paper attempts to examine the calculation of direct contact critically.

2. THEORY

The study reported in the current paper is based on numerical analysis of the mixed lubrication problem that uses a novel technique developed by the authors [4, 10]. The essence of the method is that the two fundamental equations that define the problem are fully coupled in the solution scheme. These equations are the non-Newtonian Reynolds equation relating pressure to film thickness for the hydrodynamic film and the elastic deflection equation relating the same variables through the elastic deflection of the surfaces. The case considered in the paper is that where the surface finish is oriented transverse to the rolling/sliding direction. In these circumstances Holmes et al. have shown [10] that, other than at the transverse margins of the contact, the essence of the problem is captured in a line contact analysis.

For the line contact situations discussed in the current paper these equations have the form

$$\frac{\partial}{\partial x}\left(\frac{\rho h^3}{12\eta}S\frac{\partial p}{\partial x}\right)-\frac{\partial(\rho\bar{u}h)}{\partial x}-\frac{\partial(\rho h)}{\partial t}=0 \quad (1)$$

$$\frac{\partial^2 h}{\partial x^2}=\frac{\partial^2\phi}{\partial x^2}+\frac{1}{R}+\frac{2}{\pi E'}\sum_{all\,k} f_{k-i}p_k \quad (2)$$

The deflection equation is written in this differential form in order to reap the benefits of the rapidity with which the influence coefficients $f_{i,j}$ fall to zero as the indices *i* and *j* increase from zero [14].

The non-Newtonian flow factor, *S*, is a function of viscosity, shear rate and pressure gradient and is available analytically [15] for the Eyring type behaviour used for the current paper. The viscosity is taken to follow the well known Roelands expression. The way in which equation (1) is discretised using linear finite elements with an implicit (Crank–Nicolson) time formulation, and equation (2) using central finite differences for the second derivative of *h* is described in detail in [10, 16]. They then have the form

$$A_0p_0+B_0h_0 = \hat{R}_{i,j} \quad ; \quad C_0p_0+D_0h_0 = \hat{E}_{i,j} \quad (3)$$

where A_0, B_0 C_0 and D_0 are the coefficients of the pressure and film thickness at the point in question for the two equations and the influence of the pressure and film thickness at the neighbouring points are included in the right hand sides and $\hat{R}_{i,j}$ and $\hat{E}_{i,j}$. These equations are solved in the coupled iterative scheme

$$p_{i,j}^{new} = \frac{\hat{R}_{i,j}D_0-\hat{E}_{i,j}B_0}{A_0D_0-B_0C_0} \quad ; \quad h_{i,j}^{new} = \frac{\hat{E}_{i,j}A_0-\hat{R}_{i,j}C_0}{A_0D_0-B_0C_0} \quad (4)$$

In solving transient rough surface problems with the coupled method described above, negative values of the calculated film thickness emerge at individual mesh points during a timestep calculation. These instances are usually found to occur at points which have experienced extremely thin films in the previous timestep. This is the means by which the algorithm responds to physical situations where the combined effects of the Reynolds and elasticity equations do not lead to a continuous lubricant film at all meshpoints. Negative film thicknesses have no special significance in the elasticity equation, but in the non-Newtonian Reynolds equation they represent a reversal in the direction of both the Couette and Poiseuille flow terms. Consequently the emergence of negative films in the coupled solution of the equations is an indication that flow continuity cannot be maintained with a full film.

To deal with these situations the solution algorithm was modified as follows. Whenever negative film thicknesses occur in the solution to equations (3) the Reynolds equation for that meshpoint is deleted from the overall solution scheme and replaced with a boundary condition of zero film thickness at that point. The values of pressure and film thickness are still required to satisfy the elastic equation (2), and the boundary condition $h = 0$ is imposed at the contacting point. Equations (4) are thus modified at contact points to [17]

$$h_{i,j}^{new} = 0 \; ; \qquad p_{i,j}^{new} = \hat{E}_{i,j} / C_0 \qquad (5)$$

Each mesh point is assumed to be in a full film situation at each stage in obtaining the iterative solution at each timestep unless this is found to result in a negative film thickness at that point. No assumptions are made regarding which points contact, and the coupled approach to the problem's solution ensures that h is an active variable that is not linearised during the solution scheme. This method deals effectively with the numerical occurrences of negative films.

The strategy described above deals with the numerical difficulties encountered in analysing EHL contacts between components having significant surface roughness. It is also well known from electrical contact resistance measurements [18] that such contacts can exhibit direct contact behaviour. The question of whether the numerical direct contact results can be rightfully interpreted as an indication of physical direct contact of the surfaces, or as a facet of the numerical method being used is important one to resolve. The results presented in the current paper examine the sensitivity of numerical direct contact results to mesh resolution.

3. RESULTS

A detailed comparison was made of the transient results of two rough surfaces in rolling/sliding EHL contact. The parameters used to specify the problem are given in Table 1. A series of analyses were carried out that differed only in the value assumed for the ambient viscosity, η_0. The purpose of the exercise was to compare direct contact calculation situations with full film situations so as to establish whether or not the calculated direct contact obtained using equations (5) at some points in the computing mesh was consistent with the full film results obtained using equations (4) throughout. The roughness profile illustrated in Figure 1 was used for both surfaces, and the same kinematic configuration adopted for each analysis. Thus the asperity features on each surface were in exactly the same relative configuration for each timestep of the analyses. By varying η_0 between the analyses the strength of the entrainment mechanism at each asperity interaction was varied systematically with all other factors unchanged.

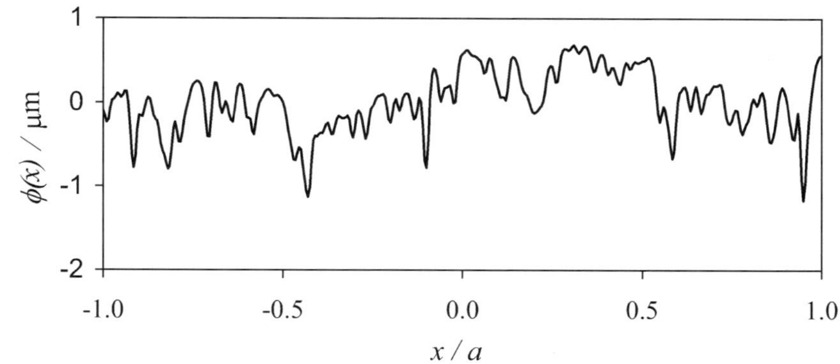

Figure 1. Roughness profile taken from FZG gear test with Ra = 0.32 μm.

Figure 2(i) shows the film thickness and pressure distributions calculated in this way for a pair of surface asperities that pass by each other during an analysis carried out with an entrainment velocity of 5 m/s and a slide/roll ratio of 0.25. Results are shown for a sequence of ambient viscosity values reducing in steps of 0.001 Pas from $\eta_0 = 0.01$, to $\eta_0 = 0.002$ Pas. The figure shows the detailed film thicknesses and pressures obtained for the surfaces at the same timestep in the analysis for each of these viscosity values. For the example shown contact develops at $x/a = -1.03$ at $\eta_0 = 0.005$ Pas, whilst the calculated film thickness at $x/a = -1.035$ is 0.001 μm. This single mesh point contact develops to include two direct contact points for the cases with $\eta_0 = 0.004$, 0.003, and 0.002 Pas. Varying η_0 systematically in this way affects the entrainment mechanism at the asperity collision while keeping other factors unchanged. The pressure carried at the contact can be seen to increase sharply as the contact becomes progressively 'heavier' as the entrainment strength is reduced. The arrows superimposed on the figure indicate the sense of viscosity decreasing which generally leads to reduction in film thickness and increase in pressure. Parameter values used for the analyses are specified in Table 1.

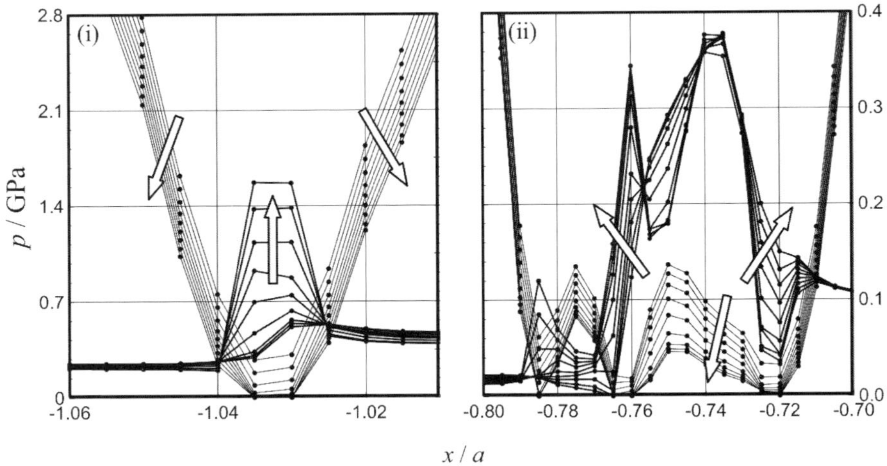

Figure 2. Pressure (thick line) and film thickness obtained at two locations where asperity collisions occur. A series of analyses are shown at $\overline{u}$ = 5 m/s, ξ = 0.25, with η_0 = 0.002, 0.003,...0.01 Pas. Arrows indicate sense of reducing viscosity.

Table 1. Parameters used for the analysis.

a	0.335 mm	E'	227 GPa	α	11.1 GPa^{-1}
R	19.1 mm	p_H	1 GPa	τ_0	10 MPa

Figure 2(ii) shows a similar comparison of calculated film thickness with viscosity for different asperity features and this particular example shows three contact positions located in close proximity. At x/a = -0.785 contact occurs for $\eta_0 \leq$ 0.004 Pas. At x/a = -0.765 contact occurs for $\eta_0 \leq$ 0.006 Pas, and at the adjacent point x/a = -0.76 contact occurs for $\eta_0 \leq$ 0.005 Pas. At x/a = -0.72 contact occurs at η_0 = 0.002 Pas. These three contacts illustrate different pressure trends as the entrainment mechanism weakens. In each direct contact case the pressure rises as the viscosity falls, but for the direct contacts at x/a = -0.785 and x/a = -0.72 the increase in pressure is from a relatively low level. The contact at x/a = -0.765 and x/a = -0.76 has an increasing pressure as seen in Figure 1 but although the contact at x/a = -0.765 first takes place at a higher viscosity, the pressure at that point remains lower than that at the neighbouring direct contact which rises to about 2.5 GPa. It is noticeable, however, that the highest pressures occur in the lubricant entrapment that forms between two of the contacts and not at one of the actual direct contacts.

The film thickness curves shown in these comparative illustrations form a consistent family of curves offset from each other as a result of the different viscosity values used. The film thickness values obtained extrapolate to suggest that contact will occur at the viscosity levels at which the negative film thicknesses first occur in the calculations. This behaviour is illustrated in Figure 3 (i) where the film thickness calculated at a direct contact point at a particular timestep is plotted against the ambient viscosity used for the calculations. In each case contact occurs at the viscosity step to be expected from the film thickness/ viscosity trend. The curves either reach $h = 0$ with reducing viscosity without any change in slope, or the slope becomes shallower in the last approach to contact. Although the figure illustrates the trend to contact for six direct contact points it is fully representative of the behaviour of a much larger number of contact points examined in this way. In none of the cases examined was there a tendency for the slope to increase in the approach to contact. This behaviour, and its consistency, gives high confidence that the method of dealing with contact detailed above is physically valid as it is entirely consistent with extrapolation from comparable full film analysis situations. Figure 3 (ii) illustrates the way in which the pressure changes at the same direct contact points as viscosity is reduced. This shows that contact is not necessarily associated with high pressures. However, whatever the pressure level at the viscosity stage prior to contact occurring, the pattern seen is of subsequent increasing pressure with decreasing viscosity and this behaviour occurs at each of the contact points.

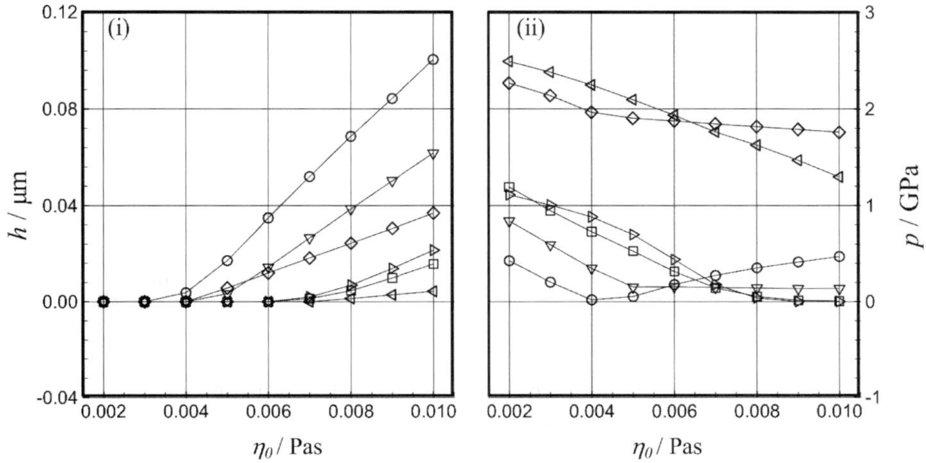

Figure 3. Variation with ambient viscosity of (i) film thickness, and (ii) pressure, at a number of direct contact points in a mixed lubrication calculation.

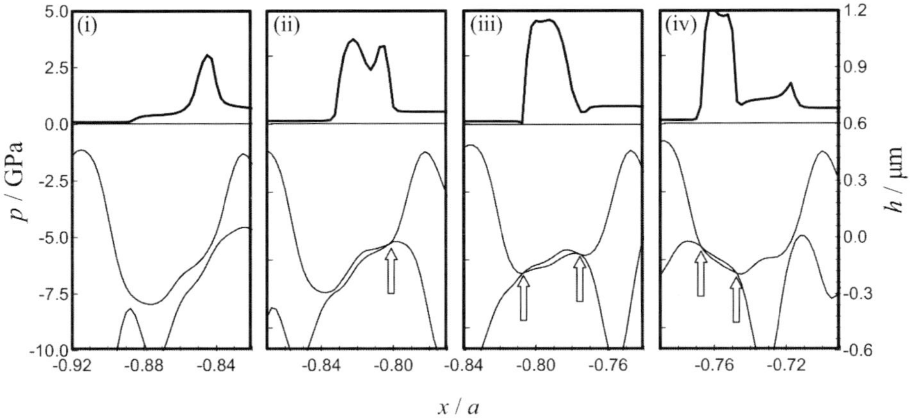

Figure 4. Sequence of timesteps showing a lubricant entrapment from an analysis with $\overline{u}$ = 0.8 m/s, ξ = 0.25, and η_0 = 0.005. Upper figures show the pressure and lower figures the contact surfaces. Each figure is separated by about 30 timesteps and arrows indicate positions where direct contact is calculated.

High pressures are not exclusively associated with direct contact. Lubricant entrapments such as seen in Figure 2 are often found to occur in these analyses as asperities collide. Figure 4 shows a sequence of timesteps that illustrate the beginning, middle and end of an asperity collision occurring within the Hertzian region where a lubricant entrapment is formed. Each of the frames shown in the figure are separated by some 30 timesteps. The entrapment is clearly visible in Figures 4 (ii–iv). In Figures 4 (iii) and 4 (iv) direct contact is calculated to occur at the limits of the entrapment (arrowed), but the pressures developed at these calculated contacts are much lower than those seen in the entrapped lubricant.

The sensitivity of results such as those shown above to the mesh spacing adopted is clearly of interest. The surface profiles are taken from profilometer traces and intermediate surface heights between the measured points are obtained using cubic splines to define the profile. The comparisons made in the paper are based on the same measured profile so that the analytical description of the surfaces for the comparisons made are the same. Figure 5 shows comparisons of direct contacts calculated at different timesteps in four different analyses. Four different mesh spacings were used to resolve the problem, namely Δx = a/100, a/200, a/300 and a/400. The appropriate mesh size for the roughness features involved in the asperity collisions illustrated in Figure 5 is probably a/200 as that analysis has captured the essence of the results obtained with finer meshes.

The timestep used is $\Delta t = u_{max}/2\Delta x$, where u_{max} is the velocity of the faster moving surface, and Δt is thus related to the mesh spacing [10]. Consequently it is advantageous to use the coarsest mesh for which meaningful results can be obtained as this minimises the computing effort.

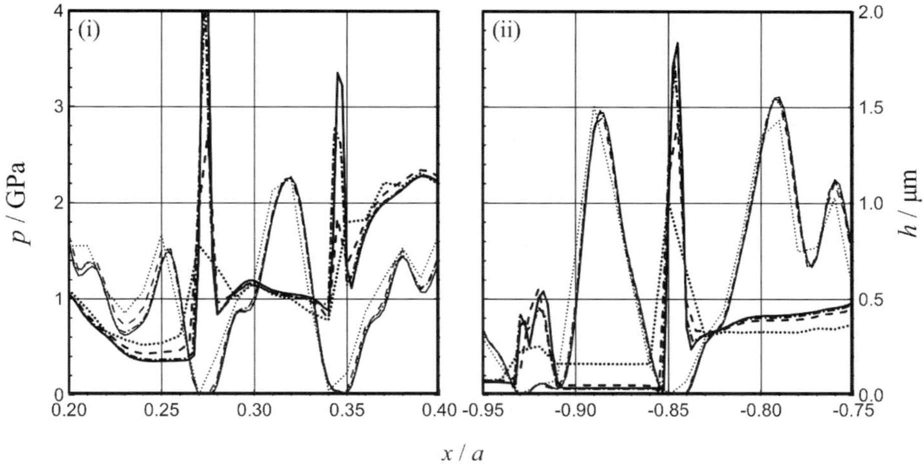

Figure 5. Pressure (heavy curves) and film thickness at two positions in the contact area where contact occurs at the timestep illustrated ($\overline{u}$ = 25 m/s, ξ = 0.25, with η_0 = 0.001). Δx=a/100 ··············, Δx=a/200 - - - - - - -, Δx=a/300 — · — · — ·, Δx=a/400 ———.

Figure 5 compares individual asperity collisions in detail, and a comparison of the calculation of direct contact over the entire contact area is given in Figure 6 for the whole of a transient analysis between two rough surfaces. The parameter Q plotted in Figure 6 is the number of direct contact occurrences at a particular mesh point during an analysis time where the faster moving surface moves through a distance of $70a$. The values of Q are scaled to be expressed in contact occurrences per millisecond. Figure 6 (i) shows the way in which Q varies for each point in the mesh for the four different mesh sizes used to analyse the same problem. The timestep used is $\Delta t = u_{max}/2\Delta x$ so that as the spatial mesh becomes finer then so too does the timestep. The same basic pattern of contact occurrences can be seen to be established from this comparison of Q, with higher values occurring as the computing mesh becomes finer. Figure 6 (ii) shows the same data, but smoothed by fitting an 8th order polynomial to it. The increase in Q seen as the mesh becomes finer may be interpreted in terms of two particular contact instances. First consider a case where contact occurs at a single mesh point and sweeps through the contact area for a fixed distance. The values of Q obtained for mesh points within this fixed distance would then be equal for each of the meshes as the number of timesteps taken for the contact to sweep through an individual mesh increment would be the same for each of the mesh sizes used. This is because the timesteps used for the analysis are proportional to the mesh spacing so that when Δx is halved, so too is Δt. Second, consider the case where contact occurs at all mesh points within a fixed distance for a fixed amount of time. In this case the values of Q obtained within this fixed distance would be in the proportions 1:2:3:4, as

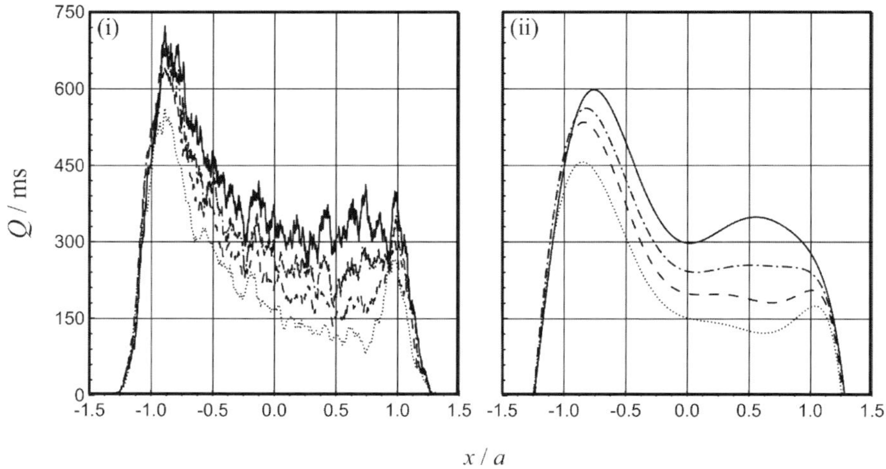

Figure 6. Variation of contact count, Q, with contact position for an analysis with $\bar{u} = 25$ m/s, $\xi = 0.25$, $\eta_0 = 0.001$. The faster moving surface has moved through a distance of $70a$. (i) shows raw data and (ii) shows a high order polynomial fit to the raw data. : $\Delta x = a/100$ ··········, $\Delta x = a/200$ - - - - - - , $\Delta x = a/300$ — · — · — , $\Delta x = a/400$ ———— .

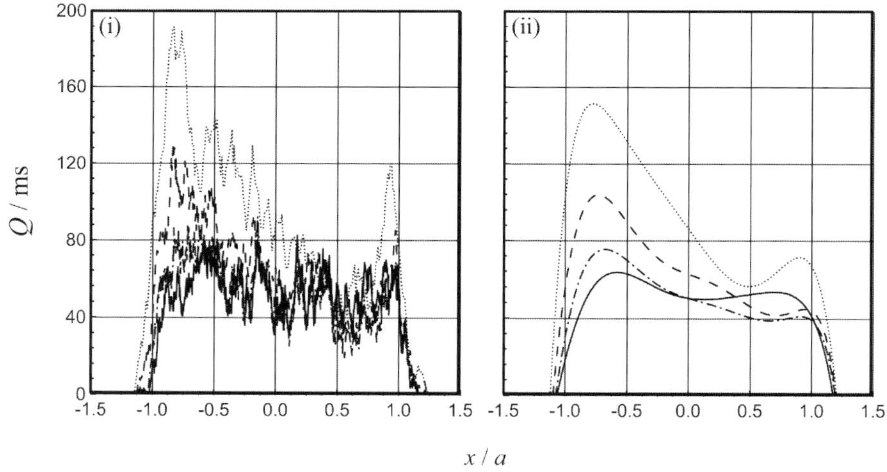

Figure 7. As Figure 6 but for case with $\eta_0 = 0.004$.

the fixed time corresponds to numbers of timesteps in these proportions. The change in Q actually seen in Figure 6 is intermediate between these two scenarios. Averaged over the contact area the change in Q with mesh resolution is in the proportions 1:1.26:1.48:1.76. This suggests that more calculated contacts occur as finer meshes are used than would correspond to simple sliding single point contacts. This is entirely plausible as the asperity tips can more easily make contact over two or three consecutive points as the mesh resolution becomes finer. This example illustrates a case with heavy asperity interaction. In examples with lighter interaction, finer meshes can lead to reduced contact count. This is illustrated in Figure 7 which is the same kinematic case analysed with a higher viscosity of 0.004 Pas. In this light contact case the finer mesh enables the algorithms to develop films at the asperity features that are calculated to be in contact with the coarser mesh. A further interesting observation is that Figures 6 and 7 suggest higher contact rates in the inlet quarter of the contact area than elsewhere.

4. DISCUSSION AND CONCLUSIONS

The results shown in this paper illustrate the process by which direct contact can be calculated in the micro EHL analysis of two rough surfaces in the fully coupled solution scheme developed by the authors [10]. The instances of calculated contact are seen to be entirely consistent with extrapolated behaviour from full film solutions obtained with higher viscosity values. This behaviour establishes that calculated contact is indeed consistent with lubricant film breakdown at asperity collisions and not a feature introduced by numerical analysis using a particular analysis technique. The sensitivity of contact calculation to mesh density is examined and discussed.

NOTATION

Parameter	Description	
a	Hertz contact dimension,	(m)
$A_k, B_k, C_k, D_k\ E_{i,j}, R_{i,j}$	coefficients in discretised equations (5) and (6)	
E'	effective modulus of elasticity	(Pa)
f_{ij}	pressure coefficient in differential deflection equation	
h	film thickness	(m)
n_c	number of neighbouring mesh points in discretisation	
p	pressure	(Pa)
p_{hz}	maximum pressure in Hertzian contact	(Pa)
Q	number of contacts occurring at a mesh point	($s^{-1}m^{-1}$)
R	radius of relative curvature	(m)
t	time	(s)
$\overline{u}$	mean surface velocity in x directions	(m/s)
u_{max}	velocity of the fastest moving surface in the x direction	(m/s)
x	co-ordinates in entrainment direction	(m)
α	pressure viscosity coefficient	(Pa^{-1})
Δx	mesh spacing	(m)
Δt	timestep	(s)
η	viscosity	(Pa s)
ξ	slide roll ratio	
ρ	density	(kg/m^3)

Parameter	Description	
τ_0	Eyring stress	(Pa)
ϕ	composite roughness height profile	(m)

ACKNOWLEDGEMENTS

The authors gratefully acknowledge the support of EPSRC (research grant GR/N33522/01) and MOD that enabled the work reported in this paper to be carried out.

REFERENCES

1. Chang, L. Webster, M.N. and Jackson, A., Trans. ASME J. of Tribology, Vol. 115, pp. 439–444, 1993.
2. Venner, C.H. and Lubrecht, A.A., Trans. ASME J. of Tribology, Vol. 116, pp. 186–193, 1994.
3. Ai, X. and Cheng, H.S., Trans. ASME J. of Tribology, Vol. 116, pp 549-558, 1994.
4. Elcoate, C.D., Evans, H.P., Hughes, T.G. and Snidle, R.W., Proc. 25th Leeds-Lyon Symp. on Tribology, Elsevier, Amsterdam, pp. 163–174, 1999.
5. Chang, L. and Zhao, Y., Trans. ASME J. of Tribology, Vol. 122, pp. 77–85, 2000.
6. Jiang, X., Hua, D.Y., Cheng, H.S., Ai, X. and Lee, S.C., Trans. ASME J. of Tribology, Vol. 121, pp. 481–491, 1999.
7. Zhao, J., Sadeghi, F. and Hoeprich, M.H., Trans. ASME J. of Tribology, Vol. 123, pp. 67–74, 2001.
8. Hu, Y-Z. and Zhu, D., Trans. ASME J. of Tribology, Vol. 122, pp. 1–9, 2000.
9. Wang, W-Z., Liy, Y-C., Wang, H. and Hu Y-Z., Trans. ASME J. of Tribology, Vol. 126, pp. 162–170, 2004.
10. Holmes, M.J.A., Evans, H.P. and Snidle, R.W., Proc. I Mech. Engrs. Part J: Journal of Engineering Tribology, Vol. 217, pp. 289–304 & 305–322, 2003.
11. Glovnea, R.P., Forrest, A.K., Olver, A.V. and Spikes, H.A., Tribology Letters, Vol. 15, pp. 217–230, 2003.
12. Jin, Z., Yang, P. and Dowson, D., Discussion on [10] to appear Proc I Mech. Engrs. Part J: Journal of Engineering Tribology, vol 218, 2004.
13. Epstein, D, Keer, L.M., Wang, Q.J., Cheng, H.S. and Zhu, D. Tribology Transactions, Vol 46, pp 506-513, 2003.
14. Evans, H. P. and Hughes, T. G., Proc. Instn. Mech. Engrs Part C, Jn of Mechanical Engineering Science, Vol 214, pp 563-584, 2000.
15. Conry, T.F., Wang, S. and Cusano, C. Trans ASME Jn. of Tribology, Vol 109, pp 648-654, 1987.
16. Elcoate, C.D., Evans, H.P., Hughes, T.G. and Snidle, R.W. Proc I Mech. Engrs. Part J: Journal. of Engineering Tribology, Vol 215, pp 319-337, 2001.
17. Holmes, M. J. A., Evans, H. P. and Snidle, R W. to appear ASME Jn. of Tribology, Vol 127, 2005.
18. Furey, M. J., ASLE Transactions Vol 4, pp 1-11, 1961.

SESSION 9

WEAR MODELLING IN WORM GEARS

K.J. Sharif, H.P. Evans and R.W. Snidle
Cardiff School of Engineering, P.O. Box 925, Newport Road, Cardiff CF24 0YF, U.K.

Abstract The paper presents a wear model for worm gears that takes into account the pressure and film thickness generated in the elastohydrodynamic oil film that separates the tooth components. Wear is calculated based on the well known Archard wear law applied to each point on the wheel tooth surface. The calculated wear is then used to modify the geometry and the calculation process is repeated in a series of wear steps. Results are presented that show the way in which the wear pattern emerges from the analysis, and that this is generally insensitive to the size of the wear step used.

Keywords: wear model, worm gears, elastohydrodynamic lubrication.

1. INTRODUCTION

Worm gears are possibly unique amongst gear power transmission systems in that the geometry of the load bearing areas of the teeth is changed continuously by wear processes during the lifetime of the gear pair. The configuration adopted for the current study is the most common one for power transmission applications: a steel worm component meshing with a phosphor bronze gear wheel. Gears of this form are usually manufactured with an "oversize" hob which gives a contact that occurs at a point rather than a line under zero load, so as to obtain benefits in lubrication and to avoid contact areas that reach the physical edges of the meshing components. The kinematic action of worm gears ensures that the wheel component is almost stationary relative to the contact point, and the gears essentially operate in a simple sliding mode. As a result, the wheel component is subject to wear due to the sliding action of the worm over its surface. The steel worm also experiences wear but this occurs at a much

R.W. Snidle and H.P. Evans (eds), IUTAM Symposium on Elastohydrodynamics and Micro-elastohydrodynamics, 371–383.

lower rate than that found in the wheel. The bronze teeth of the wheel are able to sustain this aggressive lubrication regime successfully, and their loss of material due to wear is recognised as an inevitable feature of the configuration. Indeed, the extent to which material has been lost is assessed during maintenance inspections by measuring the increase in backlash, and this change is monitored until its extent indicates that the wheel teeth are becoming too thin to sustain their duty load.

During operation of the gears there is an initial period of relatively rapid wear that is usually referred to as *bedding in* of the gears. During this period the actual contact area of the teeth under load increases so as to reduce the (average) contact stress levels. After *bedding in* has occurred the gear pair tends to wear at a reasonably steady rate for the remainder of its life cycle. This steady wear rate can be influenced by a number of factors and effectively determines the lifetime of the gear mesh.

The gear surfaces are able to sustain the high levels of sliding to which they are subjected because of elastohydrodynamic lubrication (EHL). This is the lubrication mechanism that protects the contacting surfaces of most gear pairs, but the configuration of worm gears is far from optimal as far as generating lubricant films to separate the surfaces are concerned. The contact area over which the load is carried tends to be a long thin ellipse which is distorted into a banana shape by the enveloping nature of the surfaces. Such a configuration could yield relatively thick lubricant films if lubricant entrainment was in the direction of the minor axis, i.e. across the contact area. Unfortunately, this is not the case as the sweeping action of the worm ensures that entrainment is effectively along the major axis of the contact ellipse. The authors have previously examined the film forming capability of a large number of worm gear designs [1,2] based on a full thermal EHL analysis, and this study identified locations of particularly thin films within the contact area, caused by the unfavourable kinematic action of the gears.

The current paper presents the means by which the EHL film forming analysis has been extended to include a calculation of the wear rate at each point on the wheel tooth surface. This wear pattern is then integrated over the meshing cycle to obtain the tooth wear per wheel rotation. The wear is not uniform and modifies the effective contact area. This effect of wear on wheel tooth geometry is taken into account by recalculating the EHL film and pressure distributions at the end of a series of wear stages and incorporating the calculated change in wheel tooth shape into the subsequent calculations.

2. THEORY

The shape of the surfaces of the teeth of the gears is obtained from a 'cut and mesh' worm gear contact model developed by Fish and Munro [3]. This analysis considers a series of positions of the worm gear and for each position the wheel is brought into contact with the worm in the geometric model. This establishes the point of contact under zero load and the direction of the common normal at the contact point. The position of a series of points on the surface of the worm and the wheel tooth is obtained and these are fitted with a high order polynomial of the tangent plane co-ordinates. These polynomials are taken to describe the shapes of the contacting surfaces for the subsequent EHL analysis. The EHL mechanism requires that the Reynolds equation for the lubricant's hydrodynamic action, and an elasticity equation relating the lubricant pressure to the elastic deflection are solved simultaneously. The high sliding in the contact generates heat by lubricant shear and a thermodynamic energy equation is also specified for the lubricant film and for each of the contacting solid bodies. The means by which these equations are solved is described in detail in [1,2] and is therefore not repeated in the current paper. A thermal EHL solution for each of a series of mesh positions can be obtained and these result in detailed distributions of pressure, lubricant film thickness, lubricant temperature and contacting body surface temperatures.

When the contact occurs near the root of the wheel tooth its behaviour is essentially Hertzian and the area of dry contact can be established from a semi-infinite body analysis, as is inherent in the EHL solution scheme referred to above. As the contact moves towards the tip of the wheel tooth, however, the area of contact becomes longer and narrower than that predicted on the basis of a semi-infinite body elastic analysis. This is due to the fact that tooth flexure under load influences the stiffness of the tooth in a non-uniform way with the tooth being stiffest at its centre and least stiff at the limits of its face width. A correction to the geometry is introduced to ensure that the semi-infinite body treatment of elastic deflection that is embedded in the EHL analysis calculates the contact area correctly. This correction is a curvature change in the axial direction of the tooth which is adjusted for each mesh position so as to obtain dry contact areas that coincide with those calculated based on tooth stiffness ideas described in [3, 4, 5].

The model adopted to calculate the rate of material removal by the wear process (expressed in m^3/s) is based on the Archard wear formula [6, 7]

$$Wear\ Rate = \frac{k}{H} W U_s \tag{1}$$

In this formula W is the load, U_s, is the sliding velocity, H is the hardness of the material, and the wear coefficient, k, is a dimensionless parameter that can be expected to vary according to the nature of the sliding contact, and is to be determined empirically for any given system.

For the calculation of wear on the surface of the wheel tooth it is clearly necessary to distinguish between locations that are subject to different loading and there is also a need to introduce a lubricant film thickness parameter to distinguish between regions experiencing thick and thin lubricant films in comparison with the surface roughness. The wear formula adopted for the model is

$$Wear\ Rate = \frac{k}{H} p u_s . \left(\frac{R_a}{h} \right)^n \tag{2}$$

In Eq. (2) p is the local pressure, u_s is the local sliding speed, R_a is the Centre Line Average of the composite surface roughness, h is the local film thickness, and the wear rate is the change of wear depth per unit time. Power n is a disposable parameter that controls the extent to which film thickness in the form of 'lamda' ratio h/R_a influences the wear rate. In the overall wear calculation included in relevant standards [8] this power is specified as 2.24 which causes a wear rate when $h/R_a = 0.2$ to be some 40 times that when h/R_a is unity. When $h >> R_a$ the surfaces are separated by a full lubricant film which prevents wear. As h is reduced to the order of R_a and below, localised direct interaction of the surfaces becomes increasingly likely to occur with corresponding increased wear of the surfaces. In the current paper the emphasis is on wear patterns experienced by the softer bronze wheel, and the R_a value is taken to be that of the hard steel worm component.

Figure 1 shows contours of the pressure and film thickness obtained from an EHL analysis for the worm design considered in the paper at one particular position of the worm. Also shown is the wear rate calculated by application of Eq. (2). The contact is quite heavily loaded in EHL terms with the area subject to significant hydrodynamic pressure largely restricted to the area that would carry the load in dry contact. The film thickness over this dry contact area has the characteristics of an EHL contact but these are distorted by the sweeping nature of the entrainment velocity vector as discussed at length in [2]. The central film thickness is about 0.45 μm, and there are two regions of particularly thin film where values as low as 0.15 μm occur. These regions correspond to the side constrictions seen in conventional experimental and theoretical EHL studies, but are swept across the contact by the unique entrainment characteristic of the gears.

High pressure and thin film conditions are thus restricted to the dry contact area and the wear rate obtained from Eq. (2) shows this effect clearly. The band of highest wear corresponds to the film constriction referred to above. The aspect ratio of the contact shown in figure 1 is about 18:1 and this has been distorted in the figure by a factor of two for clarity.

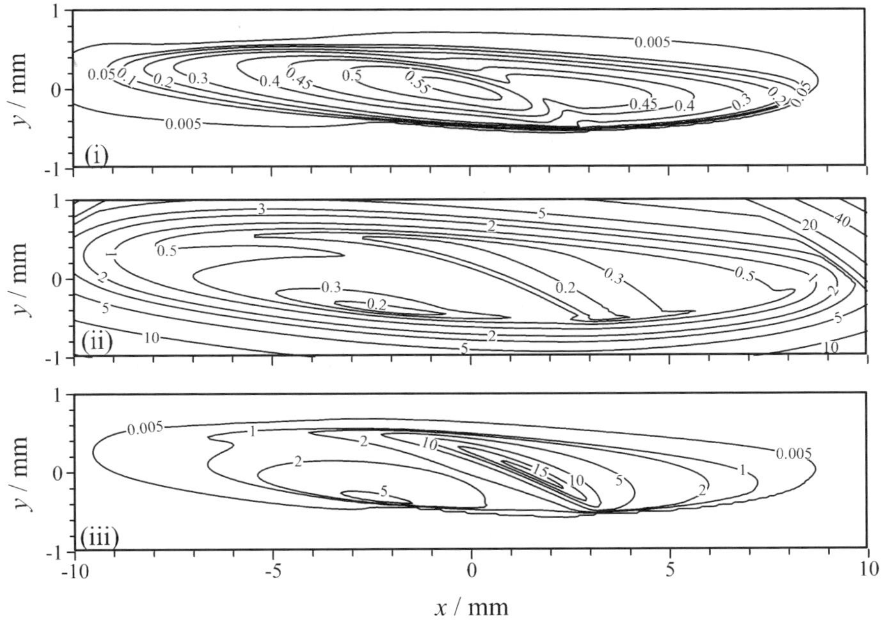

Figure 1. Contours of (i) Pressure / GPa, (ii) film thickness / μm, and (iii) calculated wear rate μm / s obtained from an EHL solution at one point in the meshing cycle of a worm gear. Note that the y axis is scaled by a factor of two to aid clarity.

During the meshing cycle of the gear the calculated wear pattern changes continuously and the area of high wear sweeps across the wheel tooth from its tip to its root. To sum the wear over the meshing cycle it is thus necessary to integrate the wear rate with respect to time over the whole tooth surface. This calculation is facilitated by considering the worm at a series of equally spaced rotational positions which are therefore equally spaced in time. EHL calculations are carried out for each of these positions and the wear rate established from Eq. (2) as described above. Figure 2 shows a series of such calculated wear rates over the wheel tooth surface. The results are presented in a projection plane perpendicular to the wheel axis. The curved boundaries indicate the worm's outer radius (top) and the wheel tooth throat radius (bottom). The transverse limits at $x = \pm\ 14.5$ mm correspond to the wheel tooth face width. The wear evaluations included here are sufficiently far apart not to influence each other as the dry contact areas for these worm positions do not overlap. The no load contact point sweeps over the wheel tooth surface during the meshing cycle as illustrated by the solid curve included in the figure. The time integration required is carried out over 10 timesteps between each of the mesh positions for which an EHL analysis is carried out. The wear at each of these timesteps is obtained by interpolating between successive wear rate calculations. In this interpolation the wear is established relative to the contact points, and used to interpolate the wear rate at each intermediate time. The interpolated wear rates are then imposed at the intermediate contact points and the total wear for the meshing cycle is calculated by summation of these interpolated wear rates.

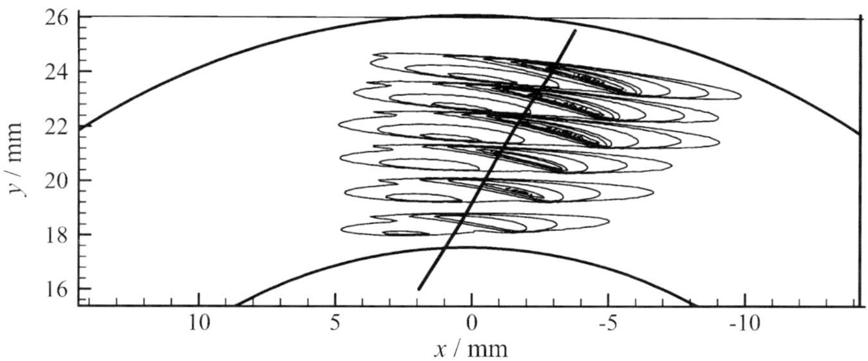

Figure 2. Contours of wear rate /μm s^{-1} calculated at a series of meshing positions on the wheel tooth surface.

The result of the calculation for the complete meshing cycle is shown in Figure 3 which shows contours of the depth of material removed per meshing cycle, δ. This shows that the wear rate is largest near the root of the wheel tooth, which occurs for two reasons. Firstly, the sliding speed is proportional to the radius from the worm's axis of rotation, and this is consequently greatest at the outer part of the worm flank which is in contact with the root of the wheel tooth. Second, the shape of the contacting bodies is such that the dry contact area is longer and narrower when contact is near the tip of the worm than when it is towards the root of the worm. This factor leads to higher pressures being generated in the contact and to thinner oil films being generated as the contact's shape becomes progressively elongated [2].

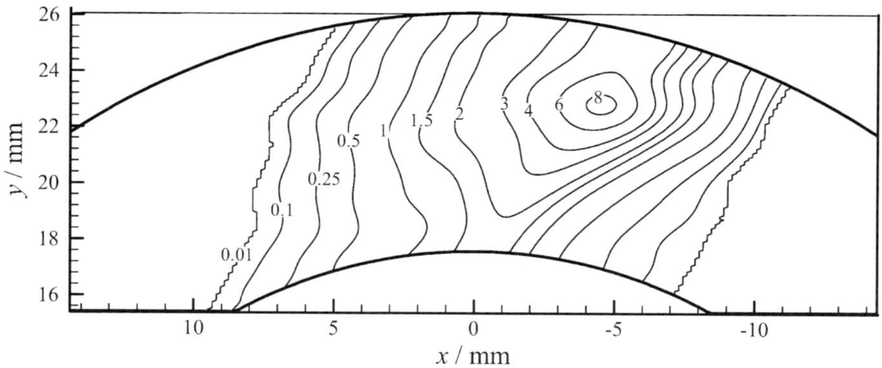

Figure 3. Contours of δ / μm showing wheel tooth wear over a single meshing cycle.

The amount of material removed during a single meshing cycle is clearly very small, but provided the basic shape of the surfaces is not changed, the amount removed during each meshing cycle will be the same and will be distributed over the tooth surface in the same way. Clearly this cannot take place indefinitely as the tooth shape will gradually change with running time and the way in which the pressure and film thickness are generated will also change. It thus becomes necessary to introduce the concept of a wear step. This is the time, ΔT, for which the underlying geometrical shape is not significantly changed, and the wear rate per meshing cycle can be assumed to also be unchanged. The contours shown in Figure 3 can therefore be scaled to show the depth of material removed during the wear stage, ΔT. The relationship between the wear step time and real time is linear as the wear coefficient, k, in Eq. (2) is an unknown that needs to be determined experimentally. For this study k was taken as 10^{-6} and H as 1 GPa. It is thus possible to scale the data so that the maximum wear depth is a particular amount, say Δw. The wear step will then correspond to the real time in which a maximum of Δw is removed from the wheel by the wear process.

The wear model then consists of the following steps.

1. Determine the "as manufactured" geometry of worm and wheel.
2. Carry out an EHL analysis at a sequence of mesh positions equally spaced in time during the meshing cycle. Determine the instantaneous wear rate for each point on the wheel tooth using Eq. (2).
3. Calculate the amount of material removed at each point on the wheel tooth during a single meshing cycle by integration in time. Scale these data by the number of meshing cycles occurring during the wear step to calculate the material removed over the tooth surface during the step.

4. Modify the geometry of the wheel tooth to take account of material removal. This completes the wear step.
5. Repeat from step 2 to determine the wear over the next wear cycle.

The question of how large the wear step can be without affecting the outcome of the wear model arises. It was found by trial and error in the example considered that the process became unstable if the wear step was such that Δw exceeded 20 μm. This is because removal of a maximum of 20 μm of material from the wheel tooth according to the pattern of Figure 3 causes significant differences to the way in which pressure is generated in the EHL film. For this reason a study was carried out with the value of Δw set to 8 μm. This determines the scaling factor σ that converts the material removal of the single meshing cycle to the wear step material removal according to $\sigma = \Delta w / \delta_{max}$. For subsequent wear stages of equal duration the same value of σ is thus used irrespective of the different pattern of δ calculated.

Trials were carried out using ΔT corresponding to Δw = 8 μm for the initial pattern of calculated wear. The wear model was run as described above for six wear stages with this value of ΔT. It was also run from the same unworn geometry for 11 stages with a wear step of $\Delta T/2$, 7 stages with a wear step of $\Delta T/4$, and 3 stages with a wear step of $\Delta T/8$.

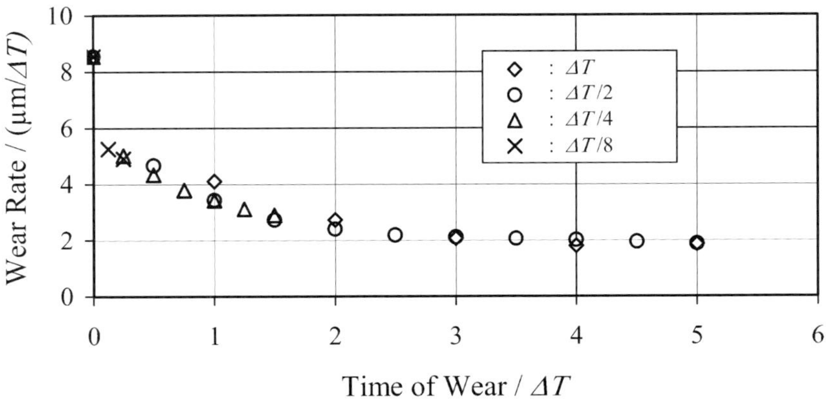

Figure 4. Variation of maximum calculated wear rate with time using wear stages of ΔT, $\Delta T/2$, $\Delta T/4$, and $\Delta T/8$.

The results of these calculations are compared in Figure 4 by plotting the maximum calculated wear rate over ΔT as a sensitive indicator of the differences between the models. The number of stages used for the smaller wear steps are those necessary for this indicator to agree on maximum wear with the next (coarser) wear step. It can be seen that the differences between the results with different wear steps are mainly apparent in the first step ΔT.

Figure 5 shows the wear rate obtained after three wear steps each of length ΔT. The maximum value for the wear rate has now fallen by a factor of four from the first calculation to 2.07 μm/ΔT The position of the highest wear rate on the tooth has also changed and is now seen in a more central position close to the worm tip. This relocation of the position of maximum wear rate reflects the changes in pressure and film thickness brought about by the calculated changes to the tooth geometry over the previous wear steps. Comparison of the differences seen in calculated wear rate after time $3\Delta T$ using wear steps of ΔT and $\Delta T/2$ show minor differences with a maximum wear rate of 2.11 μm/ΔT obtained for the finer wear step.

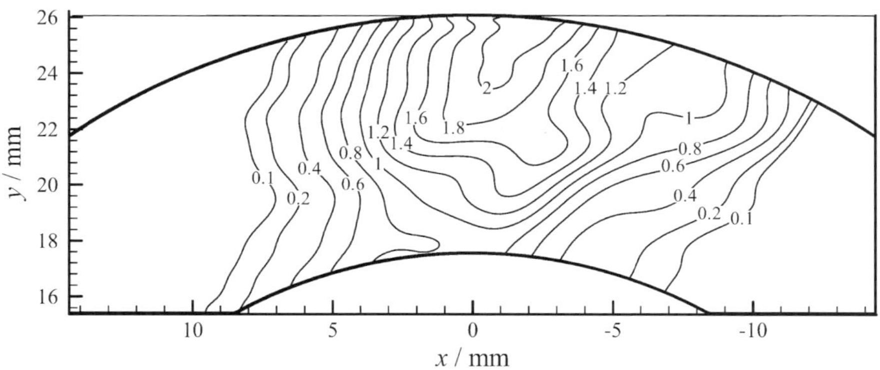

Figure 5. Contours of wear rate / μmΔT^{-1} after $3\Delta T$ calculated using wear step ΔT.

Figure 6 compares the wear rate contours calculated after $5\Delta T$ using wear steps of ΔT and $\Delta T/2$. This shows minor differences in detail but excellent general correspondence in both the values of wear rate and the way in which this varies over the tooth. This shows that in spite of the differences during the first wear stages the wear rate calculation has stabilised over this period so that wear stages ΔT are sufficiently fine to obtain the result.

Figure 7 compares the accumulated wear using wear steps of ΔT and $\Delta T/2$ after a total wear time of $6\Delta T$. This shows clearly that the differences between the models are minor with maximum calculated material removal of 13.8 and 13.2 μm, respectively.

The way in which the wear rate and accumulated wear calculated stabilises for different wear steps, as illustrated in Figures 4, 6 and 7, suggests that finer wear steps are appropriate for the initial wear from the 'as manufactured' worm geometry, whereas larger ones may well be appropriate as the wear pattern settles down. Further models were run in which the wear step was of different (temporal) duration, with the length of the wear

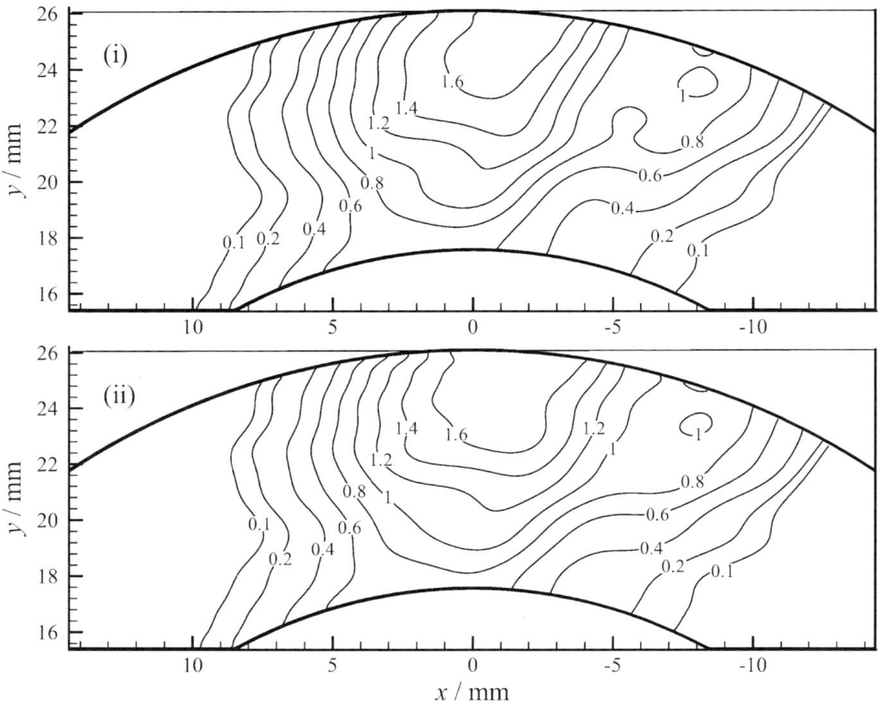

Figure 6. Contours of wear rate / $\mu m \Delta T^{-1}$ after $5\Delta T$ calculated using (i) wear step ΔT, (ii) wear step $\Delta T/2$.

step chosen so that the same maximum amount of material was removed. i.e. the scaling factor to σ was not fixed but re-calculated according to $\sigma = \Delta w / \delta_{max}$ for each individual wear step. The maximum material removal was thus Δw per wear step. A value of $\Delta w = 2$ μm was found to give the same accumulated wear after $6\Delta T$ as the fixed wear stage models.

Figure 6 compares the wear rate contours calculated after $5\Delta T$ using wear steps of ΔT and $\Delta T/2$. This shows minor differences in detail but excellent general correspondence in both the values of wear rate and the way in which this varies over the tooth. This shows that in spite of the differences during the first wear stages the wear rate calculation has stabilised over this period so that wear stages ΔT are sufficiently fine to obtain the result.

Figure 7 compares the accumulated wear using wear steps of ΔT and $\Delta T/2$ after a total wear time of $6\Delta T$. This shows clearly that the differences between the models are minor with maximum calculated material removal of 13.8 and 13.2 μm, respectively.

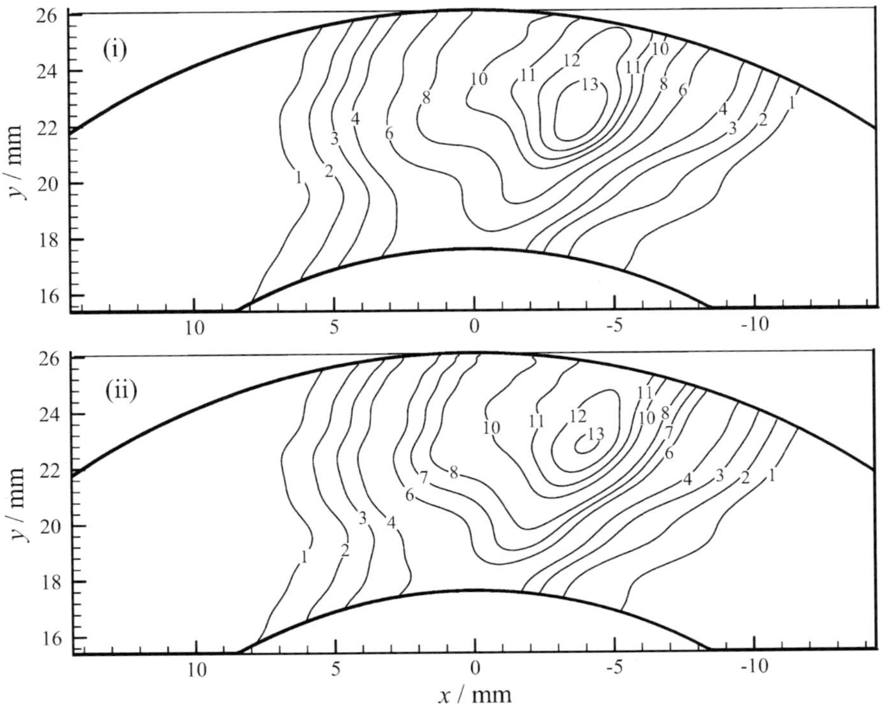

Figure 7. Contours of accumulated wear / μm after a total wear time of $6\Delta T$ calculated using (i) wear step ΔT, (ii) wear step $\Delta T/2$.

The way in which the wear rate and accumulated wear calculated stabilises for different wear steps, as illustrated in Figures 4, 6 and 7, suggests that finer wear steps are appropriate for the initial wear from the 'as manufactured' worm geometry, whereas larger ones may well be appropriate as the wear pattern settles down. Further models were run in which the wear step was of different (temporal) duration, with the length of the wear step chosen so that the same maximum amount of material was removed. i.e. the scaling factor to σ was not fixed but re-calculated according to $\sigma = \Delta w / \delta_{max}$ for each individual wear step. The maximum material removal was thus Δw per wear step. A value of $\Delta w = 2$ μm was found to give the same accumulated wear after $6\Delta T$ as the fixed wear stage models.

This mode of operation allows the calculated wear to be established over longer total wear times with fewer wear stages. The calculation of wear at a time of $14\Delta T$ obtained in this way is shown in Figure 8.

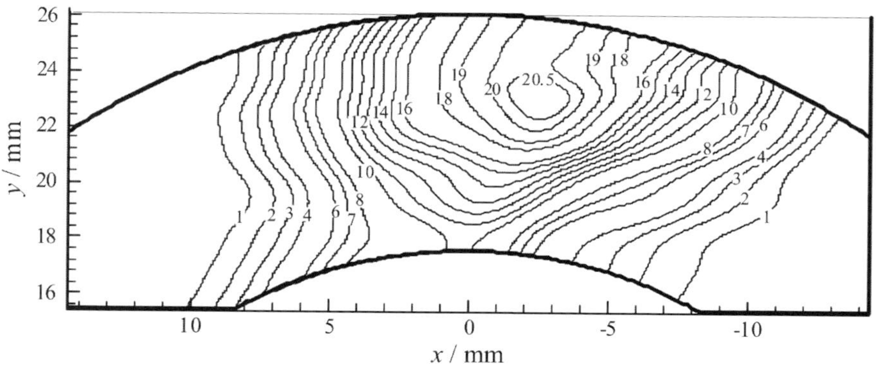

Figure 8. Contours of accumulated wear / μm after a total wear time of 14ΔT calculated using variable length wear stages with Δw = 2 μm.

3. CONCLUSIONS

A wear model for worm gears is presented based on elastohydrodynamic lubrication analyses to determine lubricant pressure and film thickness, and use of an Archard type wear model to calculate the rate of material removal from the bronze wheel tooth. The concept of a wear stage over which the wear rate is assumed not to vary due to the changes in geometry taking place is introduced, and the sensitivity of the wear rate and accumulated wear calculations to the duration of the wear stage are examined.

It is shown that the optimum approach is to use variable length wear stages characterised by the maximum amount of material removal. In this way the steady-state wear pattern can be established in relatively few overall calculation cycles.

The empirical wear model described has a film thickness dependence that can be adjusted so as to obtain correspondence with the wear patterns seen in experiments. The appropriate wear rate can be determined by consideration of the change in calculated backlash between the gears and comparison with that observed in experiments.

ACKNOWLEDGEMENTS

The authors gratefully acknowledge the support of EPSRC (grant number GR/R15429/01) and MOD that allowed this work to be completed.

REFERENCES

1. Sharif, K.J., Kong, S., Evans, H.P. and Snidle, R.W., "Contact and elastohydrodynamic analysis of worm gears: Part 1. Theoretical formulation", *Proc. Instn. Mech. Engrs. Part C, Journal of Mechanical Engineering Science*, Vol. 215, 817–830, 2001.
2. Sharif, K.J., Kong, S., Evans, H.P. and Snidle, R.W., "Contact and elastohydrodynamic analysis of worm gears: Part 2. Results", *Proc. Instn. Mech. Engrs. Part C, Jounal of Mechanical Engineering Science*. Vol. 215, 831–846, 2001.
3. Fish, M.A. and Munro, R,G., "Modelling contact characteristics in worm gear systems", British Gear Association Technical Congress, 1999.
4. Fish, M.A. and Munro, R,G., "Isolating and quantifying tooth deformation in worm gear systems operating under a torque load", Lamdamap Conference, Longhirst Hall, Northumberland, July 1999.
5. Sudoh, K., Tanaka, Y., Matsumoto, S. and Tozaki, Y., "Load distribution analysis method for cylindrical worm gear teeth", *JSME International Journal, Series C*, Vol. 39, 606–613, 1996.
6. Archard, J.F., "Contact and rubbing of flat surfaces", *J. Appl. Phys.*, Vol. 24, 981–988, 1953.
7. Peterson, M.B. and Winer, W.O. (eds), *Wear Control Handbook*, American Society of Mechanical Engineers, 1980.
8. Hőhn, B.-R. and Streingrőver, K. "DIN 3996: A New Standard for calculating the load capacity of worm gears", AGMA Technical Paper 98FTM10, 1998.

LUBRICATION MODELLING OF ARTIFICIAL HIP JOINTS

F.C. Wang and Z.M. Jin
School of Mechanical Engineering, University of Leeds, Leeds LS2 9JT, U.K.

Abstract Lubrication modelling of artificial hip joints was considered. In particular, a mixed lubrication model was developed for artificial hip joints employing ultra high molecular weight polyethylene cups, and both boundary contact and elastohydrodynamic lubrication were solved simultaneously. A full fluid-film lubrication regime was predicted in the artificial hip joint with perfectly smooth bearing surfaces under physiological conditions. However, the introduction of a simple axisymmetric sinusoidal waviness in the radial direction reduced significantly the lubricant film thickness, resulting in a mixed lubrication regime at a much higher viscosity than that of peri-prosthetic synovial fluid.

Keywords: artificial hip joint, mixed lubrication, ball-in-socket.

1. INTRODUCTION

The importance of lubrication on the successful functioning of artificial hip joints has been increasingly recognised. Lubrication plays an important role in the overall tribological performance of artificial hip joints, particularly regarding the generation of wear and wear particles, which can cause adverse tissue reactions such as osteolysis (bone-loss) and eventually lead to implant loosening [1]. Therefore, minimising wear and wear particles represents one of the most effective strategies to prolong the life span of artificial hip joints, particularly required in younger and more active patients.

In engineering, the most effective way to minimize wear is through improving lubrication. A similar principle can be applied for artificial joints [2]. The lubrication modelling of artificial hip joints should include the following factors:

(a) *Large deformation of the bearing surfaces and elastohydrodynamic lubrication (EHL) analysis*: The load experienced in artificial hip joints during normal walking can be quite high, reaching 3 to 5 times bodyweight. Therefore for a person with an average mass of 75 kg, the maximum

R.W. Snidle and H.P. Evans (eds), IUTAM Symposium on Elastohydrodynamics and Micro-elastohydrodynamics, 385–396.

load in hip joints can be as high as 3750 N, although the average load is about 1500 N.

(b) *Ball-in-socket configuration*: The bearing surfaces of artificial hip joints are closely conforming, in order to provide kinematics and stability requirements as well as to reduce contact stresses.

(c) *Anatomical position of the acetabular cup*: The acetabular cup is normally positioned with an inclination angle of approximately 45° to the horizontal plane. This may become important regarding lubricant starvation.

(d) *Transient loading and motion during walking*: This generally requires the solution of the Reynolds equation including both entraining and squeeze-film motions.

(e) *Non-Newtonian synovial fluid*: The lubricant in the replaced hip joint is periprosthetic synovial fluid, which generally experiences shear-thinning characteristics, but approaches asymptotically Newtonian behaviour under realistic shear rates encountered during walking. A representative viscosity (η) of 0.002 Pas has been estimated for the periprosthetic synovial fluid [3].

(f) *Surface topography and mixed lubrication*: The predicted lubricating film thickness from fluid-film lubrication analyses has been found to be similar or smaller than the average surface roughness of the bearing surfaces of the majority of hip implants [2]. Furthermore, there are always occasions during daily activities when potential fluid film lubrication is broken, such as during start-up and stopping.

In all previous lubrication studies of artificial hip joints, only fluid-film lubrication analyses have been conducted for various bearing surfaces, including polyethylene-on-metal, metal-on-metal and ceramics-on-ceramics [2]. Although such analyses can provide important design guidelines to promote fluid film lubrication, a mixed lubrication model is required for a better understanding of the overall tribology of artificial hip joints.

2. LUBRICATION MODEL AND GOVERNING EQUATIONS

A typical hip implant employing an ultra high molecular weight polyethylene (UHMWPE) acetabular socket against a metallic or ceramic head shown in Figure 1a was considered.

The important geometric parameters are the radius of the femoral head (R_1), and the radius and wall thickness of the UHMWPE cup (R_2, d).

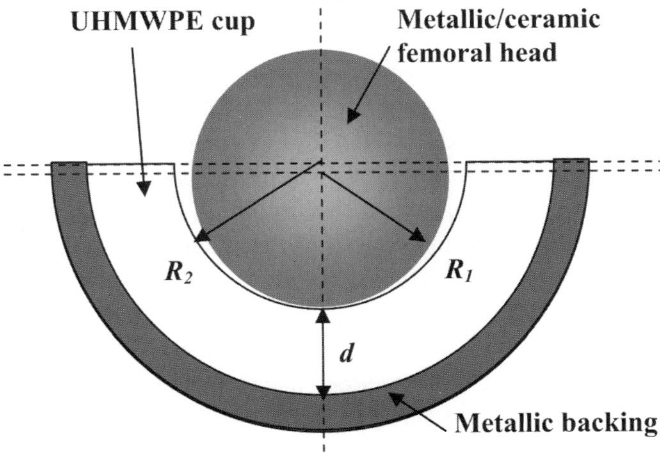

Figure 1a. A ball-in-socket model for an artificial hip joint employing an UHMWPE acetabular cup and a metallic/ceramic femoral head.

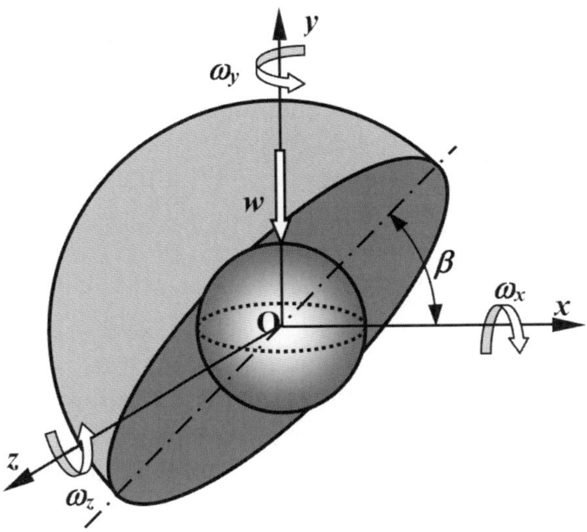

Figure 1b. Major loading and motion considered for lubrication analysis.

A ball-in-socket configuration with appropriate structural supports was used to represent the articulation between the cup and the head. Only the vertical load (w) and the flexion/extension (ω_x) as shown in Figure 1b were considered as these are generally dominant during walking.

Furthermore, only steady-state operating conditions were considered, based on the average load and speed over one walking cycle. It has been shown in previous studies that such a quasi-static estimation was very close to the average predicted lubricating film thickness from the full transient EHL analysis under normal walking [4].

Only the elastic deformation of the UHMWPE cup was considered, since the elastic modulus for either the metallic or the ceramic femoral head is at least two orders of magnitude higher. The elastic deformation of the UHMWPE bearing surface was calculated by combining both the finite element and the finite difference method incorporating the spherical fast Fourier transform (SFFT) technique [5].

Both fluid-film and boundary lubricated regions were considered simultaneously in a mixed lubrication formulation [6]. Film thickness in the mixed lubrication model was treated as either a gap in the fluid-film region or as overlapping between the two surfaces in the boundary contact problem. The pressure distribution in the boundary contact region was solved by the direct inverse method using the SFFT technique incorporating appropriate filters and smoothing functions [7]. The governing equations for the fluid-film lubricated region consisted of Reynolds equation, the film thickness equation, the elasticity equation, and the load balance equation. The following general steady-state Reynolds equation was adopted [8], relating the hydrodynamic pressure (p) and the lubricant film thickness (h):

$$\sin\theta\frac{\partial}{\partial\theta}\left(h^3\sin\theta\frac{\partial p}{\partial\theta}\right)+\frac{\partial}{\partial\phi}\left(h^3\frac{\partial p}{\partial\phi}\right)+ \\ =6\eta R_2^2\sin\theta\left[\begin{array}{l}(-\omega_x\sin\phi+\omega_y\cos\phi)\sin\theta\dfrac{\partial h}{\partial\theta}\\ \\ +(-\omega_x\cos\phi\cos\theta-\omega_y\sin\phi\cos\theta+\omega_z\sin\theta)\dfrac{\partial h}{\partial\phi}\end{array}\right]. \quad (1)$$

The corresponding spherical coordinates (ϕ, θ) and mesh grids are shown in Figures 1b and 1c, which allowed the anatomical position of the cup and the physiological direction of the load and speed to be readily considered. For the simplified load and motion assumed in the present study;

$$\begin{aligned}&\omega_y=\omega_z=0,\\&\phi=\beta\sim 180^\circ+\beta,\\&p=0\text{ at }\phi=\beta\text{ and }\phi=180^\circ+\beta,\end{aligned}$$

where (β) is the cup inclination angle and was taken as 35°.

The film thickness equation consisted of the undeformed gap between the two spherical surfaces, the roughness (r) and the corresponding elastic deformation (δ)

$$h=c-e_x\sin\theta\cos\phi-e_y\sin\theta\sin\phi-e_z\cos\theta+r+\delta, \quad (2)$$

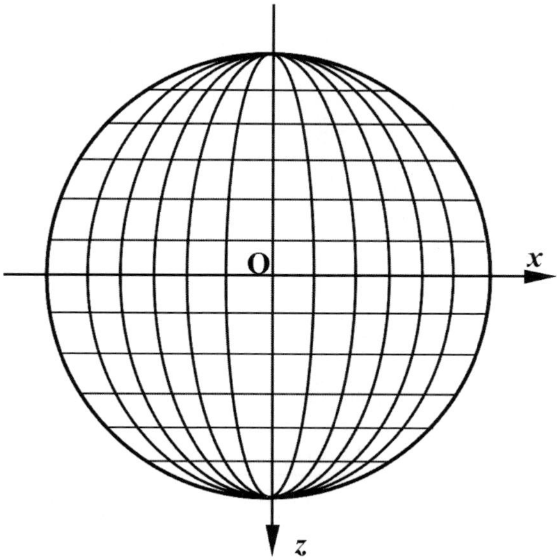

Figure 1c. Mesh grids used for the present lubrication analysis.

where (c) is the radial clearance and (e_x, e_y, e_z) are the eccentricities between the centres of the ball and cup. Finally the load balance equation requires that:

$$R_2^2 \int_{\beta}^{\beta+\pi} \int_0^{\pi} p \sin\theta \cos\phi \sin\theta \, d\theta \, d\phi = 0,$$

$$R_2^2 \int_{\beta}^{\beta+\pi} \int_0^{\pi} p \sin\theta \sin\phi \sin\theta \, d\theta \, d\phi = w,$$

$$R_2^2 \int_{\beta}^{\beta+\pi} \int_0^{\pi} p \cos\theta \sin\theta \, d\theta \, d\phi = 0. \quad (3)$$

The boundary conditions for the present lubrication problem included zero pressure at the edge of the cup and the cavitation region.

The key to the mixed lubrication formulation and solution is the criterion to be used to detect fluid-film and boundary lubricated regions. From a physical point of view, the boundary lubrication region can be assumed if the predicted lubricant film thickness from fluid-film lubrication analysis is below the boundary film thickness [6]. However, such an approach may be difficult to apply if the predicted fluid film thickness becomes inaccurate due to numerical errors and is larger than the boundary film thickness. In the present study, if the predicted fluid-film thickness became negative, boundary contact was assumed and solved accordingly. Iterations were performed to ensure the continuity

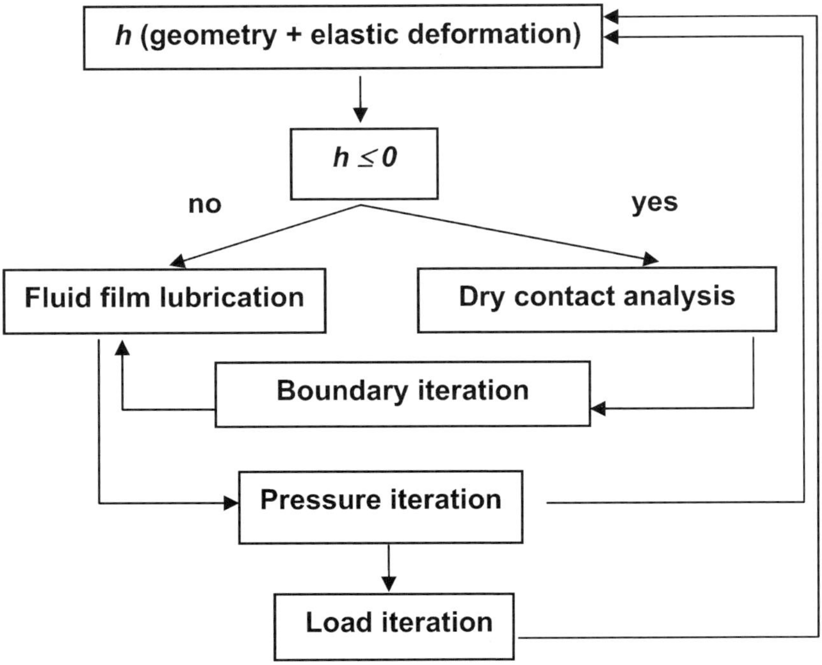

Figure 2. Flow chart for the numerical analysis of the mixed lubrication of artificial hip joints.

of the pressure at the boundary between the fluid-film and boundary contact regions. The flow chart is shown in Figure 2.

The details of the numerical methodology for solving the fluid-film and boundary (dry) contact problems developed in previous studies [4, 7] were adopted. For the present mixed lubrication problem, particularly at the boundary between fluid-film and boundary contact regions, local smoothing and averaging of the pressure was required to stabilize the iteration when the predicted fluid-film thickness became close to zero (less than 10 nm).

The mixed lubrication model was initially applied to an artificial hip joint shown in Figure 1a with perfectly smooth bearing surfaces. An axisymmetric surface topography (r) with sinusoidal waviness in the radial direction representing machining marks on the UHMWPE cup was simulated subsequently.

The following parameters were chosen in the present study:

• Femoral head radius (R_1):	14	mm
• Cup radius (R_2):	14.05	mm
• Cup wall thickness (d):	5	mm
• Elastic modulus for UHMWPE:	1.0	GPa
• Poisson's ratio for UHMWPE:	0.4	
• Angular velocity of the femoral head (ω_x)	–2.0	1/s

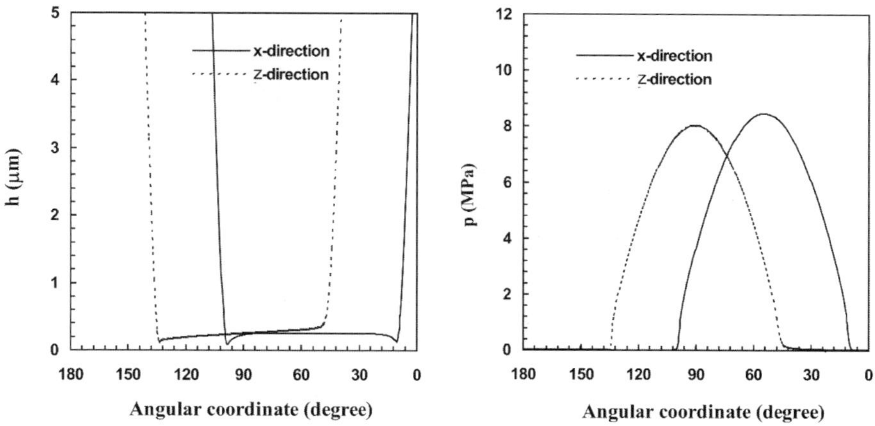

Figure 3a and 3b. Cross-sectional film (left, a) and pressure (right, b) profiles through the loading centre for a smooth surface and a viscosity of 0.005 Pas.

• Load (w)	1500	N
• Roughness amplitude	1	μm
• Roughness wavelength	9	°
• Mesh density	240×240	
• Criterion for pressure iteration (fluid-film)	10^{-4}	
• Criterion for load iteration (fluid)	10^{-3}	
• Criterion for pressure iteration (contact)	10^{-3}	

The viscosity was varied to examine the transition from full fluid-film to complete boundary lubrication regimes.

3. RESULTS

Figures 3a and 3b show the cross-sectional film thickness and pressure profiles through the centre of loading for the smooth surface and a viscosity of 0.005 Pas.

For this case, no boundary lubricated region was found and the corresponding predicted lubricant film thickness and pressure distribution exhibited typical characteristics of isoviscous soft EHL contacts reported in the literature for a ball-on-plane configuration [9]. The overall minimum film thickness occurred in the side of the contact. Consequently, initial boundary contact was observed around this region when the viscosity was decreased to about 0.001 Pas, particularly near the centre of the cup. However, as the viscosity decreased further to 0.00065 Pas, substantial contact was observed in the exit region in the entraining direction and only localized contact was observed in the side as shown in Figure 3c for the boundary contact pressure. At viscosit-

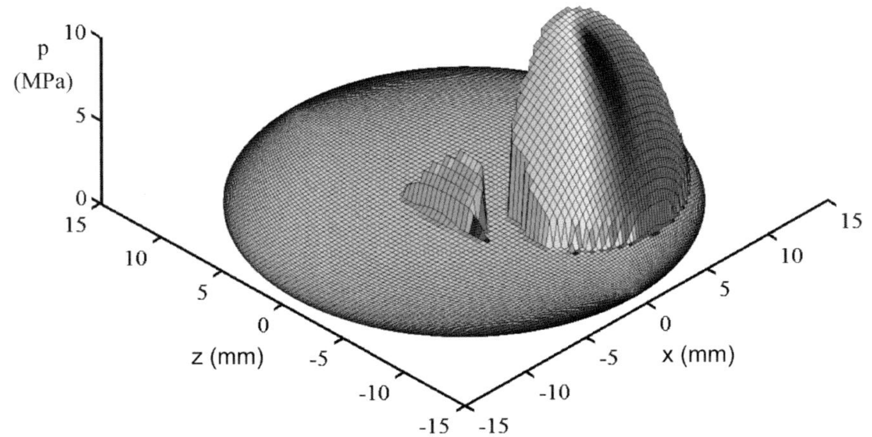

Figure 3c. Boundary contact pressure for a smooth bearing surface and a viscosity of 0.00065 Pas.

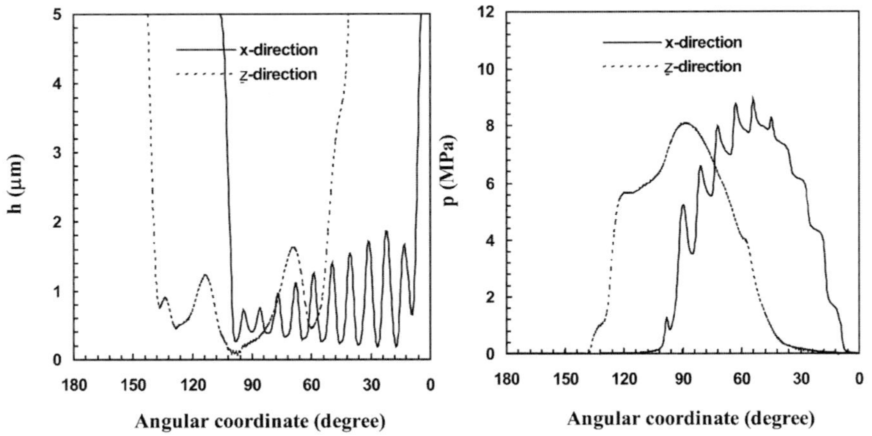

Figure 4a and 4b. Cross-sectional film (left, a) and pressure (right, b) profiles through the centre of loading for the wavy bearing surface and a viscosity of 0.1 Pas.

ies below 0.0005 Pas, almost no fluid-film lubricated area was found and the total pressure distribution was almost identical to that under boundary contact conditions.

The introduction of the sinusoidal roughness reduced the predicted lubricant film thickness as shown in Figures 4a and 4b for a viscosity of 0.1 Pas. When the viscosity decreased to 0.05 Pas, boundary contact was initiated in the side and increased substantially in the exit region along the entraining direction as the viscosity decreased further. At a viscosity of 0.001 Pas, the total pressure distribution became almost identical to that under boundary contact conditions

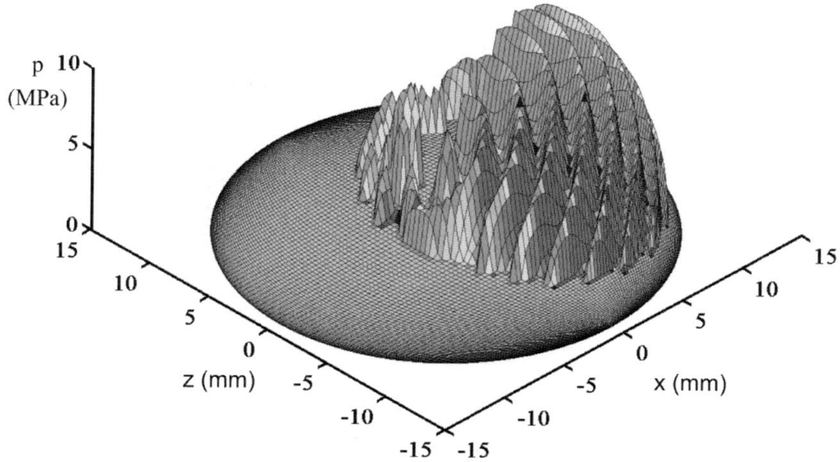

Figure 5a. Boundary contact pressure for the wavy bearing surface and a viscosity of 0.001 Pas.

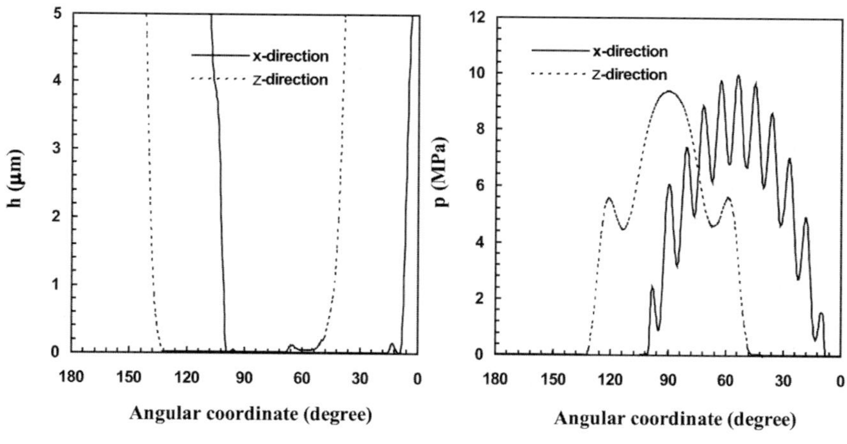

Figure 5b and 5c. Cross-sectional film (left, b) and pressure (right, c) profiles through the centre of loading for the wavy bearing surface and a viscosity of 0.001 Pas.

as shown in Figure 5a, leading to an almost complete flattening of the original roughness and a large pressure perturbation as shown in Figures 5b and 5c.

Figure 6 compares the predicted minimum film thickness between the smooth and rough surfaces.

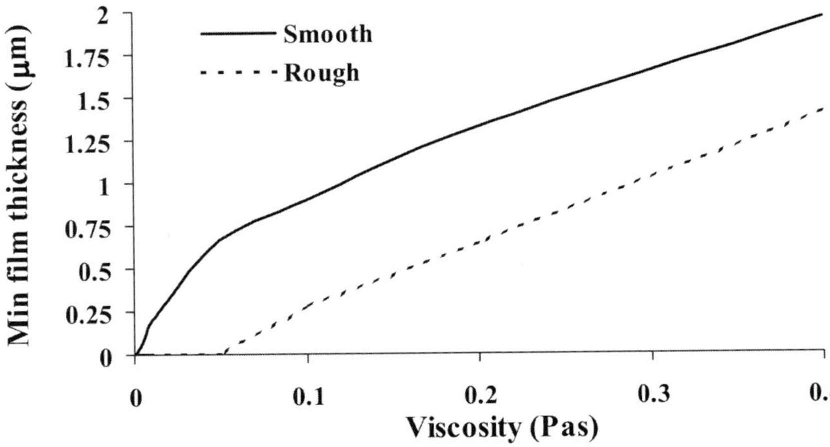

Figure 6. Comparison of the predicted minimum fluid-film thickness between smooth and wavy bearing surfaces for different viscosities.

4. DISCUSSION

It is important to estimate the lubricant film thickness and the corresponding lubrication regime in artificial hip joints employing UHMWPE cups under a realistic viscosity of 0.002 Pas representing periprosthetic synovial fluid [3]. The corresponding predicted minimum lubricating film thickness was about 21 nm for the smooth surface model, which is probably larger than the boundary film thickness due to the absorption of the proteins present in joint fluid, estimated to be 10 nm [10]. However, the introduction of a sinusoidal roughness resulted in a significant decrease in the predicted fluid-film thickness and a mixed lubrication regime under physiological conditions. Although localized micro-EHL was observed at larger viscosities, such an action was only effective near the centre of the cup. This is mainly because of the axisymmetric roughness considered in the present study. Consequently, the lay of the roughness in the centre of the cup is largely perpendicular to the direction of the entraining (transverse) while that around the centre of the contact is predominantly parallel (longitudinal). It is generally known that significant micro-EHL action is associated with the transverse roughness. The original average surface roughness of the UHMWPE cup has been reported to be of the order of microns [11]. Although these original asperities can be smoothed out significantly due to wear or creep, the average roughness of the worn cup is still of the order of 0.1 μm to 0.5 μm [12]. Therefore it is suggested that a mixed or boundary lubrication regime should be dominant in artificial hip joints employing UHMWPE cups.

5. CONCLUSION

A general mixed lubrication methodology was developed for artificial hip joints, consisting of both fluid-film and boundary lubricated regions. The general methodology was applied to a typical artificial hip joint employing an UHMWPE cup with smooth and wavy surfaces. It was found that a fluid-film lubrication regime was possible under physiological conditions. However, the lubricating film thickness was significantly reduced when the roughness was introduced, leading to a mixed lubrication regime.

ACKNOWLEDGMENT

This work was supported by the Engineering and Physical Sciences Research Council (UK).

REFERENCES

[1] E. Ingham and J. Fisher, Biological reactions to wear debris in total joint replacement, *Proc. Instn. Mech. Engrs., J. Eng. in Med.*, **214**, 21–37 (2000).

[2] Z.M. Jin, D. Dowson and J. Fisher, Analysis of fluid film lubrication in artificial hip joint replacements with surfaces of high elastic modulus, *Proc. Instn. Mech. Engrs., J. Eng. in Med.*, **211**, 247–256 (1997).

[3] J.Q. Yao, M.P. Laurent, T.S. Johnson, C.R. Blanchard and R.D. Crowinshield, The influence of lubricant and material on polymer/CoCr sliding friction, *Wear*, **255**, 780–784 (2003).

[4] D. Jalali-Vahid and Z.M. Jin, Transient elastohydrodynamic lubrication analysis of UHMWPE hip joint replacements, *Proc. Instn. Mech. Engrs., J. Mech. Eng. Sci.*, **216**, 409–420 (2002).

[5] F.C. Wang and Z.M. Jin, Prediction of elastic deformation of acetabular cup and femoral head for lubrication analysis of artificial hip joints, *Proc. IMechE J. Eng. Tribology*, **218**, 201–209 (2004).

[6] Y.Z. Hu and D. Zhu, A full numerical solution to the mixed lubrication in point contacts, *Trans. ASME J. of Tribology*, **122**, 1–9 (2000).

[7] F.C. Wang, Z.M. Jin, H.M.J. McEwen and J. Fisher, Microscopic asperity contact and deformation of ultrahigh molecular weight polyethylene bearing surface, *Proc. IMechE, J. Eng. Med.*, **217**, 477–490 (2003).

[8] P.K. Goenka and J.F. Booker, Spherical bearings: Static and dynamic analysis via the finite element method, *Trans. ASME, J. Lub. Technol.*, **102**, 308–319 (1980).

[9] B.J. Hamrock and D. Dowson, Elastohydrodynamic lubrication of elliptical contacts for materials of low elastic modulus. I: Fully flooded conjunction, *Trans. ASME, J. Lubric. Technol.*, **100**, 236–245 (1978).

[10] R.J. Green, J. Davies, M.C. Davies, C.J. Roberts and S.J. Tendler, Surface plasmon resonance for real time in situ analysis of protein adsorption to polymer surfaces, *Biomaterials*, **18**, 405–413 (1997).

[11] A. Wang, A. Essner, V.K. Polineni, C. Stark and J.H. Dumbleton, Lubrication and wear of ultra-high molecular weight polyethylene in total joint replacements, *Tribology International*, **31**(1–3), 17–33 (1998).

[12] A.P. Elfick, R.M. Hall, A. Unsworth and I.M. Pinder, Surface topography of retrieved PCA acetabular liners: Proposal for a novel wear mechanism, *J. Mater. Sci. Lett.*, **17**(13), 1085–1088 (1998).

FRACTALS AND SURFACE ROUGHNESS IN EHL

F.M. Borodich
School of Engineering, Cardiff University, Cardiff CF24 3AA, U.K.

Abstract Firstly, a brief introduction to fractals and similarity methods is given. Fractal models of rough surfaces are usually used when the spectral density function of surfaces has the power law character. It is argued that the main source for various misunderstandings in applications of fractals to mechanics is the lack of precise definitions and non-critical repetition of common statements about fractal geometry. Some key papers concerning fractal models of roughness and papers connecting EHL and fractals are reviewed. Two classes of fractal surfaces introduced by the author, namely the Cantor profile models and the parametric-homogeneous (PH) surfaces, are discussed. The well-known Weierstrass–Mandelbrot (W–M) profile is a particular case of PH-profiles. It is shown that only physical fractals (prefractals) should be attributed to real surfaces. It is argued that the Cantor profile is simple for analytical analysis. However, it has a minor drawback: all asperities of the profile have one-level character, while, as Archard showed, real roughness has a hierarchical structure. Finally, it is suggested to model rough surfaces by a multilevel prefractal model introduced by Borodich and Onishchenko.

Keywords: roughness, fractal, parametric-homogeneous surfaces, multilevel prefractal model.

1. INTRODUCTION

Surface topography plays a significant role in tribology, i.e., in problems of friction, wear, lubrication and contact. Problems of elastohydrodynamic lubrication (EHL) with rough surfaces have been actively studied. The detailed understanding of the problems has become an important issue as the ratio of film thickness to surface roughness is decreasing [1]. One has to be able to model the main features of rough surfaces and to understand the behaviour of the features at different scales of the real roughness. The first models of roughness were quite simple. It was assumed that (i) protuberances of the rough surfaces are spherical, (ii) their radii are the same and equal to R, (iii) they have various heights, and (iv) their number at a specific height increases as the level of consideration goes deeper into the rough surface (Zhuravlev [2],

R.W. Snidle and H.P. Evans (eds), IUTAM Symposium on Elastohydrodynamics and Microelastohydrodynamics, 397–408.

and Greenwood and Williamson (GW) [3]). Then Archard [4] discovered that roughness has a hierarchical structure. He presented a model where a sphere of radius R_0 was covered by spherical protuberances of radius of curvature R_1 and these in turn were covered by smaller protuberances of radius R_2. Due to its iterative construction, Archard's profile is sometimes attributed to self-similar fractals. However, Borodich and Mosolov [5] said that this is not true.

Influenced by studies of random signals, Whitehouse and Archard [6] suggested the study of roughness using the auto-correlation function $R(\delta)$ and its Fourier transform, the power spectral density $G(\omega)$

$$R(\delta) = \lim_{T\to\infty} \frac{1}{2T} \int_{-T}^{T} [z(x+\delta) - \bar{z}][z(x) - \bar{z}]\,\mathrm{d}x,$$

$$G(\omega) = \frac{2}{\pi} \int_{0}^{\infty} R(\delta) \cos \omega\delta \,\mathrm{d}\delta \quad \text{and} \quad \bar{z} = \lim_{T\to\infty} \frac{1}{2T} \int_{-T}^{T} z(x)\,\mathrm{d}x,$$

where $\bar{z}$ is the average value of the profile function $z(x)$. Now $R(\delta)$ and $G(\omega)$ are the main tools for studying statistical models of rough surfaces. The moments m_n of the spectral density $G(\omega)$ provides a useful description of the surface roughness $m_n = \int_{\omega_0}^{\infty} \omega^n G(\omega)\,\mathrm{d}\omega$, where $\omega_0 = 2p/\lambda_0$ is the wave number corresponding to the profile length λ_0. It is possible to show (see, e.g., [7]) that m_0 is the variance of heights (rms height) of the surface, m_2 is the variance of slopes (rms slope) and m_4 is the variance of curvatures (rms curvature). In 1978 Sayles and Thomas [8] showed that profiles of a large number of both natural and artificial surfaces have approximately the following form of the spectral density $G(\omega) = 2\pi\Lambda/\omega_2$ where Λ is called the topothesy of the surface. Interest in the fractal approach to surfaces started with comment by Berry and Hannay [9] to the above results. They noted that statistically isotropic surfaces which have no scale and whose profile is non-differentiable may have spectra

$$G(\omega) \propto 1/\omega^{\alpha} \tag{1}$$

with any α in the range $1 < \alpha \leq 3$; and that these surfaces are discussed in detail by Mandelbrot who calls them fractals. Indeed, if the profile $z(x)$ has spectral density of the form (1) then it is reasonable to expect that the box dimension of the graph $z(x)$ is equal to $(5 - \alpha)/2$ [10].

As I realised later, my acquaintance with examples of classic fractal sets began in 1969, when my father presented me a popular book about sets written by Vilenkin [11] for Russian high-school children. In Vilenkin's book numerous examples of sets which are more irregular than sets considered in common textbooks on Euclidean geometry were collected. Definitely, I was impressed by such sets as the Cantor staircase (devil's staircase), trajectory of Brownian motion, and the Peano curve, and by other sets that are currently attributed to self-similar fractals like the von Koch curve and snowflake, the Menger sponge

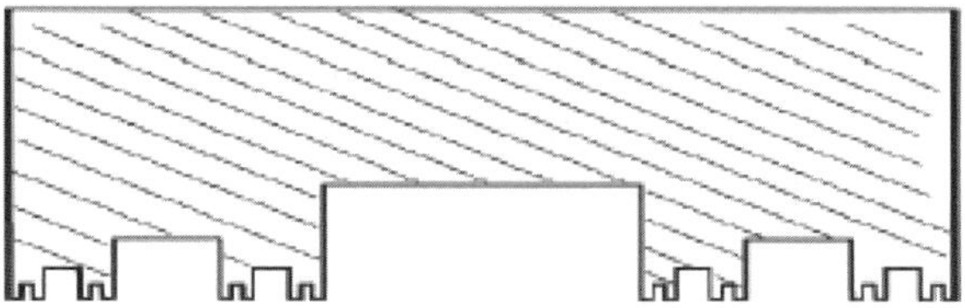

Figure 1. The C_B profile.

and the Sierpinski carpet. However, in 1969 I was too young to understand the examples in detail and naturally, I did not associate these sets with non-integer fractal dimensions. The term "fractal" itself was introduced later by B.B. Mandelbrot, who published a book concerning fractals in 1975 [12].

In 1990 Dr. A.B. Mosolov attracted my attention to the fact that in spite of considerable interest in fractal description of rough surfaces, nobody could solve any contact problem for fractal surfaces. Although I had read Vilenkin's book, I was not experienced in fractal modelling. I read the papers by Liu et al. [13] about 2D self-affine structures used for characterization of metal-electrolyte interface. I understood that the construction procedure of the Cantor–Liu bar model led to unbounded profile height in most of the considered examples. In three days, I constructed another set, namely the Cantor profile that was always bounded (Figure 1). Sometimes it is referred to in Russia as the Cantor–Borodich (C_B) profile and in the West as the Cantor set model.

I find the latter term rather confusing because a number of other constructions can be created using the 1D Cantor set. For example, the Cantor staircase, the Cantor–Liu bar model, and the Moscow university function U_M [14]. The U_M is an example of a non-fractal, nowhere differentiable, self-similar function. The C_B profile has two scaling parameters f_x and f_z such that the width L_i and height h_i of each generation of asperities are given by $L_i = L_0 f_x^i$ and $h_i = h_0 f_z^i$, respectively. I realised that the profile may be fractal for some values of the parameters: $0 < f_x < 0.5$ and $0.5 < f_z < 1$, and contact problems can be solved iteratively for the elastic foundation and the rigid plastic foundation (the Hill approach). I brought these results to Dr. Mosolov and he presented them using fractal language [15] (see also [16]). In particular, it was shown that the box dimension of the profile C_B is $\dim_B C_B = 1 - \ln(2f_z)/\ln f_x$.

There are various generalizations of the C_B profile. In particular, the profile can be multilevel [17, 18], random [19], deformable [17, 20]. It can have not only two but also several segments in each generation of its construction [21]. The hierarchical spring-connected block model considered by Schmittbuhl et al. [22] is similar to the above generalizations. The original C_B profile has a minor drawback: all asperities of the profile have one level character, while, as

Archard showed, real roughness has a hierarchical structure. The Borodich–Onishchenko hierarchical profile will be discussed later.

There is a great interest in application of fractal ideas to problems of tribology. One can see also that there are papers applying fractal language to the EHL of rough surfaces (see, e.g. [23–26] and the references therein). However, if one could collect all statements about fractals spread throughout the literature then we would get a rather confused conglomeration of various ideas of similarity theory, iterative models and theory of measure. As a result, one can meet rather sceptical views of leading researchers in the field on the applicability of fractal language to tribology (see, e.g. [27–29]). Here I present my personal point of view on application of fractal models to surface roughness description.

2. MODELS OF SURFACE ROUGHNESS AND FRACTALS

2.1 Definitions of Fractals

In papers concerning fractals, Mandelbrot usually gave examples in order to explain what is a fractal and suggested use of the term "without a pedantic definition". It seems to me that one can define fractals as *sets with noninteger fractal dimension* and split the term in two: mathematical and physical fractals. Mathematical fractal dimensions include Hausdorff dimension and box-counting dimension (this can be obtained by consideration of the limit behaviour of covers of a set by boxes of size at most δ when $\delta \to 0$ [10]), while physical (natural, empirical) fractals which are real world or numerically simulated objects exhibiting the power-law relation $N(\delta) \propto \delta^{-D}$ in a bounded region between upper (Δ_*) and lower (δ_*) cut-offs. The mathematical box dimension $\dim_{\mathrm{B}} S$ of a set S and the physical box dimension $\mathrm{D_B}$ are respectively defined as

$$\dim_{\mathrm{B}} S = \lim_{\delta \to 0+} \frac{\ln N(\delta)}{\ln(1/\delta)}, \qquad \mathrm{D_B} = \frac{\ln N(\delta)}{\ln(1/\delta)},$$

where $N(\delta)$ the number of balls, cubes or yardsticks in the cover. One often sees that these two kinds of fractals are confused. Let points A and B be connected by a tortuous fractal curve. Then the length of the curve increases with decrease of the yardstick lengths

$$L \approx N(\delta) \cdot \delta \approx (|AB|/\delta)^{\mathrm{D}} \cdot \delta \propto \delta^{l-D}, \quad -1 < 1 - D < 0.$$

However, it has a limit for a physical fractal, while the length is infinite for a mathematical fractal. In addition, a mathematical fractal curve can be continuous everywhere and differentiable nowhere. So, it has an infinite set of vertices. For this curve we cannot use, at least in the usual sense, such a common notion as a normal. In particular, it is impossible to use, for a solid with such a fractal surface, the classic formulation of boundary value problems. Hence, the physical fractal approach is more flexible than mathematical, e.g., it can take into account the change of fractal dimension with changing the scale of consideration [30].

2.2 Self-Similar Sets and Similarity Dimension

How can one construct a self-similar set? Let us consider an initial set I whose size is taken as unity, e.g. an interval, a triangle, a square or a cube, and the operation Q such that it transforms this initial set I into N similar subsets. This transformed set is called **the generator** of the fractal. Then the operation Q is applied again to each of these N smaller subsets, i.e. we replace each subset by a reduced copy of the generator. The scale factor p shows the ratio of length of the subsets $p = \varepsilon_n/\varepsilon_{n+1}$, where n is the generation number. The process is repeated infinitely. The resulting set is a self-similar fractal. A set on an intermediate level (generation) of the iterative construction is called **prefractal**. Fractal dimension of a self-similar set can be calculated without its covering using a short cut, namely the **similarity dimension** is defined simply as $D_{\text{sim}} = \ln N/\ln p$.

A random curve can be represented as a stochastic function $X(t)$ of position on the line. It is called statistically self-similar if for each $\lambda > 0$ there exists a constant $C(\lambda)$ such that all the finite dimensional distributions of $X(\lambda t)$ are identical with those of $CX(t)$ up to transition [31]. It is often claimed that the assumption of self-similarity is necessary for exhibition of fractal features. However, this is not true.

Although it was shown that the values of similarity, Hausdorff and box-counting dimensions are the same for the above self-similar fractals, the box dimension is not always equal to the Hausdorff dimension. It is only possible to prove in the general case that $\dim_{\text{H}} S \leq \dim_{\text{B}} S$. It was also claimed that for the Gaussian based models of rough surfaces, all the common definitions of fractal dimensions are applicable, and produce the same numerical value for the dimension [31]. However, no arguments were given to support the claim.

Note that it is wrong to conclude that a set is a fractal just because it can be obtained by an iterative procedure. For example, the C_{B} profile always remains rough. However, it can be both fractal and non-fractal depending on values of the structural parameters. Although the Archard profile is not a self-similar fractal [5] because one obtains overlapping regions starting with the second generation of its construction, let us apply formally the above formula. Thus, a semi-sphere of radius R_0 (the generator) is covered by m semi-spheres of radius $R_1 = R_0/p$, these in turn are covered by m^2 semi-spheres of radius $R_2 = R_1/p$, and so on. Evidently, $p = R_0/R_1 = 1/\sin(\pi/2m)$. Hence, $D_{\text{sim}} = -\ln m/\ln\sin(\pi/2m) \approx 1/(1 - \ln\pi/\ln 2m)$.

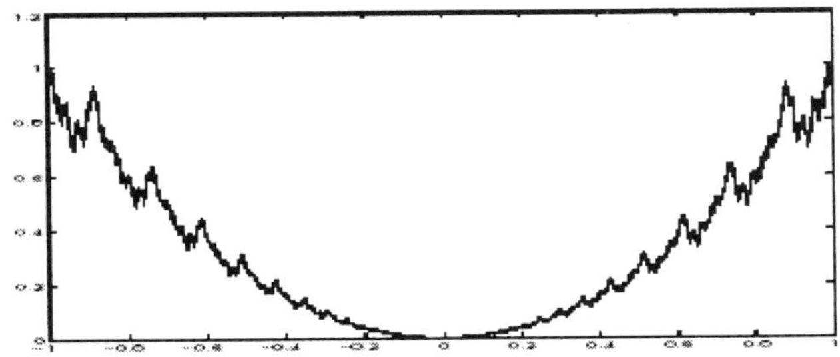

Figure 2. A graph of a fractal PH-parabola.

2.3 Parametric-Homogeneous (PH) Surfaces

I introduced parametric-homogeneous (PH) and parametric quasi-homogeneous (PQH) functions, as natural generalizations of concepts of homogeneous and quasi-homogeneous functions when the discrete (discontinuous) group of coordinate dilations $(x_1, \dots, x_l) \to (p^{k\alpha_1}x_1, \dots, p^{k\alpha_l}x_l)$ (PH-transformation), with a fixed scaling parameter $p > 0$ is considered instead of the continuous group of coordinate dilations $x_i \to \lambda^{\alpha_i}x_i$ [14, 32]. Here $k \in Z$, and Z is the set of integer numbers. The function B_{d} is called a PQH function of degree d and parameter p with weights $\boldsymbol{\alpha} = (\alpha_1, \dots, \alpha_l)$ if there exists a positive parameter p, $p \neq 1$ such that it satisfies the following identity

$$B_d(p^{k\alpha_1}x_1, \dots, p^{k\alpha_l}x_l) = p^{kd}B_d(x; p), \quad k \in Z$$

and the parameter is unique in some neighborhood. PH-functions b_d of degree d are a particular case of PQH-functions when $\alpha_1 = \dots = \alpha_l$ (Figure 2). Here the least p: $p > 1$ is taken as the parameter. Smooth sinusoidal log-periodic functions and fractal Weierstrass type functions are well-known examples of PH-functions

$$b_0(x; p) = A\cos(2\pi \ln x/\ln p + \Phi) \quad \text{or} \quad b_\gamma(x; p) = \sum_{n=-\infty}^{\infty} p^{-\gamma n}h(p^n x),$$

where A and Φ are arbitrary constants and h is an arbitrary function.

After Berry and Lewis [33], the Weierstrass type functions $C(x; p)$ with $h(p^n x) = (1 - \cos p^n x)$ or $W(x; p)$ with $h(p^n x) = \Lambda^{(D-1)}\cos 2\pi p^n x$ are called the Weierstrass–Mandelbrot (WM) functions. The WM functions with $0 < \gamma < 1$ are continuous everywhere and differentiable nowhere. In addition, the box dimension of their graphs is $D = 2 - \gamma$.

In 1986 Roques-Carmes et al. [34] started to use Weierstrass type functions for fractal modelling of surface roughness. Then a number of other researchers modelled roughness using these functions (see, e.g., [35–39]). The following truncated WM ($\widetilde{W}$) function

$$\widetilde{W}(x; p) = \Lambda^{(D-1)} \sum_{n=n_1}^{\infty} p^{(D-2)n} \cos(2\pi p^n x)$$

was suggested for fractal characterization of any surface topography (Majumdar and Bhushan [36]). Here n_1 is an integer number, which corresponds to the low cut-off frequency of the profile, and Λ is the so-called characteristic length scale of the profile. The number n_1 depends on the length L of the sample and is given by $p^{n_1} = 1/L$ and the parameter Λ determines the position of the spectral density along the $\lg G$ axis. It was often stated that both parameters Λ and D of the function W or $\widetilde{W}$ are scale-invariant characteristics of the roughness. However, later the extensive experimental studies of this fractal characterization model showed that the values of parameters Λ and D are not unique and depend on instruments or resolution of a given instrument [40].

Further, $C(x; p)$ or $W(x; p)$ have a specific trend x^{2-D}. Hence, they can be used only as particular examples of a fractal profile and they are not general fractal functional models. The assumption that the WM function gives a universal fractal representation to rough profiles can lead to wrong conclusions concerning surface roughness parameters and their distributions.

2.4 Brownian Surfaces and Self-Affine Fractals

The fractional Brownian motion (fBm) is a generalisation of the ordinary Brownian process. The fBm are widely used in creating computer-generated surfaces. For example, a profile can be constructed as a graph of 1-dimensional fBm $V_H(x)$ of index H: $z(x) = V_H(x)$, where x is taken as the time and z is the random variable of the single valued function $V_H(x)$ with

$$\left\langle [V_H(x+\delta) - V_H(x)]^2 \right\rangle \propto \delta^{2H}, \quad 0 < H < 1. \tag{2}$$

It is known [10] that both the Hausdorff and the box dimensions of $V_H(x)$ equal $2 - H$ with probability equal to 1. Usually, it is assumed that the same conclusions concerning the box dimension are valid for physical fractals as well.

The concept of self-affine fractals is often mentioned in studies of roughness of surfaces. A number of experimental investigations claim that profiles of vertical sections of real surfaces are statistically similar to themselves under repeated magnification; however, the profile should be scaled differently in the direction of nominal surface plane and in the vertical direction. Self-affine mapping on a plane means a one-to-one mapping such that images of any three collinear points are collinear in turn. However, in the literature devoted to fractals, self-affine mapping means usually its very special case, namely the quasi-homogeneous (anisotropic) coordinate dilation, $x' = \lambda x$, $z' = \lambda^H z$, where z is a graph of a surface profile and H is some scaling exponent. The concept of a self-affine fractal was introduced by Mandelbrot [42]. However, he did not give any strict definition of the concept and preferred to give some examples of self-affine fractals, such as the Brownian motion $V_{0.5}(t)$ or $V_H(t)$. Mandelbrot stated that each version of fractal dimension for self-affine fractals has a local and a global value, separated by a crossover. Hence, there are both local and global scaling for self-affine fractals. One could define *a self-affine fractal set as a set that is invariant from the statistical point of view under quasi-homogeneous (anisotropic) scaling*. However, Mandelbrot's statements about dimensions of the self-affine fractals would not follow from the definition. It is necessary to note that the above statements about self-affine fractals were not given in a rigorous manner that is traditional for mathematicians. This leads to doubts concerning the validity of the concept.

A self-affine function f is often defined as $f(x) \propto \lambda^{-H} f(\lambda x)$, where H is called the Hölder exponent and the global fractal dimension of the function is $D_f = 2 - H$. This is true for the fBm (2). However, this is not true in the general case.

Let us consider PH-functions b_0, b_1 and b_2 (Figure 2) with degrees $d = 0$, $d = 1$ and $d = 2$, respectively:

$$b_1(x; p) = x b_0(x; p), \quad b_2(x; p) = x^2 b_0(x; p), \quad b_0(x; p) = x^{-\gamma} C(x; p).$$

These functions have the same box dimension as the WM function $C(x; p)$ [14]. If one considers the trends of the functions then we could conclude that b_2 is a self-affine function with the exponent $H = 2$ and b_1 is a self-similar function. Thus, the common statement, that the scaling exponent H of self-affine fractals is always closely related to the fractal dimension, is wrong. The PH-graphs near any point x_0 are repeated in scaling form near all points $p^k x_0$, $k \in Z$. This scaling (self-affine) property was often wrongly attributed to fractal features of the graph. However, as I have just showed, the trend of a PH-function and its fractal dimension in general do not correlate. Further, anisotropic scaling is normally the so-called Lipschitz homeomorphism that preserves the box dimensions of a set [14]. Hence, the claims about self-affine fractal dimensions are not justified. It seems to me that characterization of self-affine surfaces would benefit if one avoids using fractal dimensions and employs the roughness exponent ζ [43]. This is practically the same as to use the exponent α of the power spectra (1) because $G(\omega) \propto 1/\omega^{1+2\zeta}$ [44].

2.5 Borodich–Onishchenko Prefractal Multilevel Model

As Lubrecht and Venner noted [1], even if it were possible to compute the EHL of a real rough surface, this would be of little use, as it would not allow the pre-

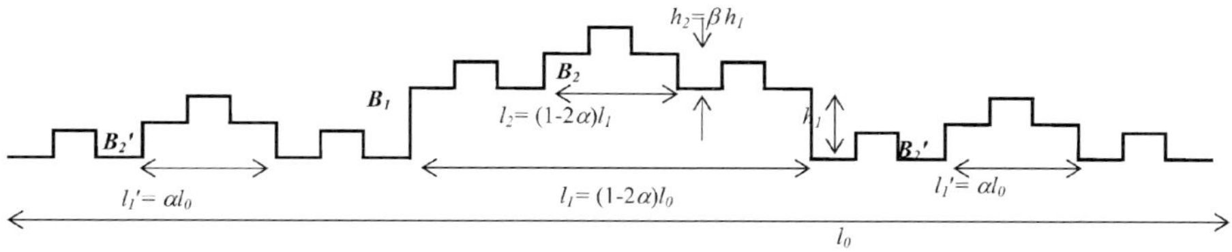

Figure 3. The iterative construction algorithm for the B–O profile.

diction of the lubrication for another profile. Incorporating a specific roughness into applied problems is much more effective if the roughness is modelled by mathematically synthesised surfaces with several structural factors. However, only a few relevant fractal models have been currently proposed. The fBm, the WM function and the C_B profile, are synthesised surfaces. However, all these models have the same drawback: although they allow surfaces to be constructed with a prescribed fractal dimension, they do not have sufficient parameters to produce a surface with both prescribed engineering characteristics, and the fractal dimension. In addition, it is difficult to handle surface models based on fBm profiles analytically, which presents a further obstacle in their study. Borodich and Onishchenko [17, 18] introduced an alternative to the C_B profile model, namely a multilevel prefractal model (the B–O model) based on iterated function algorithms. The B–O model is a hierarchical, prefractal curve that can be constructed in a step-by-step manner. Such a construction of a prefractal is a version of the iterated function system approach. In the simplest case the B–O model is characterised by two main scaling structural parameters: a horizontal factor, α, and a vertical factor, β. Their meanings are clear from Figure 3.

It is necessary to point out two further important advantages of the B–O model in comparison with other synthesised fractal curves: (i) it has an inherent multilevel structure, which allows the flexible separation of macro-, micro-, and nano- (molecular) roughness; (ii) although its shape appears jagged and irregular, it is completely regular from the viewpoint of the construction algorithm. The construction of iterated function models involves a sequence of approximations (prefractals), each obtained from its predecessor by modification in increasingly fine detail, with a pure mathematical fractal as a limiting set.

3. DISCUSSION AND CONCLUSION

It is often believed that fractal behaviour can be observed on many orders of magnitude and, hence, mathematical fractals are a good model for physical fractals. However, an extended analysis of published data [41] shows that the overwhelming majority of reported physical fractals span approximately 1.5

orders of magnitude, i.e. $\lg(\Delta_*/\delta_*) \approx 1.5$. Mathematical fractal models could be not only more difficult for study, but they also do not reflect often the physics of the phenomenon. Hence, if one models a surface using an iterative construction then prefractal models are more appropriate than mathematical fractals. It was proposed to study the bounded similarity of surface roughness on iteratively generated rough profiles, namely the B–O model [18]. It is suitable for the flexible modelling of roughness. The model can be both fractal and non-fractal depending on values of the structural parameters. Regardless of this, the model profile remains rough and possesses certain self-affine properties. The iterative regular construction of the profile allows us to analyse its prefractals of arbitrary generation. Mandelbrot noted that natural objects do not have pure shapes of classical mathematical objects, for example, coastlines are not circles, clouds are not spheres, and mountains are not cones. Borodich and Onishchenko [18] extended this list of statements saying that "roughness of real bodies is not mathematical fractal". All these geometrical objects: spheres, cones, circles as well as fractals are just mathematical idealisations of complex shapes of real physical bodies. One has to use different mathematical models to describe the same object on different scales. For example, to describe a nominally flat rough surface, one could use sequentially the model of a plain, the model of a collection of spherical asperities and the model of a fractal surface depending on the chosen magnification. One should not expect that the fractal dimension reflects details of manufacturing. However, if the considered process depends on the roughness having the power spectral density of the form (1) within some range (δ_*, Δ_*), then the model of the roughness has to have the same prefractal features.

REFERENCES

[1] Lubrecht, A.A. and Venner, C.H. (1999) Elastohydrodynamic lubrication of rough surfaces. *Proc. Instn. Mech. Engrs. Part J. J. Eng. Trib.*, **213**, 397–404.

[2] Zhuravlev, V.A. (1940) On question of theoretical justification of the Amontons–Coulomb law for friction of unlubricated surfaces. *Zh. Tekh. Fiz.*, **10**, 1447–1452.

[3] Greenwood, J.A. and Williamson, J.B.P. (1966) Contact of nominally flat surfaces. *Phil. Trans. Roy. Soc. Lond.*, **A295**, 300–319.

[4] Archard, J.F. (1957) Elastic deformation and the laws of friction. *Proc. Roy. Soc. Lond.*, **A243**, 190–205.

[5] Borodich, F.M. and Mosolov, A.B. (1992) Fractal roughness in contact problems. *J. Appl. Math. Mech.*, **56**, 786–795.

[6] Whitehouse, D.J. and Archard, J.F. (1970) The properties of random surfaces of significance in their contact. *Proc. Roy. Soc. Lond.*, **A316**, 97–121.

[7] Brown, S.R. (1995) Simple mathematical model of rough fracture. *J. Geophys. Res.*, **100**(B4), 5941–5952.

[8] Sayles, R.S. and Thomas, T.R. (1978) Surface topography as a nonstationary random process. *Nature*, **271**, 431–434.

[9] Berry, M.V. and Hannay, J.H. (1978) Topography of random surfaces. *Nature*, **273**, 573.

[10] Falconer, K.J. (1990) *Fractal Geometry: Mathematical Foundations an Applications*, John Wiley, Chichester.

[11] Vilenkin, N.Ya. (1968) *Stories about Sets*, Academic Press, New York.

[12] Mandelbrot, B.B. (1975) *Les Objects Fractals: Forme, Hasard et Dimension*, Flammarion, Paris.

[13] Liu, S.H., Kaplan T. and Gray, L.J. (1986) Theory of the AC response of rough interfaces. In: *Fractals in Physics*, L. Pietronero and E. Tosatti (eds), North Holland, Amsterdam, pp. 383–392.

[14] Borodich, F.M. (1998) Parametric homogeneity and non-classical self-similarity. II. Some applications. *Acta Mechanica*, **131**, 47–67.

[15] Borodich, F.M. and Mosolov, A.B. (1991) Fractal contact of solids. *Sov. Phys.-Tech. Phys.*, **61**, 50–54.

[16] Argatov, I.I. and Dmitriev, N.N. (2003) *Fundamentals of Theory of Elastic Discrete Contact*, Politekhnika Press, St. Petersburg.

[17] Borodich, F.M. and Onishchenko, D.A. (1993) Fractal roughness for problem of contact and friction (the simplest models). *J. Frict. Wear*, **14**, 14–19.

[18] Borodich, F.M. and Onishchenko, D.A. (1999) Similarity and fractality in the modelling of roughness by multilevel profile with hierarchical structure. *Int. J. Solids Struct.*, **36**, 2585–2612.

[19] Warren, T.L. and Krajcinovic, D. (1996) Random Cantor set models for the elastic perfectly plastic contact of rough surfaces. *Wear*, **196**, 1–15.

[20] Plesha, M.E. and Ni, D. (2001) Scaling of geological discontinuity normal load-deformation response using fractal geometry. *Int. J. Num. Anal. Meth. Geomech.*, **25**, 741–756.

[21] Warren, T.L., Majumdar, A. and Krajcinovic, D. (1996) A fractal model for the rigid-perfectly plastic contact of rough surfaces. *J. Appl. Mech.*, **63**, 47–54.

[22] Schmittbuhl, J., Vilotte, J.-P. and Roux, S. (1996) Velocity weakening friction: A renormalization approach. *J. Geophys. Res.*, **101**(B6), 13911–13917.

[23] Sugimura, J. (1995) Fractal surfaces and hydrodynamic lubrication. *J. Japan Soc. Trib.*, **40**, 549–554 (in Japanese).

[24] Kennedy, F.E., Brown, C.A., Kolodny, J. and Sheldon, B.M. (1999) Fractal analysis of hard disk surface roughness and correlation with static and lowspeed friction. *Trans. ASME J. Tribology*, **121**, 968–974.

[25] Jhon, M.S. and Choi, H.J. (2001) Lubricants in future data storage technology. *J. Ind. Eng. Chem.*, **7**, 263–275.

[26] Plouraboue, F., Prat, M. and Letalleur, N. (2001) Sliding lubricated anisotropic rough surfaces. *Phys. Rev. E*, **64**, Art. No. 011202.

[27] Whitehouse, D.J. (2001) Fractal of fiction. *Wear*, **249**, 345–353.

[28] Greenwood, J.A. and Wu, J.J. (2001) Surface roughness and contact: An apology. *Meccanica*, **36**, 617–630.

[29] Bhushan, B. (2001) *Modern Tribology Handbook*, CRC Press, Boca Raton, p. 1430.

[30] Borodich, F.M. (1999) Fractals and fractal scaling in fracture mechanics. *Int. J. Fracture*, **95**, 239–259.

[31] Davies, S. and Hall, P. (1999) Fractal analysis of surface roughness by using spatial data. *J. R. Statist. Soc. B*, **61**, 3–37.

[32] Borodich, F.M. (1993) Similarity properties of discrete contact between a fractal punch and an elastic medium. *C.R. Acad. Sci. (Paris), Ser. 2*, **316**, 281–286.

[33] Berry, M.V. and Lewis, Z.V. (1980) On the Weierstrass–Mandelbrot fractal functions. *Proc. Roy. Soc. Lond.*, **A370**, 459–484.

[34] Roques-Carmes, C., Wehbi, D., Quiniou, J.F. and Tricot, C. (1988) Modelling engineering surfaces and evaluating their non-integer dimension for application in material science. *Surface Topography*, **1**, 435–443.

[35] Sun, X. and Jaggard, D.L. (1990) Wave scattering from non-random fractal surfaces. *Optics Communications*, **78**, 20–24.

[36] Majumdar, A. and Bhushan, B. (1990) Role of fractal geometry in roughness characterization and contact mechanics of surfaces. *J. Tribology*, **112**, 205–216.

[37] Moreira, J.G., Kamphorst Leal da Silva, J. and Oliffson Kamphorst, S. (1994) On the fractal dimension of self-affine profiles. *J. Phys. A: Math. Gen.*, **27**, 8079–8089.

[38] Lopez, J., Hansali, G., Zahouani, Le Bosse, J.C. and Mathia, T. (1995) 3-D fractal-based characterization for engineered surface topography. *Int. J. Mach. Tools Manufacture*, **35**, 211–217.

[39] Blackmore, D. and Zhou, J.G. (1996) A general fractal distribution function for rough surface profiles. *SIAM J. Appl. Math.*, **56**, 1694–1719.

[40] Bhushan, B. (1995) A fractal theory of the temperature distribution at elastic contacts of fast sliding surfaces. Discussion. *J. Tribology*, **117**, 214–215.

[41] Malcai, O., Lidar, D.A., Biham, O. and Avnir, D., (1997) Scaling range and cutoffs in empirical fractals. *Phys. Rev. E*, **56**, 2817–2828.

[42] Mandelbrot, B.B. (1985) Self-affine fractals and fractal dimension. *Phys. Scripta.*, **32**, 257–260.

[43] Hansen, A., Hinrichsen, E.L. and Roux, S. (1991) Roughness of crack interfaces. *Phys. Rev. Lett.*, **66**, 2476–2479.

[44] Måløy, K.J., Hansen, A., Hinrichsen, E.L. and Roux, S. (1992) Experimental measurements of the roughness of brittle cracks. *Phys. Rev. Lett.*, **68**, 2266–2269.

SESSION 10

ROUGHNESS ATTENUATION AND PRESSURE RIPPLING IN EHL CONTACTS

C.J. Hooke
School of Engineering, The University of Birmingham, U.K.

Abstract Surface roughness tends to reduce the life of EHL contacts and from the early work of Tallian onwards it has been recognized that the mechanism is via the production of pressure fluctuations in the conjunction.

Long wavelength components of the roughness are flattened as they pass through the contact, shorter components are largely unmodified. The transition occurs, approximately, where the roughness wavelength is equal to the length of the inlet pressure sweep and it is possible to produce a single master curve relating the attenuation of the roughness to this ratio. This paper examines pressure rippling in EHL contacts and shows that the maximum amplitude occurs at the transition from short to long wavelength behaviour and that, for both short and long wavelengths, the amplitude tends to zero.

Keywords: rough EHL, pressure.

1. INTRODUCTION

Increases in surface roughness generally tend to reduce the life of the contacting components. A mechanism for the life reduction was suggested by Tallian [1–3] who argued that the surface roughness produces fluctuations or ripples in the pressure distribution and it is these, as they modify the smooth pressure profile that increase the cyclic stresses and reduce the fatigue life. Tallian also suggested that there might be a simple, empirical relationship between the ratio of roughness amplitude to film thickness and life reduction. The well known, but widely discredited, statement that component life is reduced if the film thickness is less than three times the combined surface roughness arises from this source.

With the availability of more powerful numerical techniques [4] detailed analyses of rough surfaces became possible and although Tallian's basic premise, that roughness affects the stresses in the surfaces, appears sound it is clear that details of the roughness distribution affect the contact's behaviour. As an example, Figure 1 shows the pressures and clearances, at one instant

R.W. Snidle and H.P. Evans (eds), IUTAM Symposium on Elastohydrodynamics and Micro-elastohydrodynamics, 411–422.

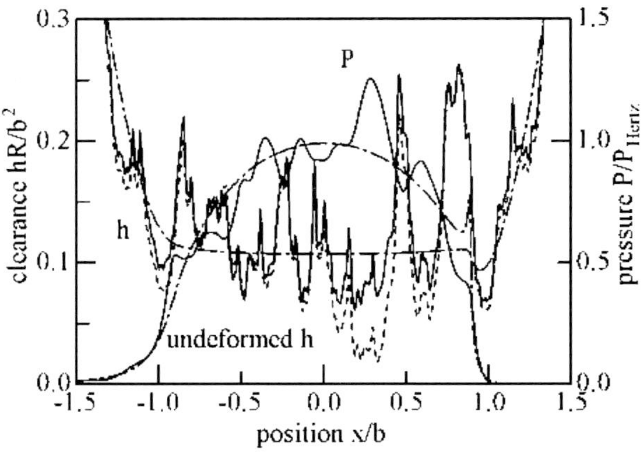

Figure 1. Rough EHL contact. $P = 20$, $S = 10$, $P_{\text{Hertz}} = 1$ GPa.

of time, under a rolling EHL contact where the surface roughness is relatively large compared to the film thickness. The surface profile is that used by Evans and Snidle [5] in their analysis of roughness in gears but for comparison with later results in this paper, the contact is assumed to be rolling without slip and the results correspond, therefore, to conditions on the pitch line of the gears. It may be seen that the deformation of the surfaces has increased the minimum film thickness substantially above that predicted using the smooth film thickness and the undeformed surface profile and that there is substantial rippling of the pressure profile.

A study of the behaviour of low amplitude, sinusoidal roughness in EHL contacts [6] showed that the roughness wavelength has a major effect on the way the surface behaves. For short wavelengths the roughness passes through the contact unchanged; with long wavelengths the roughness is substantially reduced in amplitude. The authors were also able to produce a master equation for the reduction in amplitude. Later studies [7–10] showed that the amplitude reduction was, in fact, related to the ratio of the wavelength to the length of the inlet pressure sweep and that this applied whatever the orientation of the roughness. However, details of the relationship do depend on the orientation. Finally a detailed study [11] showed that the same relationship between amplitude reduction and wavelength ratio applied to line, point and elliptic contacts allowing a single relationship to be used for all conjunctions.

Most of these studies have concentrated on the reduction in amplitude of the surface roughness rather than its effect on the pressure distribution. It is the purpose of this paper to extend the analysis to pressure rippling and it will be shown that there is a simple relationship between amplitude reduction and the magnitude of the pressure perturbations. However, it will also be shown that the magnitude of the pressure ripple depends on both the ratio of the wavelength to the length of the inlet pressure sweep and the ratio to the Hertz contact width.

2. ANALYSIS

If low amplitude surface roughness is introduced into an otherwise smooth EHL contact, the changes in pressure and clearance will be small allowing the conjunction to be treated as a linear system. This means that, for example, doubling the amplitude of the roughness will double the changes in the pressures and clearances. It also means that the responses to different surface profiles may be super-positioned. Thus an undeformed surface roughness such as that shown in Figure 1 may be split into its Fourier components, the response to each component calculated, and the results added to obtain the response of the contact to the original profile.

Where both surfaces are rough, the response may, of course, be obtained by superimposing the results for the two separate surfaces.

There are obviously limitations to this process when the roughness becomes large compared to the film thickness but even here this approach enables the main features of the contact's response to be obtained.

Accordingly, this paper will concentrate on the response of EHL contacts to low amplitude, sinusoidal roughness. The analysis starts with the calculation of the pressures and clearances under a smooth EHL line contact – a rapid process using a multi-grid solver. Then low amplitude, sinusoidal roughness is introduced and the changes in pressures and clearances calculated. Full details of this perturbation analysis are given in [10] and only a brief outline will be given here.

It is assumed that only one surface is rough and that the undeformed surface profile is given by:

$$R = A\cos(\omega x - \omega v t)\cos(\xi y), \tag{1}$$

where A is the amplitude of the roughness, ω and ξ the wave numbers in the x and y directions respectively (x is measured through the conjunction, y along it) and it is assumed that the rough surface is moving into the conjunction with a velocity v.

This equation may be rewritten, in complex form, as

$$R = \mathrm{Re}[\underline{R}\,\mathrm{e}^{-i\varphi}]\cos(\xi y), \tag{2}$$

where $\varphi = \omega v t$ and $\underline{R} = A\,\mathrm{e}^{i\omega x}$.

It may be noted that while the roughness is a function of x, y and t, the complex variable $\underline{R}$ is a function of x alone, with the variations with y and t being incorporated via the $\cos(\xi y)$ and $\mathrm{e}^{-i\varphi}$ terms.

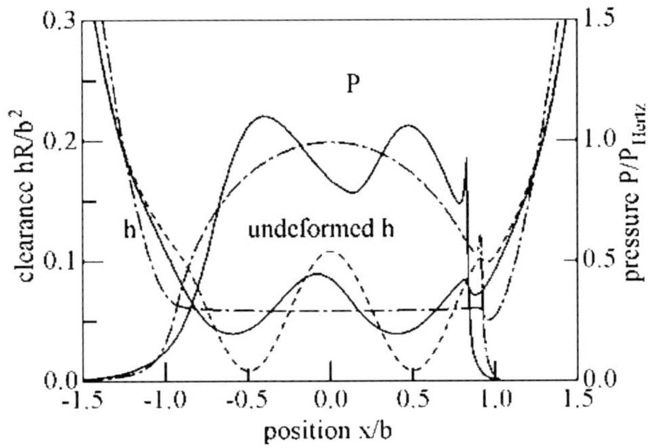

Figure 2. Sinusoidal roughness under a line contact obtained from a multi-grid analysis, $\lambda/b = 1$, $P = 20$, $S = 8$, $P_{\mathrm{Hertz}} = 1$ GPa.

This roughness will produce changes δP in the pressure and δh in the clearance and, since the system is linear, these must have the form

$$\delta P = \mathrm{Re}(\underline{\delta P}\, \mathrm{e}^{-i\varphi}) \cos(\xi y), \tag{3a}$$

$$\delta h = \mathrm{Re}(\underline{\delta h}\, \mathrm{e}^{-i\varphi}) \cos(\xi y). \tag{3b}$$

Similar expressions will apply for the changes in deformation, density and viscosity. This enables a series of complex, linear perturbation equations to be obtained from which the responses $\underline{\delta P}$ and $\underline{\delta h}$ can be calculated.

It will be noted that the inclusion of the $\cos(\xi y)$ term in the analysis allows two dimensional roughness to be examined even though the smooth contact is one dimensional. However, for the examples given in this paper only transverse roughness will be considered and the $\cos(\xi y)$ term is, therefore, taken as 1.

2.1 Sinusoidal Roughness

As an example of the results from this perturbation analysis, Figure 2 shows the pressures and clearances under an EHL contact as a surface with a sinusoidal, transverse roughness passed through it. This result was obtained using a standard multi-grid solver and shows the pressures and clearances under pure rolling conditions as the peak of the original roughness reaches the centre of the contact.

The equivalent result from the perturbation analysis is shown in Figure 3. This plots the undeformed profile and the changes in clearances and pressures at time, $t = 0$. The chain dashed lines show the magnitude of $\underline{\delta P}$ and $\underline{\delta h}$.

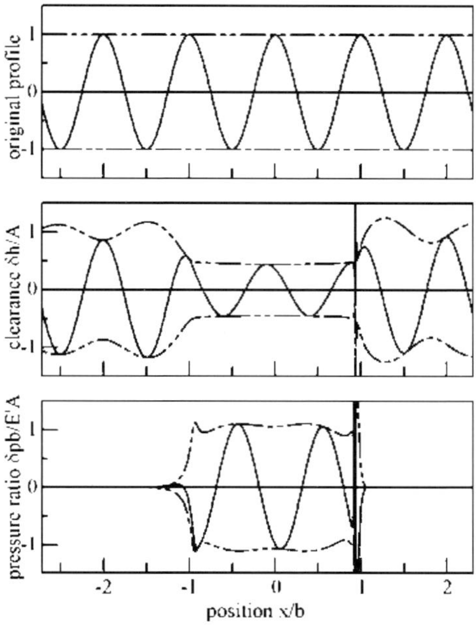

Figure 3. Results from the perturbation analysis for the contact shown in Figure 2.

The oscillation of the δh magnitude outside the conjunction arises because the separation, h_0, varies cyclically as the roughness passes through the contact.

It may be seen that, inside the contact, the magnitude of the clearance perturbation is almost constant and that its waveform is sinusoidal with a wavelength equal to that of the original roughness. There is, however, a difference in phase between the original waveform and the clearance under the contact. The pressures do vary slightly across the contact but for practical purposes the central pressures give a reasonable representation of the distribution. Under the contact, therefore, both the clearances and pressures can be treated as having a sinusoidal waveform moving at the velocity of the rough surface and this can be used to simplify the analysis of general, low amplitude, rough surfaces.

It may also be noted that both the clearance and pressure perturbations show large excursions at the exit to the conjunction. Essentially, these arise because the roughness changes the position of the exit pressure spike slightly, producing large, localized changes in both the clearance and pressure as it does so. These changes are not, however, associated with major changes in the actual height of the pressure spike or in the clearance and can, therefore be ignored.

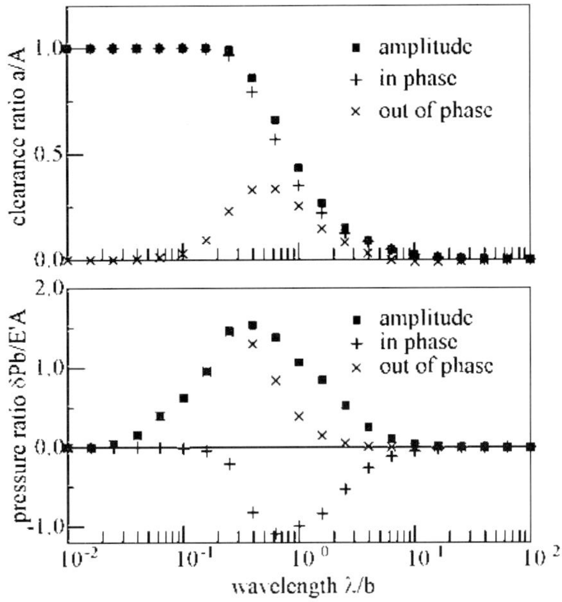

Figure 4. Attenuation of sinusoidal, transverse roughness and magnitude of the pressure perturbation. $P = 20$, $S = 8$, $P_{\text{Hertz}} = 1$ GPa.

Calculation of the results shown in Figure 3 is rapid and repetition for a range of different wavelengths enables the clearance and pressure perturbations at the centre of the contact to be determined. These are shown in Figure 4.

The figure shows the real (in phase) and imaginary (out of phase) parts of the response plus its amplitude. The out of phase component lags 90° behind the original surface roughness.

The attenuation of the roughness amplitude follows the well established pattern with no change occurring for short wavelengths and with the roughness being completely flattened at long wavelengths. The transition between the two is centred, approximately, on $\lambda/b = 0.8$ but extends over the range $\lambda/b = 0.08$ to 8.

The magnitude of the pressure perturbation is zero at short wavelengths where the surface is undeformed and also drops to zero at long wavelengths. It reaches a peak around $\lambda/b = 0.3$, but is significant over wavelengths from $\lambda/b = 0.03$ to around 8.

At low wavelengths the out of phase component of the pressure is dominant and this, essentially, corresponds to a pressure peak in advance of each clearance minimum, as shown in Figures 2 and 3. However, the amount of surface distortion is low because of the high wave number of the pressure ripple. As the wavelength increases, the magnitude of the in phase component of the pres-

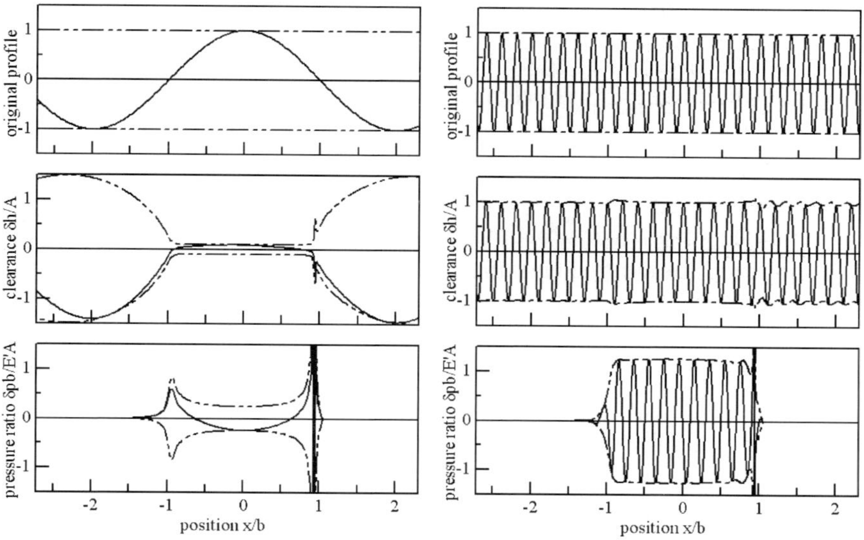

Figure 5. Perturbation analysis, $\lambda/b = 4$ and $\lambda/b = 0.2$.

sure also increases. The negative values in Figure 4 arise because the pressure is at a maximum where the clearance is lowest. This pressure reduces the amplitude of the clearance perturbation giving rise to the familiar curve shown in the upper figure.

At still higher wavelengths, the wavelength exceeds the contact width and the effect is to produce a sinusoidal variation in the effective contact radius. Here the pressure and clearance perturbations correspond to the changes that this would produce in a smooth conjunction.

This behaviour may be seen more clearly in Figure 5 which shows the results of the perturbation analysis for a long and a short wavelength. In the right hand curves, the wavelength is 20% of the semi-contact width and there is little change in the amplitude of the roughness as it passes through the contact. However, there is a substantial pressure ripple approximately 90° out of phase with the roughness.

The left hand curves show the behaviour at long wavelengths. Here the wavelength is twice the total contact width and the main effect of the roughness is to generate a cyclic change in effective contact radius and a cyclic variation in the slope of the contacting surfaces. The changes in pressure are essentially equal to the changes in Hertz contact pressure produced by the radius variation while the clearance variation reflects the effect that changing the radius has on the smooth film thickness.

2.2 Non-Dimensional Results

Results similar to those of Figure 4 were obtained for line contacts with a range of operating conditions. In previous studies it has been found that the clearance attenuation can be reduced to a single master curve by plotting the amplitude ratio against the parameter, $\lambda/bP^{1.5}S^{-2}$. This has been done for the present results in Figure 6 which shows the clearance and pressure perturbations. Only the magnitudes of the perturbations are given in the figure but similar curves can be obtained for the real and imaginary components. The solid symbols are for the pressure perturbations, the hollow for the roughness attenuation.

The parameter $\lambda/bP^{1.5}S^{-2}$ is proportional to the length of the inlet pressure sweep and it is clear that the clearance perturbations fall, closely, onto a single curve, confirming that it is the ratio of the roughness wavelength to the length of the inlet sweep that controls the clearance variation.

The pressure perturbations are more complex and these have been plotted in the form $\delta p\,\lambda/E'A$. It should be noted that the inclusion of the wavelength, λ, in this non-dimensional grouping has the effect of exaggerating the pressure perturbations at the longer wavelengths.

At the lower non-dimensional wavelengths, all the pressure perturbations initially follow a single master curve that increase from zero reaching a maximum value of $\pi/2$ at a value of $\lambda/bP^{1.5}S^{-2}$ around 80. In interpreting this curve it may be noted that the original roughness would be completely flattened by a pressure perturbation, in non-dimensional form, of $\delta p\,\lambda/E'A = \pi/2$. The close correlation between the amount the original profile is flattened and the pressure magnitude is, therefore, to be expected. What deviation there is arises from the fact that both the pressure and clearance perturbations are complex and it is the distortion, not the clearance that is related to the pressure perturbation. The actual relationship between pressure and amplitude ratio is given, later, in Equation (6).

At longer wavelengths the pressure perturbations deviate from this curve falling to zero at long wavelengths. The drop occurs when the wavelength exceeds the contact width, producing the type of behaviour seen in the left hand curves of Figure 5. To investigate the behaviour at longer wavelengths it may be noted that, for a sinusoidal waveform of amplitude, A, and wavelength, λ, the surface profile has a maximum change of curvature of $4\pi^2 A/\lambda^2$. Then noting that the Hertz pressure is related to the effective contact radius, R, by

$$P_{\text{Hertz}} = \sqrt{\frac{WE'}{2\pi R}} \tag{4}$$

the pressure perturbation becomes

$$\frac{\delta p\,\lambda}{E'A} = \frac{\pi^2 b}{2\lambda}. \tag{5}$$

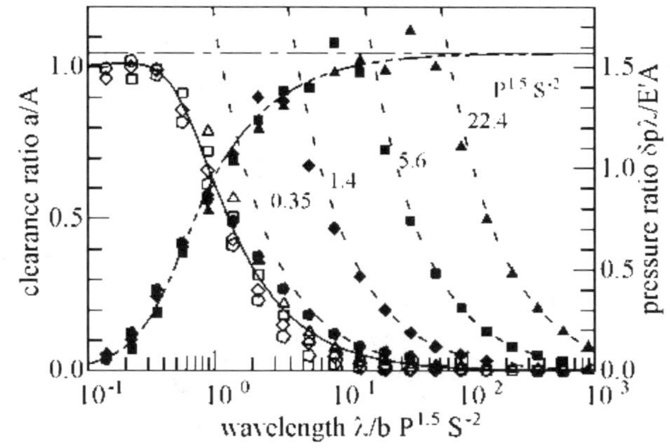

Figure 6. Clearance attenuation and pressure perturbation. The hollow symbols are for clearance attenuation; the solid for the pressure perturbation.

The curves corresponding to this equation are shown dashed in the figure for the four operating conditions presented. It may be seen that, at long wavelengths, the pressure perturbations follow Equation (5) closely until the curves cross the low wavelength curve after which they switch to that curve. For the most heavily loaded contacts the transition between the two curves occurs with the pressure perturbation approximately equal to $\pi/2$ and here the wavelength at the change over is given by $\lambda/b = \pi$. For the less heavily loaded contacts the transition occurs at a lower pressure perturbation and a slightly longer wavelength.

The pressure and clearance distributions given in Figure 6 may be approximated, for short wavelength behaviour where the pressures follow the master curve in the figure, to a reasonable accuracy by:

$$\frac{\delta h}{A} = 1 + \frac{2}{\pi} wp \quad \text{and} \quad \frac{\delta P\ b}{E'A} = P^{1.5} S^{-2} p, \tag{6}$$

where

$$p = -\frac{\pi}{2} \frac{1}{c_0/w + c_1 + w}, \qquad w = \frac{\lambda}{b} P^{1.5} S^{-2},$$

and

$$c_0 = -0.22 + 0.40i, \quad c_1 = 0.55 + 0.55i.$$

At longer wavelengths Equation (5) applies. The easiest way to deal with the transition from short to long wavelength behaviour is to calculate the pressure perturbation using both equations. The one with the lowest magnitude is then selected.

It may be noted that Equations (5) and (6) were used to plot the trend lines in Figure 6 and it may be seen that the simple relationships of these equations give a reasonable fit to the detailed numerical data.

3. ANALYSIS OF ROUGH CONTACTS

The results for low amplitude sinusoidal roughness, summarized by Equations (5) and (6), can be used to obtain rapid estimates of the pressures and clearances under rough EHL contacts with transverse roughness.

It was noted earlier that, for sinusoidal roughness, both the pressures and clearances under the conjunction consist, to a reasonable approximation, of sinusoidal curves of the same wavelength as the original roughness moving at the velocity of the rough surface. If it is assumed that the measured surface profile, $r(x)$ represents the profile at time $t = 0$, then, at any other time the profile will be given by $r(x - vt)$ since the surface is moving through the conjunction with a velocity, v. Similarly, the moving pressure and clearance variations under the conjunction will have the form $\delta P\ (x - vt)$ and $\delta h\ (x - vt)$.

Because of this, the values of δP and δh at $t = 0$ may be found by taking a Fourier transform of $r(x)$, scaling this by the coefficients given in Equations (5) and (6) and taking an inverse transform of the result. The distributions at any other time may then be found by applying a shift, vt.

As an example, Figure 7 shows the variation of the pressures and clearances at the centre of the contact shown, for one particular instant of time, in Figure 1. Also shown, for comparison, are the results obtained from the multi-grid analysis. Clearly the results from the FFT analysis provide a good estimate of the general behaviour of the conjunction. There are differences, particularly where the clearances are low, with the FFT method predicting somewhat lower clearances and pressures. These arise because the FFT approach assumes that the system is linear whereas the actual contact is not. In spite of this, the FFT analysis provides an extremely rapid and easy way of obtaining an estimate of the likely behaviour of any transverse roughness in a rolling EHL conjunction.

4. CONCLUSIONS

Surface roughness produces pressure rippling in EHL conjunctions and this rippling may affect the life of the contacting surfaces.

For low amplitude roughness it is possible to treat the contact as a linear system and examine each of the roughness components separately. A perturbation analysis shows that for short wavelengths it is the ratio of the wavelength to the contact's inlet length that controls both the roughness attenuation and the magnitude of the pressure ripple. For longer wavelengths, above $\lambda/b = \pi$, the roughness simply alters the effective radius of the contacting surfaces and the pressure variation can be calculated from modifications to the Hertz pressure.

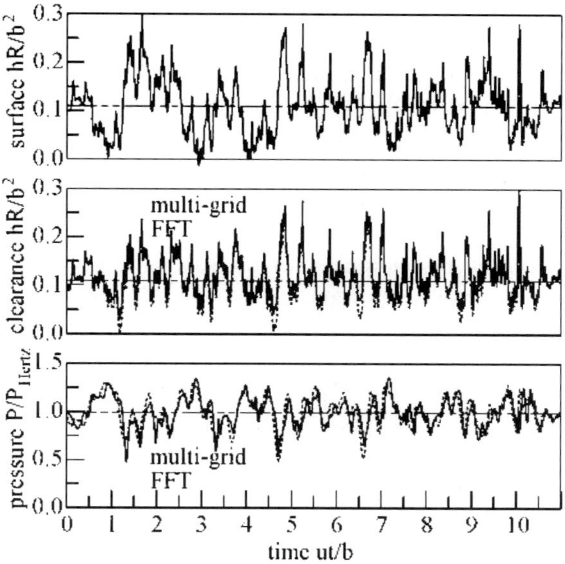

Figure 7. Comparison of the pressures and clearances at the centre of a rough EHL contact calculated using a multi-grid process and from a FFT analysis, $P = 20$, $S = 10$.

For transverse roughness, the only case considered in detail here, it is possible to obtain simple approximations for the pressure ripple and hence the roughness attenuation.

This enables rapid estimates to be obtained of the likely pressure and clearance variations in EHL conjunctions with transverse roughness by using a FFT analysis of the measured rough surface. Although there are differences between these results and those obtained using a multi-grid solver as the roughness amplitude becomes large compared to the film thickness the process does allow a real time assessment of the effect of particular roughnesses on EHL contacts.

Although the analysis in this paper has been limited to transverse roughness, the method may be readily extended to three dimensional profiles.

REFERENCES

[1] Chiu, Y.P., Tallian, T.E. and McCool, J.I., "An engineering model of spalling fatigue failure in rolling contact – The subsurface model", *Wear*, **17**, 1971, 433–446.

[2] Tallian, T.E. and McCool, J.I., "An engineering model of spalling fatigue failure in rolling contact – The surface model", *Wear*, **17**, 1971, 447–461.

[3] Tallian, T.E., "Pressure and traction rippling in elastohydrodynamic contact of rough surfaces", *Trans. ASME, J. Lubric. Technol.*, **96**, 1974, 399–405.

[4] Venner, C.H., "Multi-level solution of the elastohydrodynamic line and point contact problems", Proefschrift, Universiteit Twente, the Netherlands, 1991.

[5] Tao, J., Hughes, T.G., Evans, H.P. and Snidle, R.W., "Elastohydrodynamic response of transverse ground gear teeth", in *Proc. 28th Leeds–Lyon Symposium on Tribology*, Vienna, 2001, pp. 447–458.

[6] Venner, C.H., Couhier, F., Lubrecht, A.A. and Greenwood, J.A., "Amplitude reduction of waviness in transient EHL line contacts", in *Proc. 1996 Leeds–Lyon Symposium on Tribology*, Leeds, 1996, pp. 103–112.

[7] Hooke, C.J., "Surface roughness modification in elastohydrodynamic line contacts operating in the elastic piezoviscous regime", *Proc. Instn. Mech. Engrs., J. of Eng. Trib.*, **212**, 1998, 145–162.

[8] Hooke, C.J., "The behaviour of heavily loaded line contacts with transverse roughness", *Proc. Instn. Mech. Engrs., Part C*, **213**, 1999, 309–320.

[9] Hooke, C.J., "The behaviour of low amplitude surface roughness under line contacts", *Proc. Instn. Mech. Engrs., J. of Eng. Trib.*, **213**, 1999, 275–286.

[10] Hooke, C.J., "The behaviour of low amplitude surface roughness under line contacts: Non-Newtonian fluids", *Proc. Instn. Mech. Engrs., Part C*, **214**, 2000, 253–265.

[11] Hooke, C.J. and Venner, C.H., ?Surface roughness attenuation in line and point contacts", *Proc. Instn. Mech. Engrs., J. of Eng. Trib.*, **214**, 2000, 439–444.

EFFECT OF FILM THICKNESS RATIO ON GEARING CONTACT FATIGUE LIFE

C.K. Gao[1], X.M. Qi[1], R.W. Snidle[2] and H.P. Evans[2]
[1]*Taiyuan University of Technology, Taiyuan 030024, P.R. China*
[2]*School of Engineering, Cardiff University, Cardiff CF24 0YF, U.K.*

Abstract The paper considers the effect of the ratio of film thickness to composite surface roughness, λ, on the contact fatigue life, N, in gearing applications. More than 50 sets of thermal non-Newtonian EHL numerical calculations and six sets of disc fatigue tests under the conditions ranging from mixed to full film lubrication are considered. The results show that in seeking to improve gearing fatigue life it is generally useful to increase the value of the film thickness ratio, λ. But it is not simply the case that the higher the film thickness ratio, the better. When $\lambda < 5$, a quantitative relation of $N \propto \lambda^{1/2}$ is seen in the test results. When $\lambda > 5$, however, increasing the λ ratio is found to reduce the fatigue life, so that $\lambda \approx 5$ gives the maximum value for fatigue life. Based on this result, the paper raises questions concerning the accuracy of the lubrication factor recommended by the International Standard for Computing Cylindrical Gearing Strength (ISO/423E).

Keywords: elastohydrodynamic lubrication, fatigue test, film thickness ratio, gearing.

1. INTRODUCTION

The means of lubrication of gear tooth contacts is elastohydrodynamic lubrication (EHL). For EHL contacts the ratio of the central film thickness h_c to the composite root mean square (RMS) surface roughness, σ, is defined as the film thickness ratio, or λ ratio. This ratio is known to be influential in determining the fatigue life of rolling element bearings, and its effect on contact fatigue life in gearing has been the focus of much research over the years. In 1962, Dawson [1] studied the effect of metallic contact on the pitting of lubricated rolling surface. He proposed a power law relation between the fatigue life N and λ ratio. Onions and Archard [2] conducted fatigue tests on a gear test rig and revealed for the first time the quantitative relation, $N \propto \lambda^{2.21}$. Subsequently Czyzewski [3] carried out seven sets of fatigue tests with rollers and obtained a parabolic relation between N and λ. In particular, he emphasized that excessive λ ratio

R.W. Snidle and H.P. Evans (eds), IUTAM Symposium on Elastohydrodynamics and Microelastohydrodynamics, 423–434.

values would be detrimental to fatigue life. In 1976, the Lubrication Research Group [4] of the ASME obtained a result of the form $N \propto \lambda^{0.53}$ on the basis of 433 contact fatigue tests using rollers. Simultaneously Wellaur and Holloway [5] presented a relation curve between the failure probability of gear tooth surfaces and the λ ratio, based on both test results and field data of gear-set fatigue failures. The curve indicated that the failure probability decreased with increasing λ ratio.

The studies referred to above were based on experimental investigations. Only in recent years have related theoretical analyses begun to emerge for boundary and mixed lubrication. Zhu and co-workers [6–8] were among the first to develop a mixed lubrication EHL model for elliptical contacts and have investigated the contact ratio, frictional forces, flash temperatures and subsurface stresses developed in rough surface EHL contacts. Jiang et al. [9] made a theoretical analysis of mixed lubrication and presented the average gap, oil-film pressure and contact pressure. Recently Wang et al. [10] provided mixed lubrication solutions in which a curve of average pressure against λ ratio was given.

Different numerical methods have been adopted by researchers, but the same result is obtained, namely that the average contact pressures and subsurface stresses decrease monotonically with increasing λ ratio. This implies that fatigue life can be expected to increase monotonically with λ ratio, which is generally felt to be a reasonable and acceptable finding for the case of boundary or mixed lubrication. However, whether or not this same principle is applicable to full film EHL contacts is not established and the current paper seeks to throw some light on this question.

2. GOVERNING EQUATIONS

As is stated above the objective of the current paper is to investigate the effect of λ ratio on fatigue life under the conditions of full film lubrication where the composite surface roughness is small compared to the film thickness. Consequently surface roughness effects were not included in the EHL numerical analyses. Furthermore, some previous results [11] have shown that a line contact analysis is sufficient to identify the EHL response of gear tooth contacts. Therefore, it should suffice to use a model contact problem of two elastic, smooth-surface cylinders.

For the thermal non-Newtonian EHL line contact between smooth surfaces, the governing equations can be expressed as follows.

2.1 Reynolds Equation

The Reynolds equation is formulated using Yang and Wen's [12] approach:

$$\frac{\partial}{\partial x}\left[\left(\frac{\rho}{\eta}\right)_e h^3 \frac{\partial p}{\partial x}\right] = 12u_r \frac{12(\rho^* h)}{\partial x}. \tag{1}$$

Here

$$\left(\frac{\rho}{\eta}\right)_e = 12\left(\frac{\eta_e \rho'_e}{\eta'_e} - \rho''_e\right), \quad \rho^* = [\rho'_e \eta_e (u_2 - u_1) + \rho_e u_1]/u_r,$$

$$\rho_e = \frac{1}{h}\int_0^h \rho\, \mathrm{d}z, \quad \rho'_e = \frac{1}{h^2}\int_0^h \rho\left(\int_0^z \frac{1}{\eta^*}\mathrm{d}z\right)\mathrm{d}z,$$

$$\rho''_e = \frac{1}{h^3}\int_0^h \rho\left(\int_0^z \frac{z}{\eta^*}\mathrm{d}z\right)\mathrm{d}z,$$

$$\frac{1}{\eta_e} = \frac{1}{h}\int_0^h \frac{1}{\eta^*}\mathrm{d}z, \quad \frac{1}{\eta'_e} = \frac{1}{h^2}\int_0^h \frac{z}{\eta^*}\mathrm{d}z.$$

An effective viscosity η^* has been introduced in Equation (1) to describe the non-Newtonian lubricant properties. For any non-Newtonian model whose constitutive equation is $\partial u/\partial z = f(\tau, \eta)$, the η^* is defined as:

$$\tau = \eta^* \frac{\partial u}{\partial z}. \tag{2}$$

For the Eyring type fluid model used in this analysis proposed by Johnson and Tevaarwerk [13], $f(\tau, \eta) = (\tau_0/\eta)\sinh(\tau/\tau_0)$, then

$$\frac{1}{\eta^*} = \frac{1}{\eta_e}\frac{\tau_0}{\eta}\sinh\left(\frac{\tau}{\tau_0}\right).$$

The pressure boundary conditions for Equation (1) are $p(x_i) = 0$ and $p(x_e) = 0$ at the inlet boundary, x_i and the exit cavitation position, x_e. The inlet boundary is located at $x_i = -4b$ and the outlet location, x_e, is determined in the solution process.

The Roelands relation [14] is used to define the change in viscosity with pressure and temperature:

$$\eta = \eta_0 \exp\left\{(\ln \eta_0 + 9.67)\left[(1 + 5.1 \times 10^{-9} p)^Z \left(\frac{T - 138}{T_0 - 138}\right)^{-s_0}\right] - 1\right\} \tag{3}$$

and the density variation after Wang et al. [15] as

$$\rho = \rho_0[1 + Ap/(1 + Bp) - D_0(T - T_0)]. \tag{4}$$

2.2 Film Thickness Equation

$$h(x) = h_0 + \frac{x^2}{2R} - \frac{2}{\pi E}\int_{x_1}^{x_e} p \ln\left(\frac{x - s}{s}\right)^2 \mathrm{d}s. \tag{5}$$

2.3 Energy Equation

$$uc_f\rho\frac{\partial T}{\partial x} - k_f\frac{\partial^2 T}{\partial z^2} + \frac{T}{\rho}\frac{\partial \rho}{\partial T}u\frac{\partial p}{\partial x} = \tau\frac{\partial u}{\partial z}. \tag{6}$$

The boundary conditions for Equation (6) are

$$\begin{aligned} T(x,0) &= \frac{k_f}{\sqrt{(\pi k_1\rho_1 c_1 u_1)}}\int_{-\infty}^{x}\left.\frac{\partial T}{\partial z}\right|_{z=0}\frac{\mathrm{d}s}{\sqrt{x-s}} + T_0, \\ T(x,h) &= \frac{k_f}{\sqrt{(\pi k_2\rho_2 c_2 u_2)}}\int_{-\infty}^{x}\left.\frac{\partial T}{\partial z}\right|_{z=h}\frac{\mathrm{d}s}{\sqrt{x-s}} + T_0. \end{aligned} \tag{7}$$

2.4 Load Equation

$$P_0 = \int_{x_1}^{x_e} p(s)\,\mathrm{d}s. \tag{8}$$

3. NUMERICAL APPROACH

Equations (1–8) were solved simultaneously to obtain the pressure, film thickness and temperature distributions within the lubricant film. The hybrid forward and inverse method [16] was extended to allow solutions for the nonlinear viscous fluid rheology model. The extension is mainly involved with the evaluation of the effective viscosity η^*, which can be determined by considering fluid element force equilibrium as follows [12]:

$$\frac{\partial \tau}{\partial z} = \frac{\partial p}{\partial x}. \tag{9}$$

Integrating with respect to z across the film gives

$$\tau = \tau_1 + z\frac{\partial p}{\partial x}. \tag{10}$$

Substituting Equation (2) into (9), we get

$$\frac{\partial}{\partial z}\left(\eta^*\frac{\partial u}{\partial z}\right) = \frac{\partial p}{\partial x}. \tag{11}$$

From Equations (2), (10) and (11), the shear strain rate which will be used in solving the energy equation (6) can be written as

$$\frac{\partial u}{\partial z} = f\left(\tau_1 + z\frac{\partial p}{\partial x}, \eta\right). \tag{12}$$

Integrating with respect to z, for the Eyring type fluid model, gives

$$u_2 - u_1 = \int_0^h \frac{\tau_0}{\eta} \sinh\left(\frac{\tau_1 + z\frac{\partial p}{\partial x}}{\tau_0}\right) \mathrm{d}z$$

$$= \int_0^h \frac{\tau_0}{\eta}\left[\sinh\left(\frac{\tau_1}{\tau_0}\right)\cosh\left(\frac{z}{\tau_0}\frac{\partial p}{\partial x}\right) + \cosh\left(\frac{\tau_1}{\tau_0}\right)\sinh\left(\frac{z}{\tau_0}\frac{\partial p}{\partial x}\right)\right]\mathrm{d}z,$$

which can be used to determine the value of τ_1.

With the value of τ_1 known, Equations (10) and (12) give τ and $\partial u/\partial z$ at each nodal point across lubricant film. Thus, the film velocity profile can be obtained by quadrature from

$$u = u_1 + \int_0^z \frac{\partial u}{\partial z}\,\mathrm{d}z.$$

After the values of η^*, u and τ have been determined, it is feasible to conduct the numerical calculations of EHL for the non-Newtonian fluid.

When the value of τ_1 has been determined the coefficient of friction, μ, on the pinion tooth surface can be obtained from

$$\mu = F/P_0 = \left(\int_{x_i}^{x_e} \tau_1\,\mathrm{d}x\right) \Big/ P_0. \tag{13}$$

When the film pressure and coefficient of friction are known it is possible to compute the subsurface stress components at any point $m(x, z)$ with Equations (14). For the current work this is done on the simplifying assumption that the surface shear stress is μp at all points so that

$$\sigma_x = \frac{-2}{\pi}\int_{x_1}^{x_e} p(\xi)\frac{z(x-\xi)^2 + \mu(x-\xi)^3}{((x-\xi)^2 + z^2)}\,\mathrm{d}\xi,$$

$$\sigma_x = \frac{-2}{\pi}\int_{x_1}^{x_e} p(\xi)\frac{z^3 + \mu z^2 + (x-\xi)}{((x-\xi)^2 + z^2)}\,\mathrm{d}\xi,$$

$$\tau_{xz} = \frac{-2}{\pi}\int_{x_1} x_e p(\xi)\frac{z^2(x-\xi) + \mu z(x-\xi)^2}{((x-\xi)^2 + z^2)}\,\mathrm{d}\xi. \tag{14}$$

4. THEORETICAL RESULTS

For a given load parameter $W = 16.6 \times 10^{-5}$ and material parameter $G = 5000$, a total of fifty-four sets of numerical calculations are considered for variation of the speed parameter U from 2.09×10^{-11} to 2.3×10^{-9}, and the computational results are used as the basis for the relation curve between the

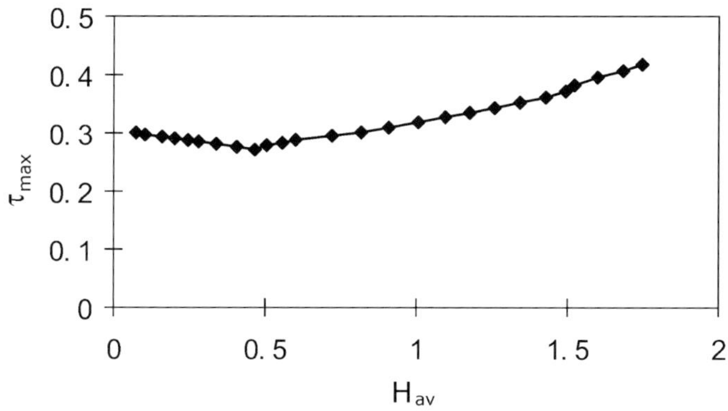

Figure 1. The relation between maximum shear stress, $\tau_{\max}$, and average film thickness, H_{av}, from EHL computations.

dimensionless maximum shear stress, $\tau_{\max}$, and the dimensionless average film thickness, H_{av}, shown in Figure 1. The conditions examined correspond to the fatigue test conditions specified in Section 5.2 with the product $\eta_0 u_r$ varied systematically between 0.0617 and 6.79 Pa·m. This shows that the maximum subsurface shear stress does not decrease monotonically with increasing film thickness. From this it can be deduced that the fatigue life does not necessarily increase monotonically with increasing λ ratio. To examine this conjecture it is essential to carry out some fatigue tests.

5. FATIGUE SIMULATION TESTS

5.1 Test Principle

Fatigue tests were conducted on a commercially available contact fatigue test rig which is comprised of four systems, namely, the drive, loading, lubrication and electrical system. The electrical system includes a relay trip and a cycle counter. The relay is activated by the level of vibration detected at the contact, which, in general, increases systematically with the level of pitting. The level is set so as to correspond to pitting distributed around the circumference of the discs as shown in the examples of Figures 2 and 3 which show enlarged views of examples of two of the pitted surfaces. When the trip is activated the relay cuts off the electric power supply and the cycle counter records automatically the cycle number at this stage which is viewed as a value of the fatigue life N corresponding to this test condition. If, on inspection, the surface damage was seen to be localised then the test was discarded so that the results correspond only to cases where pitting was developed uniformly around the disc circumference. Localised pitting occurred in about 5% of the tests carried out and was attributed to the defects in the test discs.

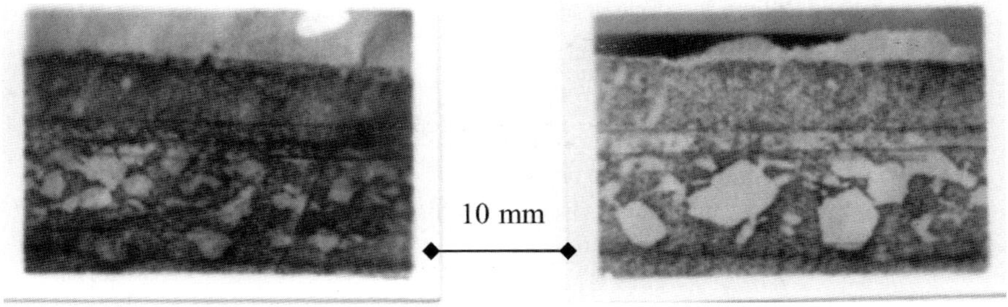

Figure 2. Photo of a fatigue disc (Test 1).

Figure 3. Photo of a fatigue disc (Test 5).

5.2 Test Method

Weck et al. [17] established, based on many tests, that the contact states between gear teeth can be effectively simulated by a pair of rollers when their parameters are equivalent to those of gears. Steel discs were chosen as fatigue testing specimens, with inner and outer diameters of 32 mm and 50 mm respectively. The hardness of HB240-260 was produced by quenching and tempering heat treatment, and the representative surface finish of $R_a \leq 0.32\ \mu$m gained by means of circumferential grinding. The tests were carried out under the conditions specified in Table 1. Before testing, the surface topography of each disc was measured by a mechanical stylus profilometer. The composite RMS roughness of the surface was then determined by $\sigma = \sqrt{\sigma_1^2 + \sigma_2^2}$, which remained about the same for all tests. In order to achieve different λ ratios it was thus necessary to vary film thickness. To this end, for a given testing load, rotation and dimensions of the discs, six mineral lubricants of the same type but having different viscosities were used. The nominal kinematic viscosities used in the tests were 10, 20, 30, 50, 70, 80 cSt. Table 2 gives the measured viscosity of the test lubricants as quoted by the suppliers at temperatures of 50°C and 40°C.

During the tests the oil temperature was monitored and then the kinematic viscosity, ν, corresponding to this monitored temperature can be determined by

$$\lg \lg(\nu + 0.6) = C_1 + C_2 \lg(T + 273).$$

Here, C_1 and C_2 are constants whose magnitudes can be obtained by the least square method for a specific mineral oil. For the current test lubricants the corresponding values have been calculated and are also given in Table 2.

Table 1. Operating conditions for fatigue tests.

Entrainment velocity	5.7 m/s
Slide roll ratio	0 and 0.22
Load	490 kN/m
Maximum Hertzian pressure	1.2 GPa
Lubricant pressure viscosity coefficient	21.2 GPa^{-1}
Lubricant Eyring stress	18 MPa

Table 2. Values of viscosity (cSt) and C_1, C_2 for test lubricants.

No.	1	2	3	4	5	6
ν_{50}	9.8	19.2	30.8	48.8	69.5	79.3
ν_{40}	12.3	28.3	47.7	80.8	109.8	138.4
C_1	9.52	9.71	9.61	9.85	9.34	9.77
C_2	3.77	3.82	3.76	3.84	3.62	3.78

Table 3. Fatigue test results obtained at pure rolling and at $\xi = 0.22$.

No. of test	ξ	Lubricant temp. T/°C	Lubricant viscosity ν/cSt	Central film thickness h_c/μm	Composite RMS roughness σ/μm	λ ratio	Fatigue life $N/10^3$
1	0	36.7	22.2	0.821	0.526	1.56	2020
2	0	38.3	30.5	1.20	0.541	2.21	2310
3	0	39.6	48.7	1.51	0.546	2.77	2670
4	0	40.0	80.8	2.15	0.541	3.97	3040
5	0	41.2	104	2.62	0.526	4.97	3250
6	0	42.1	116	2.91	0.539	5.41	2700
7	0.22	38.9	16.1	0.758	0.528	1.44	487
8	0.22	40.1	28.3	1.01	0.537	1.88	545
9	0.22	41.5	44.5	1.40	0.535	2.62	686
10	0.22	43.1	69.1	1.04	0.545	3.56	767
11	0.22	43.5	91.1	2.36	0.533	4.43	849
12	0.22	44.3	105.4	2.67	0.544	4.92	758

5.3 Test Results

With the chosen lubricants twelve sets of fatigue tests were carried out, six at pure rolling and six with a slide roll ratio $\xi = 0.22$. The results are presented in Table 3. For each set between ten and fifteen fatigue tests were conducted and a Weibull plot was used to determine the 50% fatigue life which is the value of N given in Table 3. The variation of N with λ is shown in Figure 4

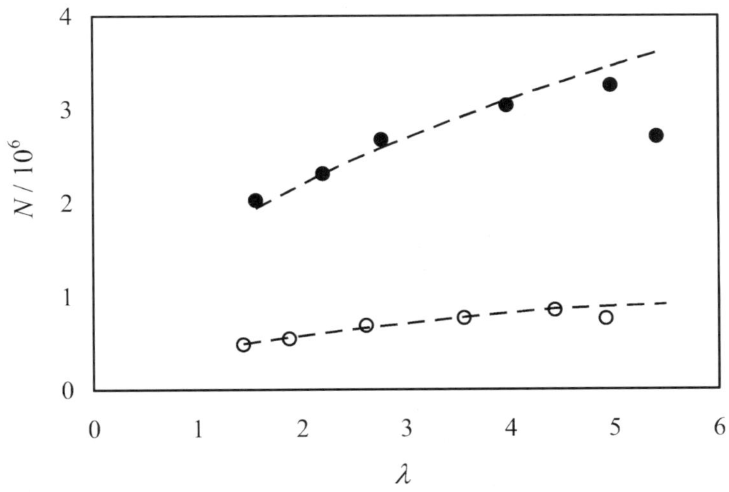

Figure 4. The relation between fatigue life, N, and λ ratio from disc fatigue tests; closed symbols give pure rolling results, open symbols correspond to results at $\xi = 0.22$. Broken curves indicate $N \propto \lambda^{1/2}$ curve fits.

for the two ξ values in the tests. The film thickness calculations were obtained from the theoretical analysis at the nominal test conditions. The results suggest that the fatigue life N is not a monotonic function of the film thickness ratio λ. For the set of fatigue results in the current work when $\lambda < 5$, the fatigue life N is approximately proportional to the $\lambda^{0.5}$ as shown by the broken lines in the figure. When $\lambda > 5$, however, increasing the λ ratio is seen to reduce the fatigue life for the current test conditions. However, departure of the fatigue life from the $\sqrt{\lambda}$ characteristic may well occur sooner for the pure rolling tests at $\lambda > 4$.

5.4 Analysis of Test Results

It can be seen from Figure 4 that the current tests cover the regions ranging from mixed to full film lubrication. For mixed lubrication the load is shared between the asperity contact and lubricant film. As the λ ratio increases the contribution of the lubricant film to the total load increases and this results in the decline of the maximum subsurface shear stress. When $\lambda > 3$, according to the previous experimental results [18], the frictional force decreases drastically and reaches its minimum at $\lambda \approx 5$. Thereafter it goes up again with the λ ratio, mainly due to the higher lubricant viscosity associated with thicker oil film. In addition it may be that as the λ values become large the contacts may become sufficiently lightly loaded (in EHL terms) for them to develop significantly larger pressure spikes at the exit of the contact. This could cause increased subsurface shear stress and stress concentration with a consequent decrease in fatigue life.

6. CONCLUSIONS

Increasing the film thickness ratio λ can help to improve contact fatigue life N, but the life N does not increase monotonically with increasing λ ratio. For the test conditions considered in the current paper when $\lambda < 5$ a quantitative relation of the form $N \propto \sqrt{\lambda}$ is detected, which is in good agreement with the result [4] of the ASME where the maximum λ value considered was 2. When $\lambda > 5$, however, increasing the λ ratio is found to reduce the fatigue life in the current tests.

As it is well known that the film thickness is strongly related to the lubricant viscosity, from the above result it can be deduced that it is not the case that the higher the lubricant viscosity, the better. However, in the International Standard for Computing Cylindrical Gearing Strength (ISO/423E), the lubrication factor Z_L increases monotonically as the viscosity rises. Therefore, it is reasonable to suggest that the lubrication factor Z_L recommended by ISO/423E has some limitations and can only be used under specific conditions.

NOTATION

A, B	pressure-density coefficients	m^2/N
B	half Hertzian contact width	m
c_1, c_2	specific heat of pinion and gear	J/kg K
c_f	specific heat of lubricant	J/kg K
D_0	temperature-density coefficient	K^{-1}
E	equivalent Young's modulus	N/m^2
F	frictional force per unit tooth length	N/m
G	dimensionless material parameter $G = \alpha E$	
h	lubricant film thickness	m
H_{av}	dimensionless average film thickness $H_{av} = h_{av}/bR^2$	
h_c	central film thickness	m
h_0	film thickness constant	m
k_f	thermal conductivity of lubricant	W/mK
k_1, k_2	thermal conductivities of pinion and gear	W/mK
p	pressure	N/m
P_0	applied load per unit tooth length	N/m
R	equivalent radius, $1/R = 1/R_1 + 1/R_2$	m
R_1, R_2	radii of curvature of pinion and gear	m
s_0	temperature-viscosity index	
T	temperature	K
T_0	inlet lubricant temperature	K

T_1, T_2	temperatures on pinion and gear surface	K
u	X-component of lubricant velocity	m/s
u_1, u_2	pinion and gear teeth surface velocities	m/s
u_r	contact rolling velocity, $u_r = (u_1 + u_2)/2$	m/s
U	dimensionless velocity parameter $U = (\eta_0 u_r)/(ER)$	
W	dimensionless load parameter $W = P_0/(ER)$	
x	coordinate in surface motion direction	m
z	coordinate across lubricant	m
Z	viscosity index of Roelands' model	
α	pressure viscosity coefficient	m^2/N
η_0	ambient viscosity of lubricant	Ns/m^2
η^*	effective viscosity	Ns/m^2
η	viscosity of lubricant	Ns/m^2
λ	film thickness ratio $\lambda = h_c/\sigma$	
ξ	slide/roll ratio $= 2(u_2 - u_1)/(u_2 + u_1)$	
ρ_0	ambient density of lubricant	kg/m^3
ρ	density of lubricant	kg/m^3
ρ_1, ρ_2	densities of pinion and gear	kg/m^3
σ_x, σ_z	normal stresses on pinion subsurface	N/m^2
σ	composite RMS roughness $\sigma = (\sigma_1^2 + \sigma_2^2)^{1/2}$	m
τ	shear stress within lubricant film	N/m^2
τ_0	Eyring stress	N/m^2
τ_1	shear stress of film on pinion tooth surface	N/m^2
τ_{xz}	shear stress on pinion subsurface	N/m^2

ACKNOWLEDGEMENT

The authors would like to express their sincere gratitude for the support from provincial Natural Science Foundation of Shanxi of China (20041057). The paper was written during a sabbatical visit of the first author to Cardiff University.

REFERENCES

[1] Dawson, P.H., Effect of metallic contact on the pitting of lubricated rolling surface. *J. of Mechanical Engineering Science*, **7**(1), 147–155, 1962.

[2] Onions, R.A. and Archard, J.F., Pitting of gears and disks. *Proc. Instn. Mech. Engrs.*, **188**, 673–682, 1974.

[3] Czyzewski, T., Changes in the stress field in the EHD contact zone due to some operating factors and their possible role in the rolling contact fatigue of cylindrical surfaces. *Wear*, **31**(2), 119–140, 1975.

[4] Bhattacharyya, S., Bock, F.C., Howes, M.A.H. and Parikh, N.M., Chemical effects of lubrication in contact fatigue, Part 2: The statistical analysis, summary, and conclusions. *J. Lubr. Tech., Trans. ASME*, **98**(2), 299–307, 1976.

[5] Wellaur, E.J. and Holloway, G.A., Application of EHD oil film theory to industrial gear drive. *Trans. ASME, Series B*, **98**(2), 317–321, 1976.

[6] Zhu, D. and Hu, Y.Z., The study of transition from full film elastohydrodynamic to mixed and boundary lubrication. In: *Proc. STLE/ASME*, H.S. Cheng (ed.), Tribology Surveillance, 1999, pp. 150–156.

[7] Zhu, D. and Hu, Y.Z., A computer program package for the prediction of EHL and mixed lubrication characteristics, friction, subsurface stresses and flash temperature based on measured 3-d surface roughness. *Tribology Transactions*, **44**(3), 383–390, 2001.

[8] Zhu, D., Effect of surface roughness on mixed EHD lubrication characteristics. *Tribology Transactions*, **46**(1), 44–48, 2003.

[9] Jiang, X., Cheng, H.S. and Hua, D.Y., A theoretical analysis of mixed lubrication by macro-micao approach: Part I – Results in a gear surface contact. *Tribology Transactions*, **43**(4), 689–699, 2000.

[10] Wang, Q.J., Zhu, D. and Cheng, H.S., Mixed lubrication analysis by a macro-micro approach and a full-scale mixed EHL model. *Trans. ASME, J. Tribology*, **126**, 81–91, 2004.

[11] Snidle, R.W., Evans, H.P. and Alanou M.P., Gearing tribology. In: *Proceedings of the 29th Leeds–Lyon Symposium on Tribology*, Leeds, 2002, Elsevier, pp. 575–558, 2003.

[12] Yang, P. and Wen, S., A generalized Reynolds equation for non-Newtonian thermal elastohydrodynamic lubrication. *Trans. ASME, J. Tribology*, **112**, 631–636, 1990.

[13] Johnson, K.L. and Tevaarwerk, L.L., Shear behaviour of elastohydrodynamic oil films. *Proc. Roy. Soc. Lond.*, **A356**, 215–236, 1977.

[14] Roelands, C.J.A., Correlational aspects of the viscosity-temperature-pressure relationship of lubricating oil. PhD Thesis, Delft, The Netherlands, 1966.

[15] Wang, W.Z., Liu, Y.C., Wang, H. and Hu, Y.Z., A computer thermal model of mixed lubrication in point contacts. *Trans. ASME, J. Tribology*, **126**(1), 162–169, 2004.

[16] Dowson, D. and Higginson G.R., A numerical solution to the elastohydrodynamic problems. *J. of Mech. Engng. Sci.* **1**(1), 6–15, 1959.

[17] Weck, M., Kruse, A. and Gohritz, A., Determination of surface fatigue of gear material by roller tests. *J. of Mechanical Design*, **100**, 433–439, 1978.

[18] Gao, C.K. and Qi, X.M., Experimental study to the effect of frictional force on gearing contact stress. *J. of Mechanical Strength*, **25**(6), 642–645, 2003 [in Chinese].

AUTHOR INDEX

SUBJECT INDEX

Mechanics

SOLID MECHANICS AND ITS APPLICATIONS

Series Editor: G.M.L. Gladwell

Aims and Scope of the Series

The fundamental questions arising in mechanics are: *Why?*, *How?*, and *How much?* The aim of this series is to provide lucid accounts written by authoritative researchers giving vision and insight in answering these questions on the subject of mechanics as it relates to solids. The scope of the series covers the entire spectrum of solid mechanics. Thus it includes the foundation of mechanics; variational formulations; computational mechanics; statics, kinematics and dynamics of rigid and elastic bodies; vibrations of solids and structures; dynamical systems and chaos; the theories of elasticity, plasticity and viscoelasticity; composite materials; rods, beams, shells and membranes; structural control and stability; soils, rocks and geomechanics; fracture; tribology; experimental mechanics; biomechanics and machine design.

1. R.T. Haftka, Z. Gürdal and M.P. Kamat: *Elements of Structural Optimization*. 2nd rev.ed., 1990 ISBN 0-7923-0608-2
2. J.J. Kalker: *Three-Dimensional Elastic Bodies in Rolling Contact.* 1990 ISBN 0-7923-0712-7
3. P. Karasudhi: *Foundations of Solid Mechanics.* 1991 ISBN 0-7923-0772-0
4. *Not published*
5. *Not published.*
6. J.F. Doyle: *Static and Dynamic Analysis of Structures.* With an Emphasis on Mechanics and Computer Matrix Methods. 1991 ISBN 0-7923-1124-8; Pb 0-7923-1208-2
7. O.O. Ochoa and J.N. Reddy: *Finite Element Analysis of Composite Laminates.* ISBN 0-7923-1125-6
8. M.H. Aliabadi and D.P. Rooke: *Numerical Fracture Mechanics.* ISBN 0-7923-1175-2
9. J. Angeles and C.S. López-Cajún: *Optimization of Cam Mechanisms.* 1991 ISBN 0-7923-1355-0
10. D.E. Grierson, A. Franchi and P. Riva (eds.): *Progress in Structural Engineering.* 1991 ISBN 0-7923-1396-8
11. R.T. Haftka and Z. Gürdal: *Elements of Structural Optimization.* 3rd rev. and exp. ed. 1992 ISBN 0-7923-1504-9; Pb 0-7923-1505-7
12. J.R. Barber: *Elasticity.* 1992 ISBN 0-7923-1609-6; Pb 0-7923-1610-X
13. H.S. Tzou and G.L. Anderson (eds.): *Intelligent Structural Systems.* 1992 ISBN 0-7923-1920-6
14. E.E. Gdoutos: *Fracture Mechanics.* An Introduction. 1993 ISBN 0-7923-1932-X
15. J.P. Ward: *Solid Mechanics.* An Introduction. 1992 ISBN 0-7923-1949-4
16. M. Farshad: *Design and Analysis of Shell Structures.* 1992 ISBN 0-7923-1950-8
17. H.S. Tzou and T. Fukuda (eds.): *Precision Sensors, Actuators and Systems.* 1992 ISBN 0-7923-2015-8
18. J.R. Vinson: *The Behavior of Shells Composed of Isotropic and Composite Materials.* 1993 ISBN 0-7923-2113-8
19. H.S. Tzou: *Piezoelectric Shells.* Distributed Sensing and Control of Continua. 1993 ISBN 0-7923-2186-3
20. W. Schiehlen (ed.): *Advanced Multibody System Dynamics.* Simulation and Software Tools. 1993 ISBN 0-7923-2192-8
21. C.-W. Lee: *Vibration Analysis of Rotors.* 1993 ISBN 0-7923-2300-9
22. D.R. Smith: *An Introduction to Continuum Mechanics.* 1993 ISBN 0-7923-2454-4
23. G.M.L. Gladwell: *Inverse Problems in Scattering.* An Introduction. 1993 ISBN 0-7923-2478-1

Mechanics

SOLID MECHANICS AND ITS APPLICATIONS

Series Editor: G.M.L. Gladwell

24. G. Prathap: *The Finite Element Method in Structural Mechanics.* 1993 ISBN 0-7923-2492-7
25. J. Herskovits (ed.): *Advances in Structural Optimization.* 1995 ISBN 0-7923-2510-9
26. M.A. González-Palacios and J. Angeles: *Cam Synthesis.* 1993 ISBN 0-7923-2536-2
27. W.S. Hall: *The Boundary Element Method.* 1993 ISBN 0-7923-2580-X
28. J. Angeles, G. Hommel and P. Kovács (eds.): *Computational Kinematics.* 1993 ISBN 0-7923-2585-0
29. A. Curnier: *Computational Methods in Solid Mechanics.* 1994 ISBN 0-7923-2761-6
30. D.A. Hills and D. Nowell: *Mechanics of Fretting Fatigue.* 1994 ISBN 0-7923-2866-3
31. B. Tabarrok and F.P.J. Rimrott: *Variational Methods and Complementary Formulations in Dynamics.* 1994 ISBN 0-7923-2923-6
32. E.H. Dowell (ed.), E.F. Crawley, H.C. Curtiss Jr., D.A. Peters, R. H. Scanlan and F. Sisto: *A Modern Course in Aeroelasticity.* Third Revised and Enlarged Edition. 1995 ISBN 0-7923-2788-8; Pb: 0-7923-2789-6
33. A. Preumont: *Random Vibration and Spectral Analysis.* 1994 ISBN 0-7923-3036-6
34. J.N. Reddy (ed.): *Mechanics of Composite Materials.* Selected works of Nicholas J. Pagano. 1994 ISBN 0-7923-3041-2
35. A.P.S. Selvadurai (ed.): *Mechanics of Poroelastic Media.* 1996 ISBN 0-7923-3329-2
36. Z. Mróz, D. Weichert, S. Dorosz (eds.): *Inelastic Behaviour of Structures under Variable Loads.* 1995 ISBN 0-7923-3397-7
37. R. Pyrz (ed.): *IUTAM Symposium on Microstructure-Property Interactions in Composite Materials.* Proceedings of the IUTAM Symposium held in Aalborg, Denmark. 1995 ISBN 0-7923-3427-2
38. M.I. Friswell and J.E. Mottershead: *Finite Element Model Updating in Structural Dynamics.* 1995 ISBN 0-7923-3431-0
39. D.F. Parker and A.H. England (eds.): *IUTAM Symposium on Anisotropy, Inhomogeneity and Nonlinearity in Solid Mechanics.* Proceedings of the IUTAM Symposium held in Nottingham, U.K. 1995 ISBN 0-7923-3594-5
40. J.-P. Merlet and B. Ravani (eds.): *Computational Kinematics '95.* 1995 ISBN 0-7923-3673-9
41. L.P. Lebedev, I.I. Vorovich and G.M.L. Gladwell: *Functional Analysis.* Applications in Mechanics and Inverse Problems. 1996 ISBN 0-7923-3849-9
42. J. Menčik: *Mechanics of Components with Treated or Coated Surfaces.* 1996 ISBN 0-7923-3700-X
43. D. Bestle and W. Schiehlen (eds.): *IUTAM Symposium on Optimization of Mechanical Systems.* Proceedings of the IUTAM Symposium held in Stuttgart, Germany. 1996 ISBN 0-7923-3830-8
44. D.A. Hills, P.A. Kelly, D.N. Dai and A.M. Korsunsky: *Solution of Crack Problems.* The Distributed Dislocation Technique. 1996 ISBN 0-7923-3848-0
45. V.A. Squire, R.J. Hosking, A.D. Kerr and P.J. Langhorne: *Moving Loads on Ice Plates.* 1996 ISBN 0-7923-3953-3
46. A. Pineau and A. Zaoui (eds.): *IUTAM Symposium on Micromechanics of Plasticity and Damage of Multiphase Materials.* Proceedings of the IUTAM Symposium held in Sèvres, Paris, France. 1996 ISBN 0-7923-4188-0
47. A. Naess and S. Krenk (eds.): *IUTAM Symposium on Advances in Nonlinear Stochastic Mechanics.* Proceedings of the IUTAM Symposium held in Trondheim, Norway. 1996 ISBN 0-7923-4193-7
48. D. Ieşan and A. Scalia: *Thermoelastic Deformations.* 1996 ISBN 0-7923-4230-5

Mechanics

SOLID MECHANICS AND ITS APPLICATIONS

Series Editor: G.M.L. Gladwell

49. J.R. Willis (ed.): *IUTAM Symposium on Nonlinear Analysis of Fracture*. Proceedings of the IUTAM Symposium held in Cambridge, U.K. 1997 ISBN 0-7923-4378-6
50. A. Preumont: *Vibration Control of Active Structures*. An Introduction. 1997 ISBN 0-7923-4392-1
51. G.P. Cherepanov: *Methods of Fracture Mechanics: Solid Matter Physics*. 1997 ISBN 0-7923-4408-1
52. D.H. van Campen (ed.): *IUTAM Symposium on Interaction between Dynamics and Control in Advanced Mechanical Systems*. Proceedings of the IUTAM Symposium held in Eindhoven, The Netherlands. 1997 ISBN 0-7923-4429-4
53. N.A. Fleck and A.C.F. Cocks (eds.): *IUTAM Symposium on Mechanics of Granular and Porous Materials*. Proceedings of the IUTAM Symposium held in Cambridge, U.K. 1997 ISBN 0-7923-4553-3
54. J. Roorda and N.K. Srivastava (eds.): *Trends in Structural Mechanics*. Theory, Practice, Education. 1997 ISBN 0-7923-4603-3
55. Yu.A. Mitropolskii and N. Van Dao: *Applied Asymptotic Methods in Nonlinear Oscillations*. 1997 ISBN 0-7923-4605-X
56. C. Guedes Soares (ed.): *Probabilistic Methods for Structural Design*. 1997 ISBN 0-7923-4670-X
57. D. François, A. Pineau and A. Zaoui: *Mechanical Behaviour of Materials*. Volume I: Elasticity and Plasticity. 1998 ISBN 0-7923-4894-X
58. D. François, A. Pineau and A. Zaoui: *Mechanical Behaviour of Materials*. Volume II: Viscoplasticity, Damage, Fracture and Contact Mechanics. 1998 ISBN 0-7923-4895-8
59. L.T. Tenek and J. Argyris: *Finite Element Analysis for Composite Structures*. 1998 ISBN 0-7923-4899-0
60. Y.A. Bahei-El-Din and G.J. Dvorak (eds.): *IUTAM Symposium on Transformation Problems in Composite and Active Materials*. Proceedings of the IUTAM Symposium held in Cairo, Egypt. 1998 ISBN 0-7923-5122-3
61. I.G. Goryacheva: *Contact Mechanics in Tribology*. 1998 ISBN 0-7923-5257-2
62. O.T. Bruhns and E. Stein (eds.): *IUTAM Symposium on Micro- and Macrostructural Aspects of Thermoplasticity*. Proceedings of the IUTAM Symposium held in Bochum, Germany. 1999 ISBN 0-7923-5265-3
63. F.C. Moon: *IUTAM Symposium on New Applications of Nonlinear and Chaotic Dynamics in Mechanics*. Proceedings of the IUTAM Symposium held in Ithaca, NY, USA. 1998 ISBN 0-7923-5276-9
64. R. Wang: *IUTAM Symposium on Rheology of Bodies with Defects*. Proceedings of the IUTAM Symposium held in Beijing, China. 1999 ISBN 0-7923-5297-1
65. Yu.I. Dimitrienko: *Thermomechanics of Composites under High Temperatures*. 1999 ISBN 0-7923-4899-0
66. P. Argoul, M. Frémond and Q.S. Nguyen (eds.): *IUTAM Symposium on Variations of Domains and Free-Boundary Problems in Solid Mechanics*. Proceedings of the IUTAM Symposium held in Paris, France. 1999 ISBN 0-7923-5450-8
67. F.J. Fahy and W.G. Price (eds.): *IUTAM Symposium on Statistical Energy Analysis*. Proceedings of the IUTAM Symposium held in Southampton, U.K. 1999 ISBN 0-7923-5457-5
68. H.A. Mang and F.G. Rammerstorfer (eds.): *IUTAM Symposium on Discretization Methods in Structural Mechanics*. Proceedings of the IUTAM Symposium held in Vienna, Austria. 1999 ISBN 0-7923-5591-1

Mechanics

SOLID MECHANICS AND ITS APPLICATIONS

Series Editor: G.M.L. Gladwell

69. P. Pedersen and M.P. Bendsøe (eds.): *IUTAM Symposium on Synthesis in Bio Solid Mechanics*. Proceedings of the IUTAM Symposium held in Copenhagen, Denmark. 1999 ISBN 0-7923-5615-2
70. S.K. Agrawal and B.C. Fabien: *Optimization of Dynamic Systems*. 1999 ISBN 0-7923-5681-0
71. A. Carpinteri: *Nonlinear Crack Models for Nonmetallic Materials*. 1999 ISBN 0-7923-5750-7
72. F. Pfeifer (ed.): *IUTAM Symposium on Unilateral Multibody Contacts*. Proceedings of the IUTAM Symposium held in Munich, Germany. 1999 ISBN 0-7923-6030-3
73. E. Lavendelis and M. Zakrzhevsky (eds.): *IUTAM/IFToMM Symposium on Synthesis of Nonlinear Dynamical Systems*. Proceedings of the IUTAM/IFToMM Symposium held in Riga, Latvia. 2000 ISBN 0-7923-6106-7
74. J.-P. Merlet: *Parallel Robots*. 2000 ISBN 0-7923-6308-6
75. J.T. Pindera: *Techniques of Tomographic Isodyne Stress Analysis*. 2000 ISBN 0-7923-6388-4
76. G.A. Maugin, R. Drouot and F. Sidoroff (eds.): *Continuum Thermomechanics*. The Art and Science of Modelling Material Behaviour. 2000 ISBN 0-7923-6407-4
77. N. Van Dao and E.J. Kreuzer (eds.): *IUTAM Symposium on Recent Developments in Non-linear Oscillations of Mechanical Systems*. 2000 ISBN 0-7923-6470-8
78. S.D. Akbarov and A.N. Guz: *Mechanics of Curved Composites*. 2000 ISBN 0-7923-6477-5
79. M.B. Rubin: *Cosserat Theories: Shells, Rods and Points*. 2000 ISBN 0-7923-6489-9
80. S. Pellegrino and S.D. Guest (eds.): *IUTAM-IASS Symposium on Deployable Structures: Theory and Applications*. Proceedings of the IUTAM-IASS Symposium held in Cambridge, U.K., 6–9 September 1998. 2000 ISBN 0-7923-6516-X
81. A.D. Rosato and D.L. Blackmore (eds.): *IUTAM Symposium on Segregation in Granular Flows*. Proceedings of the IUTAM Symposium held in Cape May, NJ, U.S.A., June 5–10, 1999. 2000 ISBN 0-7923-6547-X
82. A. Lagarde (ed.): *IUTAM Symposium on Advanced Optical Methods and Applications in Solid Mechanics*. Proceedings of the IUTAM Symposium held in Futuroscope, Poitiers, France, August 31–September 4, 1998. 2000 ISBN 0-7923-6604-2
83. D. Weichert and G. Maier (eds.): *Inelastic Analysis of Structures under Variable Loads*. Theory and Engineering Applications. 2000 ISBN 0-7923-6645-X
84. T.-J. Chuang and J.W. Rudnicki (eds.): *Multiscale Deformation and Fracture in Materials and Structures*. The James R. Rice 60th Anniversary Volume. 2001 ISBN 0-7923-6718-9
85. S. Narayanan and R.N. Iyengar (eds.): *IUTAM Symposium on Nonlinearity and Stochastic Structural Dynamics*. Proceedings of the IUTAM Symposium held in Madras, Chennai, India, 4–8 January 1999 ISBN 0-7923-6733-2
86. S. Murakami and N. Ohno (eds.): *IUTAM Symposium on Creep in Structures*. Proceedings of the IUTAM Symposium held in Nagoya, Japan, 3-7 April 2000. 2001 ISBN 0-7923-6737-5
87. W. Ehlers (ed.): *IUTAM Symposium on Theoretical and Numerical Methods in Continuum Mechanics of Porous Materials*. Proceedings of the IUTAM Symposium held at the University of Stuttgart, Germany, September 5-10, 1999. 2001 ISBN 0-7923-6766-9
88. D. Durban, D. Givoli and J.G. Simmonds (eds.): *Advances in the Mechanis of Plates and Shells The Avinoam Libai Anniversary Volume*. 2001 ISBN 0-7923-6785-5
89. U. Gabbert and H.-S. Tzou (eds.): *IUTAM Symposium on Smart Structures and Structonic Systems*. Proceedings of the IUTAM Symposium held in Magdeburg, Germany, 26–29 September 2000. 2001 ISBN 0-7923-6968-8

Mechanics

SOLID MECHANICS AND ITS APPLICATIONS

Series Editor: G.M.L. Gladwell

90. Y. Ivanov, V. Cheshkov and M. Natova: *Polymer Composite Materials – Interface Phenomena & Processes*. 2001 ISBN 0-7923-7008-2
91. R.C. McPhedran, L.C. Botten and N.A. Nicorovici (eds.): *IUTAM Symposium on Mechanical and Electromagnetic Waves in Structured Media*. Proceedings of the IUTAM Symposium held in Sydney, NSW, Australia, 18-22 Januari 1999. 2001 ISBN 0-7923-7038-4
92. D.A. Sotiropoulos (ed.): *IUTAM Symposium on Mechanical Waves for Composite Structures Characterization*. Proceedings of the IUTAM Symposium held in Chania, Crete, Greece, June 14-17, 2000. 2001 ISBN 0-7923-7164-X
93. V.M. Alexandrov and D.A. Pozharskii: *Three-Dimensional Contact Problems*. 2001 ISBN 0-7923-7165-8
94. J.P. Dempsey and H.H. Shen (eds.): *IUTAM Symposium on Scaling Laws in Ice Mechanics and Ice Dynamics*. Proceedings of the IUTAM Symposium held in Fairbanks, Alaska, U.S.A., 13-16 June 2000. 2001 ISBN 1-4020-0171-1
95. U. Kirsch: *Design-Oriented Analysis of Structures*. A Unified Approach. 2002 ISBN 1-4020-0443-5
96. A. Preumont: *Vibration Control of Active Structures*. An Introduction (2nd Edition). 2002 ISBN 1-4020-0496-6
97. B.L. Karihaloo (ed.): *IUTAM Symposium on Analytical and Computational Fracture Mechanics of Non-Homogeneous Materials*. Proceedings of the IUTAM Symposium held in Cardiff, U.K., 18-22 June 2001. 2002 ISBN 1-4020-0510-5
98. S.M. Han and H. Benaroya: *Nonlinear and Stochastic Dynamics of Compliant Offshore Structures*. 2002 ISBN 1-4020-0573-3
99. A.M. Linkov: *Boundary Integral Equations in Elasticity Theory*. 2002 ISBN 1-4020-0574-1
100. L.P. Lebedev, I.I. Vorovich and G.M.L. Gladwell: *Functional Analysis*. Applications in Mechanics and Inverse Problems (2nd Edition). 2002 ISBN 1-4020-0667-5; Pb: 1-4020-0756-6
101. Q.P. Sun (ed.): *IUTAM Symposium on Mechanics of Martensitic Phase Transformation in Solids*. Proceedings of the IUTAM Symposium held in Hong Kong, China, 11-15 June 2001. 2002 ISBN 1-4020-0741-8
102. M.L. Munjal (ed.): *IUTAM Symposium on Designing for Quietness*. Proceedings of the IUTAM Symposium held in Bangkok, India, 12-14 December 2000. 2002 ISBN 1-4020-0765-5
103. J.A.C. Martins and M.D.P. Monteiro Marques (eds.): *Contact Mechanics*. Proceedings of the 3rd Contact Mechanics International Symposium, Praia da Consolação, Peniche, Portugal, 17-21 June 2001. 2002 ISBN 1-4020-0811-2
104. H.R. Drew and S. Pellegrino (eds.): *New Approaches to Structural Mechanics, Shells and Biological Structures*. 2002 ISBN 1-4020-0862-7
105. J.R. Vinson and R.L. Sierakowski: *The Behavior of Structures Composed of Composite Materials*. Second Edition. 2002 ISBN 1-4020-0904-6
106. Not yet published.
107. J.R. Barber: *Elasticity*. Second Edition. 2002 ISBN Hb 1-4020-0964-X; Pb 1-4020-0966-6
108. C. Miehe (ed.): *IUTAM Symposium on Computational Mechanics of Solid Materials at Large Strains*. Proceedings of the IUTAM Symposium held in Stuttgart, Germany, 20-24 August 2001. 2003 ISBN 1-4020-1170-9

Mechanics

SOLID MECHANICS AND ITS APPLICATIONS

Series Editor: G.M.L. Gladwell

109. P. Ståhle and K.G. Sundin (eds.): *IUTAM Symposium on Field Analyses for Determination of Material Parameters – Experimental and Numerical Aspects.* Proceedings of the IUTAM Symposium held in Abisko National Park, Kiruna, Sweden, July 31 – August 4, 2000. 2003 ISBN 1-4020-1283-7
110. N. Sri Namachchivaya and Y.K. Lin (eds.): *IUTAM Symposium on Nonlinear Stochastic Dynamics.* Proceedings of the IUTAM Symposium held in Monticello, IL, USA, 26 – 30 August, 2000. 2003 ISBN 1-4020-1471-6
111. H. Sobieckzky (ed.): *IUTAM Symposium Transsonicum IV.* Proceedings of the IUTAM Symposium held in Göttingen, Germany, 2–6 September 2002, 2003 ISBN 1-4020-1608-5
112. J.-C. Samin and P. Fisette: *Symbolic Modeling of Multibody Systems.* 2003 ISBN 1-4020-1629-8
113. A.B. Movchan (ed.): *IUTAM Symposium on Asymptotics, Singularities and Homogenisation in Problems of Mechanics.* Proceedings of the IUTAM Symposium held in Liverpool, United Kingdom, 8-11 July 2002. 2003 ISBN 1-4020-1780-4
114. S. Ahzi, M. Cherkaoui, M.A. Khaleel, H.M. Zbib, M.A. Zikry and B. LaMatina (eds.): *IUTAM Symposium on Multiscale Modeling and Characterization of Elastic-Inelastic Behavior of Engineering Materials.* Proceedings of the IUTAM Symposium held in Marrakech, Morocco, 20-25 October 2002. 2004 ISBN 1-4020-1861-4
115. H. Kitagawa and Y. Shibutani (eds.): *IUTAM Symposium on Mesoscopic Dynamics of Fracture Process and Materials Strength.* Proceedings of the IUTAM Symposium held in Osaka, Japan, 6-11 July 2003. Volume in celebration of Professor Kitagawa's retirement. 2004 ISBN 1-4020-2037-6
116. E.H. Dowell, R.L. Clark, D. Cox, H.C. Curtiss, Jr., K.C. Hall, D.A. Peters, R.H. Scanlan, E. Simiu, F. Sisto and D. Tang: *A Modern Course in Aeroelasticity.* 4th Edition, 2004 ISBN 1-4020-2039-2
117. T. Burczyński and A. Osyczka (eds.): *IUTAM Symposium on Evolutionary Methods in Mechanics.* Proceedings of the IUTAM Symposium held in Cracow, Poland, 24-27 September 2002. 2004 ISBN 1-4020-2266-2
118. D. Ieşan: *Thermoelastic Models of Continua.* 2004 ISBN 1-4020-2309-X
119. G.M.L. Gladwell: *Inverse Problems in Vibration.* Second Edition. 2004 ISBN 1-4020-2670-6
120. J.R. Vinson: *Plate and Panel Structures of Isotropic, Composite and Piezoelectric Materials, Including Sandwich Construction.* 2005 ISBN 1-4020-3110-6
121. *Forthcoming*
122. G. Rega and F. Vestroni (eds.): *IUTAM Symposium on Chaotic Dynamics and Control of Systems and Processes in Mechanics.* Proceedings of the IUTAM Symposium held in Rome, Italy, 8–13 June 2003. 2005 ISBN 1-4020-3267-6
123. E.E. Gdoutos: *Fracture Mechanics. An Introduction. 2nd edition.* 2005 ISBN 1-4020-3267-6
124. M.D. Gilchrist (ed.): *IUTAM Symposium on Impact Biomechanics from Fundamental Insights to Applications.* 2005 ISBN 1-4020-3795-3
125. J.M. Huyghe, P.A.C. Raats and S. C. Cowin (eds.): *IUTAM Symposium on Physicochemical and Electromechanical Interactions in Porous Media.* 2005 ISBN 1-4020-3864-X
126. H. Ding and W. Chen: *Elasticity of Transversely Isotropic Materials.* 2005ISBN 1-4020-4033-4
127. W. Yang (ed): *IUTAM Symposium on Mechanics and Reliability of Actuating Materials.* Proceedings of the IUTAM Symposium held in Beijing, China, 1–3 September 2004. 2005 ISBN 1-4020-4131-6
128. J.-P. Merlet: *Parallel Robots.* 2006 ISBN 1-4020-4132-2

Mechanics

SOLID MECHANICS AND ITS APPLICATIONS

Series Editor: G.M.L. Gladwell

129. G.E.A. Meier and K.R. Sreenivasan (eds.): *IUTAM Symposium on One Hundred Years of Boundary Layer Research.* Proceedings of the IUTAM Symposium held at DLR-Göttingen, Germany, August 12–14, 2004. 2006 ISBN 1-4020-4149-7
130. H. Ulbrich and W. Günthner (eds.): *IUTAM Symposium on Vibration Control of Nonlinear Mechanisms and Structures.* 2006 ISBN 1-4020-4160-8
131. L. Librescu and O. Song: *Thin-Walled Composite Beams.* Theory and Application. 2006 ISBN 1-4020-3457-1
132. G. Ben-Dor, A. Dubinsky and T. Elperin: *Applied High-Speed Plate Penetration Dynamics.* 2006 ISBN 1-4020-3452-0
133. X. Markenscoff and A. Gupta (eds.): *Collected Works of J. D. Eshelby.* Mechanics and Defects and Heterogeneities. 2006 ISBN 1-4020-4416-X
134. R.W. Snidle and H.P. Evans (eds.): *IUTAM Symposium on Elastohydrodynamics and Micro-elastohydrodynamics.* Proceedings of the IUTAM Symposium held in Cardiff, UK, 1–3 September, 2004. 2006 ISBN 1-4020-4532-8